Geomorphology and Soils

TITLES OF RELATED INTEREST

Geomorphology and Soils

Edited by
K. S. RICHARDS
R. R. ARNETT
S. ELLIS

Department of Geography, University of Cambridge

and

Department of Geography, University of Hull

Published in association with a conference of the British Geomorphological Research Group, University of Hull, 28−30 September 1984

London
GEORGE ALLEN & UNWIN
Boston   Sydney

George Allen & Unwin (Publishers) Ltd,
40 Museum Street, London WC1A 1LU, UK

George Allen & Unwin (Publishers) Ltd,
Park Lane, Hemel Hempstead, Herts HP2 4TE, UK

Allen & Unwin Inc.,
9 Winchester Terrace, Winchester, Mass. 01890, USA

George Allen & Unwin Australia Pty Ltd,
8 Napier Street, North Sydney, NSW 2060, Australia

First published in 1985

ISBN 0 04551093 8

Printed in Great Britain by
Mackays of Chatham, Kent

Preface

This collection of essays has been published in association with a conference of the British Geomorphological Research Group, held at the University of Hull from 28–30 September 1984. The conference theme was chosen to encourage integration between geomorphologists concerned with short-term process studies and those dealing with investigation of longer-term landform evolution, especially in the context of Quaternary studies. It was also felt that since this is an area in which common interests can be identified between geomorphologists, soil scientists and Quaternary and engineering geologists, interdisciplinary research could be fostered. Soils provide a key to integrated and collaborative research since they influence and reflect contemporary processes, preserve evidence of past processes and landform development, and provide a basis for both relative and absolute chronologies.

The book was planned as a collection of invited essays gathered *before* the conference, rather than as a 'proceedings' volume compiled *after* the event on the basis of submitted papers. It was hoped that this would allow the imposition of a clearer structure on the contents, with consequent improvement in the overall unity of the volume. Contributions were solicited covering a range of themes, a wide spatial and temporal context, and a balanced mixture of review and research. The timeliness of the project was perhaps best illustrated by the willingness of the multidisciplinary authorship to contribute: 21 chapters finally materialised − more or less on time − after 25 contributors had been initially contacted.

The editors would like to thank the authors for their co-operation throughout. No one was excessively tardy, so that the manuscript was prepared by 5 March 1984, after initial contacts had been made in March 1983 and a contract signed with the publishers on 7 July 1983. Our hope is that publication will be at or soon after the date of the conference, so that the momentum towards the co-operative research is not lost through a long time-lag. Thanks are also due to Roger Jones and Geoff Palmer of George Allen & Unwin for their continuing support for the venture. Finally, we thank the secretarial staff in the Department of Geography at Hull, and the draughtsman Keith Scurr, for their assistance in the preparation of the manuscript.

Keith Richards
Roger Arnett
Stephen Ellis
Department of Geography, University of Hull
10 April 1984

Acknowledgements

The following individuals and organisations have kindly given permission for the reproduction of copyright material:

Figure 1.3 reproduced from *The science of speleology*, T.D. Ford and C. H. D. Cullingford (eds), copyright © Academic Press Inc. (London) Ltd; *Field Studies* (1.7b & c); Figures 1.8 and 1.9 reproduced from *Hillslope form and process*, M. A. Carson and M. J. Kirkby, Cambridge University Press; M. Summerfield (2.1, 2.5); J. A. A. Jones, J. Dalrymple and Gebrüder Borntraeger (6.1); Pageant Books (7.4); *Journal of Soil Science* (7.5); *Canadian Journal of Soil Science* (7.6); H. J. Mücher (7.7); *Geotechnique* (8.1, 8.8); Elsevier (8.3, 8.4); Geological Society (8.5); A. W. Skempton (8.6); Norwegian Geotechnical Institute (8.7); Figure 8.9 reproduced from Chandler, R., *Earth Surface Processes* 7, 427–38, copyright © 1982 Wiley. Reprinted by permission of John Wiley & Sons, Ltd; Figure 13.2a reproduced from *Water Res. Res.* 1979, **15**, 212, by permission of F. Oldfield and American Geophysical Union, copyright © American Geophysical Union; Ministère de la Recherche et de l'Industrie (13.2d); Figure 13.4 reprinted by permission from *Nature* **281**, 111. Copyright © 1979 Macmillan Journals Limited; Society of Economic Paleontologists and Mineralogists (13.6a); *Boreas* (13.6c); American Society of Limnology and Oceanography (13.7b); Figure 13.7c reprinted by permission from Elsevier and *Nature*, **274**, 548–53. Copyright © 1978 Macmillan Journals Limited; Elsevier (13.7f, 13.9); Figure 13.8 reproduced from F. Oldfield in *Background to palaeohydrology*, K. J. Gregory (ed.), copyright © 1983 Wiley. Reprinted by permission of John Wiley & Sons, Ltd; *Geologiska Föreningens i Stockholm Förhandlingar* (14.1); John G. Evans and Geologists Association (18.6); Figures 20.1 and 20.2 reproduced with permission from *J. Terramechanics* **19**, 31–53, copyright © 1982 Pergamon Press Ltd; Figure 20.3 reproduced with permission from *J. Terramechanics* **13**, 45–55, copyright © 1976 Pergamon Press Ltd; Butterworth (20.5); Table 20.1 reproduced with permission from *J. Terramechanics* **4**, 39–45, copyright © 1967 Pergamon Press Ltd.

Contents

PART IV   SOILS AND DATING

List of tables

List of contributors

Anderson, M. G., Department of Geography, University of Bristol, Bristol, UK

Arnett, R. R., Department of Geography, The University, Hull, UK

Boardman, J., Department of Humanities, Brighton Polytechnic, Brighton, UK

Burt, T. P., School of Geography, University of Oxford, Oxford, UK

Catt, J. A., Soils and Plant Nutrition Department, Rothamsted Experimental Station, Harpenden, UK

Dearing, J. A., Department of Geography, Coventry (Lanchester) Polytechnic, Priory Street, Coventry, UK

Dembroff, G. R., US Geological Survey, Menlo Park, California, USA

De Ploey, J., Laboratory of Experimental Geomorphology, University of Leuven, Belgium

Ellis, S., Department of Geography, The University, Hull, UK

Eyles, N., Department of Geology, University of Toronto, Toronto, Canada

Finlayson, B. L., Department of Geography, University of Melbourne, Victoria, Australia

Gerrard, A. J., Department of Geography, University of Birmingham, Birmingham, UK

Goudie, A. S., School of Geography, University of Oxford, Oxford, UK

Green, C. P., Department of Geography, Bedford College, London, UK

Harris, C., Geography Section, Department of Geology, University College, Cardiff, UK

Johnson, D. L., Department of Geography, University of Illinois, Urbana, USA

Keller, E. A., Department of Geological Sciences, University of California, Santa Barbara, USA

Kemp, R. A., Department of Geography, Birbeck College, London University, London, UK

McCaig, M   Hydrologist, Geomorphological Services Limited, Marlow, Bucks, UK

Maher, B. A. Department of Geophysics, University of Edinburgh, UK

Matthews, J. A., Geography Section, Department of Geology, University College, Cardiff, UK

Mellor, A., Department of Mineral Soils, The Macaulay Institute for Soil Research, Craigiebuckler, Aberdeen, UK

Morgan, R. P. C., Department of Agricultural Engineering, Silsoe College, Bedford, UK

Oldfield, F., Department of Geography, University of Liverpool, Liverpool, UK

Poesen, J., Laboratory of Experimental Geomorphology, University of Lueven, Belgium, Research Associate, National Fund for Scientific Research, Belgium

Reading, A., Department of Geography, University College of Wales, Swansea, UK

Richards, K. S., Department of Geography, University of Cambridge, Cambridge, UK

Rockwell, T. K., Department of Geological Sciences, San Diego State University, San Diego, California, USA

Rose, J., Department of Geography, Birbeck College, London University, London, UK

Rouse, C., Department of Geography, University College of Wales, Swansea, UK

Russell, D. J., Engineering and Terrain Geology Section, Ontario Geological Survey, Toronto M5S 1B3, Ontario, Canada

Trudgill, S. T., Department of Geography, University of Sheffield, Sheffield, UK

Whalley, W. B., Department of Geography, The Queen's University of Belfast, Belfast, UK

Whiteman, C. C., Department of Geography, Birbeck College, London University, London, UK

Wilson, M. J., Department of Mineral Soils, The Macaulay Institute for Soil Research, Craigiebuckler, Aberdeen, UK

Introduction

K. S. Richards, R. R. Arnett and S. Ellis

In commenting on 'the unfulfilled geomorphological promise of earth surface process investigations', Douglas (1982, p. 101) echoes the scepticism of Gage (1970) concerning the ability of short-term process observations to provide a basis for explaining long-term landform evolution. This pessimistic diagnosis may reflect the result of two reactions to Davisian geomorphology, one shifting emphasis radically from structure and time towards process, the other replacing cyclic temporal considerations with the denudation chronology approach. A consequence of these reactions has been that geomorphologists have tended to interpret landform development in terms of erosional, or denudational, processes rather than an interrelated set of weathering, transportational and depositional mechanisms. This legacy has, until recently, weakened contact between geomorphologists, sedimentologists and pedologists, to the detriment of geomorphological explanation. Thus Hack's (1960) classic geomorphological statement of dynamic equilibrium concepts postdates Nikiforoff's (1942) pedological exposition by almost 20 years, and is perhaps significantly concerned with 'erosional topography'. A denudation chronological interpretation of landform development was traditionally morphological, but the mechanisms and chronology of evolution are much clarified by parallel investigations of weathering evidence preserved in soils and depositional environments displayed by sediments. Thus Costa and Cleaves (1984) have re-evaluated the Maryland piedmont landscape in terms of both equilibrium and episodic landform development over different temporal and spatial scales, using evidence of weathering residues and sediments as well as surface morphology.

The soil is defined and investigated by different specialists in different ways. Soil scientists traditionally view the soil as a plant-growing medium and are concerned with the biochemical reactions involved in soil−plant relationships. Pedologists and agricultural engineers are soil scientists broadly concerned with pure and applied aspects respectively, the former with soil-forming processes and their environmental controls, and the latter with maintenance of the soil's physical character and its associated fertility. For all soil scientists, nevertheless, the soil is surface-weathered material affected by processes of organic matter accumulation and decomposition, vertical and lateral translocation in solution, suspension and *en masse*, and the development of horizons subparallel to the earth's surface. The engineering geologist, however, uses the term 'soil' to refer to easily workable, surface, unlithified or unconsolidated material of low strength, ranging from weathered regolith to colluvium and sedimentary deposits, but also including weak clay-rich sedimentary rocks. Geomorphological perspectives on the soil may encompass both of these definitions, as soil investigations are exploited to aid explanation of land-surface morphology, process and evolution. The relationships between geomorphology and soils defined by these perspectives are both static and dynamic, and the dynamic relationships reflect both continuous, or equilibrium, conditions and episodic processes.

The static relationships arise from the spatial correspondence between landforms and soils, well exemplified by those apparent in fluvial, coastal and glacial depositional environments (Gerrard 1981). The simplest and most general static landform−soil relationship is the catena or toposequence. This has been represented as a two-dimensional quantitative solution of Jenny's (1941) state-factor equation, in the form of a topofunction (Yaalon 1975) in which soil properties are related to local slope gradient or distance down slope (e.g. Ruhe & Walker 1968, Furley 1971). However, such regression-based empirical models encounter serious autocorrelation problems, and conflicts also occur between continuous and discontinuous soil−slope association. Relationships between some soil and slope variables are continuous only within partial slope units. Thus qualitative, theoretical models such as the nine-unit land-surface model (Dalrymple *et al.* 1968) provide an alternative summary of discrete soil−slope associations. Arnett and Conacher (1973) extended this approach into three dimensions, showing that spatial variations occur in slope form and soil catena with basal stream order within drainage basins. First order streams have valley side slopes of simple form in which only three of the nine units are represented, while sixth order streams have more complex valley side soil−slope relations. Such spatially associated soil−slope relations can be mapped using canonical correlation of soil and slope variate sets (Roy *et al.* 1980). This mapping is of particular importance in the context of reduction of variability in soil mapping units. Field variability of soil properties has been investigated extensively by Webster and Beckett, and is reviewed by Trudgill (1983). Depending on the soil property being measured, sample sizes of up to 180 may be necessary to reduce the confidence limits to 20% (for some soil moisture variables, field variability is particularly high). The problem of sampling error and associated misclassification of soils can, however, be reduced by using landform units to define sampling patterns. Thus Chartres (1982) shows that parent material−landform−soil texture relationships are strong in certain Nigerian soil surveys. However, soil chemistry does not appear to vary as systematically between landform units except where ionic concentrations are sufficiently high to influence the physical properties markedly, as in saline environments (Parakshin 1982).

Both physical and chemical denudation processes are influenced by soil properties, and therefore constitute a set of dynamic soil−geomorphology interrelationships. When landform development is transport-limited and occurs beneath a continuous regolith cover, erosion is controlled by soil properties inherited from the parent material rather than by the underlying rock lithology, since the erosional processes operate on and within the regolith. Soil texture, density and moisture content affect the soil mechanical properties (Chorley 1959), which in turn influence resistance to entrainment by runoff or transfer down slope by mass movement. Thus gross topographic features such as the Lower Greensand escarpment in east-central England reflect structural influences but also regolith properties which vary spatially in response to facies changes in the parent material (Chorley 1969). The role of soil mechanical behaviour in relation to slope development is now well established in geomorphological theory, and the interaction between geomorphological and pedological process is exemplified by Carson and Petley's (1970) model of discontinuous slope decline in which weathering of stable slope regoliths eventually alters the cohesion and frictional strength so that renewed instability occurs. Chemical denudation within and beneath a soil cover is a dominant influence on surface lowering in limestone areas, and karst geomorphologists have also demonstrated the importance of CO_2 production in the soil atmosphere on non-carbonate rocks (Gunn & Trudgill 1982). The equilibrium models of soil−landform development initially expounded by Nikiforoff (1942) and elaborated by Jahn (1968) require a balance between surface erosion and weathering at the soil−rock interface. Slope development is weathering-limited when surface removal rates exceed weathering rates and only a thin debris cover exists. Rapid weathering of the bedrock relative to surface removal of soil results in a thick soil cover and transport-limited evolution. Three-dimensional models of soil−slope−landform development, such as those of Ahnert (1977), are critically dependent on establishing the relationships between soil thickness and weathering rate, in addition to the weathering−erosion balance. In thin soils, water transmission is rapid and chemical weathering rates low. Weathering is impeded under thick soils by slow water circulation and chemical saturation in soil moisture. Thus optimal weathering conditions occur for soils of intermediate thickness, although the nature of the relationship between soil thickness and weathering rate still requires investigation both in process terms and in terms of its mathematical representation for modelling purposes (Cox 1980). Process investigations of structural bypassing by rapid, transient soil water flow in macropores (and its implications for solute uptake in soils) may be increasingly incompatible with the latter objective. Dynamic three-dimensional models of soil−landform relationships have emphasised the positive feedback implicit in the convergence and divergence of water and sediment flows into hollows and over spurs. Perhaps the most explicit link between land-surface form and pedogenesis is Huggett's (1975) simulation of lateral downslope translocation of soil plasma through the soil skeleton into a topographic hollow. The net accumulation of plasma in the hollow thus simulated may, however, reflect the boundary conditions imposed and the feedback processes specified; a natural equilibrium system may exist when basal removal and reduced hydraulic conductivity in the hollow both occur.

Between periods of equilibrium, landform evolution is commonly episodic, and soil studies inform the geomorphologist of this fact through use of the soil stratigraphic principles elaborated by Finkl (1980). Episodic development is characteristic of hillslopes affected by K-cycles (Butler 1967), and a pedological approach is central to interpretation of upper hillslope erosion and basal accumulation. However, phases of instability may reflect external controls (e.g. climatic change or vegetation clearance) or intrinsic thresholds (e.g. a critical loss of strength on weathering), and external controls may be local (e.g. a random storm effect) or regional. Site-specific evidence of episodic development is therefore potentially misleading, and the accumulation of absolute dating evidence of widespread phases of landform development is necessary; for example, radiocarbon dating of valley aggradation in the mid west USA from buried organic material (Knox 1975) indicates synchroneity and widespread spatial occurrence. The absolute and relative chronologies of episodic landform development may use radiometric (e.g. ^{14}C and uranium series), thermoluminescence and amino acid racemisation analyses, and chronosequences of buried and surface soils arranged in stratigraphic order. In the longer term, Quaternary episodic development has frequently been identified by soil stratigraphic methods. Early reconstruction of the Quaternary history of the USA followed recognition of soil stratigraphic units in association with geological surveys conducted in the 19th century (Follmer 1978), and soil scientists have contributed significantly to the interpretation of the Clay-with-flints and its relationship to surfaces on the Chalk dip-slopes of south-east England (Catt and Hodgson 1976). The frequency of environmental oscillation in the Quaternary, and the potential of soil studies in unravelling environmental history, are clearly illustrated by the correlations between terrestrial palaeosol evidence in the loess sequences of Czechoslovakia and Austria and the oceanic oxygen isotope record (Fink & Kukla 1977, Kukla 1977). They also indicate that continuous environmental oscillations yield discrete terrestrial responses. Palaeopedological evidence of this episodic behaviour of landforms is exemplified by Vreeken's (1984) study of late Quaternary loess accumulation, erosion and redeposition in the Netherlands. Here it is possible to demonstrate, by mapping the spatial patterns of buried soils and erosional unconformities, that specific sites have 'migrated' from hilltop to hillside to valley bottom locations (and vice versa) as landscapes have been blanketed by loess and then subjected to renewed valley formation. This example also very clearly illustrates the benefit to geomorphological explanation of applying pedological techniques such as micromorphology.

An integration of geomorphology and soils should therefore improve interpretation of landforms in terms of the weathering, erosional and depositional processes that have combined to produce them, and should help to link the process and evolutionary aspects of geomorphology. There are approximately 1.5×10^{12} ways of grouping 20 papers into 5 sets of 4, and fortunately not enough geomorphologists and soil scientists to find a logical explanation for all of them! We hope that the structure we have chosen for this collection both illustrates the breadth of potential in integrated soil−geomorphological studies, and focuses on certain key areas for integration. These are (a) the improved understanding of landform development, particularly in the context of the processes involved; (b) the improved understanding of hillslope processes, the operation of which is dependent on soil characteristics both in terms of the process type and intensity; (c) the interpretation of past processes from evidence preserved in soils; (d) the dating of episodes of landform development in relative and absolute terms using soil properties; and (e) the solution of land management problems in an applied pedogeomorphological context.

Several perspectives on soil–landform relationships are presented in the first four chapters (Part I). Burt and Trudgill discuss the process of solute uptake in soil water, and the resulting spatial patterns of chemical denudation which can be experimentally investigated and which relate to two-dimensional (slope profile) and three-dimensional landform evolution. Long-term consequences and evidence of chemical weathering processes are then reviewed by Goudie and Green. Goudie summarises the geomorphological role of duricrusts, pedogeomorphological phenomena whose early association with subhorizontal land surfaces has been progressively re-evaluated, as understanding of the pedological and sedimentological processes of duricrust formation has developed. Green demonstrates the evidence for similar pre-Quaternary tropical and subtropical weathering (also analysed by Wilson in Chapter 12) in southern Britain, and exemplifies the role of weathering residues and sediments in improving denudation chronological studies. The same principles of stratigraphic investigation are implicit in Gerrard's analysis of recent, largely man-induced landscape change in Iceland, caused by physical erosion processes. Soil erosional products in this landscape have been transferred to sediment sinks, preserving the evidence of erosional episodes, and buried soils and tephra layers provide the means of establishing the chronology and lateral correlation of events. This chapter thus illustrates the landform consequences of the agricultural soil erosion discussed by Morgan (Chapter 19), and the application of dating methods outlined in Part IV.

Parts II, III and IV include a series of papers dealing with contemporary slope processes, past processes and the dating of past events – the framework for the interpretation of soil–landform relationships. The four chapters in Part II consider the ways in which certain soil properties influence those slope processes which cause progressive downslope movement of soil materials. The sequence from splash and wash, through throughflow and lateral translocation, and soil creep, to shallow landsliding reflects a continuum from sediment *transport* to sediment *transfer* processes as the water : solids ratio decreases and the depth of material mobilised increases, from the surface particles to the full regolith depth. Aggregate stability and soil surface crustability are investigated in relation to splash, wash and rill erosion by De Ploey and Poesen, providing the physical background to the assessment of soil erosion in an agricultural engineering context (Chapter 19). McCaig discusses the soil properties that influence the nature of subsurface hydrology, and the resulting balance of chemical and mechanical transport and deposition in slope soils, providing a complement to the chapter by Burt and Trudgill. Finlayson attacks the conventional geomorphological wisdom on soil creep and emphasises a pedogeomorphic approach in which micromorphological techniques can be exploited to assess random soil turbation processes. An engineering definition of the soil is then adopted by Rouse and Reading, who illustrate the problems of interpreting the mechanisms of threshold slope development from the morphological (slope angle) and soil mechanical evidence. This involves reconstruction of process rather than direct observation, and thus provides a link with the methodology of Part III.

This third section includes five contributions on process reconstruction based on evidence contained within soils or sediments. Such evidence includes grain surface texture, particle size distribution, soil micromorphological characteristics, optical and clay mineralogy and magnetic mineral properties. It therefore embraces the range from textural and other physical characteristics of soils to the chemical and mineralogical, with a parallel increase in the complexity of the processes responsible for the evidence and the techniques required for its identification and interpretation. The papers also illustrate an important distinction between incorporation of evidence into soils, preservation within soils and maintenance of soil properties in sediments formed after erosion. Whalley and Catt discuss, respectively, grain surface texture and particle size and mineralogy in the context of evidence within soils of past geomorphological activity, indicated by incorporation of deposited materials within the soil. Catt, for example, considers the evidence presented by the silt fractions of soils in England and Wales for incorporation of Quaternary loess. Harris deals with the modification of soil micromorphology by various periglacial processes including ice segregation and solifluction, identified in both present and fossil slope deposits. Preservation of evidence of past processes is also exemplified by Wilson's analysis of clay mineralogical development during *in situ* weathering in Scottish soils under preglacial, subtropical climatic conditions. This complements Green's review in Chapter 3. Finally, Dearing, Maher and Oldfield summarise research into the formation in soils of magnetic minerals by pedological processes, which give characteristic magnetic signatures to topsoil and subsoil that are distinct from those of unweathered rock. These signatures permit the identification of the changing source areas producing sediments accumulating in sediment sinks such as lakes after catchment soil erosion.

Part IV provides examples of the establishment of a chronological framework for the identification of past processes, based on soil studies. Matthews reviews the problems and potential of absolute ^{14}C dating of organic material in palaeosols, and Mellor describes the construction of a relative dating method based on chronofunctions for surface soils on Neoglacial moraine ridges in Norway. The following chapters by Rockwell *et al.*, Ellis and Richards, and Rose *et al.* provide illustrations of chronosequence and palaeosol investigations in dating a range of landform types over varying timescales. Mellor's chronosequence study shows how a geomorphological environment with systematically dated landforms can be used to examine temporal variation of soil properties and processes. The chapter by Rockwell, Johnson, Keller and Dembroff uses variation in soil properties to date geomorphic surfaces of fluvial origin, but displaced by tectonic activity. These two chapters contrast chronosequences based on dynamic soil properties operating over timescales of, respectively, a few hundred years (organic matter accumulation and decomposition, incipient podzolisation) and several thousand years (reddening and clay illuviation). Rockwell *et al.* have constructed a chronosequence for soils in an area affected by Neotectonic folding and faulting, and soil stratigraphic methods have proved invaluable in interpreting the age of fault-block surfaces, correlating displaced surfaces and estimating rates of Neotectonic deformation. The remaining papers in this part illustrate episodic landform development on two timescales. Ellis and Richards use both chronofunction and absolute dating methods − the latter encountering the problem of soil-organic matter residence time discussed in Chapter 14 by Matthews − to assess phases of Postglacial slope development in Norway. This study suggests that soil mechanical properties (see Rouse & Reading, Chapter 8) influence the occurrence of slope instability, but that periods of slope stability and pedogenesis may have a significant influence in preparing the slope soils for instability by altering their drainage characteristics as podzolisation progresses. Finally, Rose, Boardman, Kemp and Whiteman discuss research which is unique in Britain and also rare in western Europe, in which palaeosolic evidence is synthesised in a stratigraphic context to interpret long-term landform development in the British Quaternary.

Applied pedogeomorphology can clearly be developed in terms of the static identification of soil landform units for the purpose of improving soil mapping strategies. However, the examples of applied soil−geomorphology studies in the final part are more concerned with dynamic process-related integration of soil and surface processes. Morgan discusses soil degradation and erosion as an interaction of fluid stresses and soil resistance, both of which have been evaluated previously (Chapters 5 and 8). Much research has focused on the intersection of geomorphology, pedology and hydrology, as the chapters by Burt and Trudgill and McCaig testify. In Part V, Anderson describes a practical application of this integration by summarising the development of a simulation model which uses atmospheric conditions to predict near-surface evaporation of soil moisture. This information is then incorporated into a dynamic model of soil trafficability, which has both agricultural and military applications. However, applied investigations are rarely as distinct from 'pure' investigations as the adjectives imply, and the final contribution by Russell and Eyles effectively squares the circle. They adopt an engineering geological approach to geotechnical weathering profiles in overconsolidated clays (Palaeozoic mudrocks being compared with Quaternary lodgement tills). The emphasis is not, however, on practical engineering significance of surface weathering, but on the application of investigation of such weathering profiles to the interpretation of Quaternary geomorphology.

This collection of 21 chapters includes contributions by geomorphologists, soil scientists, Quaternary geologists and engineering geologists, and exemplifies an interdisciplinary approach to common problems concerning soils and landforms. Many of the chapters indicate profitable areas for future collaborative research. To these may be added the more general issues discussed above: the rationalisation of field variability of soil properties by geomorphologically defined stratified sampling, the identification of relationships between soil depth and weathering rates for modelling purposes, and the use of soil stratigraphic procedures to illuminate the relative contribution of continuous and episodic processes to landform development over different temporal and spatial scales. The various specialists concerned with soils in somewhat different contexts have much to gain from collaboration, and solutions to these problems may be reached more rapidly if this can be fostered. It is hoped, therefore, that this collection can mark a significant step in the direction of collaborative research.

References

Ahnert, F. 1977. Some comments on the quantitative formulation of geomorphological process in a theoretical model. *Earth Surf. Proc.* **2**, 191–201.

Arnett, R. R. and A. J. Conacher 1973. Drainage basin expansion and the nine-unit landsurface model. *Austral. Geogr.* **12**, 237–49.

Butler, B. E. 1967. Soil periodicity in relation to landform development in south-eastern Australia. In *Landform studies from Australia and New Guinea*, J. Jennings and J. A. Mabbut (eds), 231–55. Cambridge: Cambridge University Press.

Carson, M. A. and D. J. Petley 1970. The existence of threshold hillslopes in the denudation of the landscape. *Trans Inst. Brit. Geogs* **49**, 71–95.

Catt, J. A. and J. M. Hodgson 1976. Soils and geomorphology of the chalk in SE England. *Earth Surf. Proc.* **1**, 181–93.

Chartres, C. J. 1982. The use of landform–soil associations in irrigation soil surveys in northern Nigeria. *J. Soil Sci.* **33**, 317–28.

Chorley, R. J. 1959. The geomorphic significance of some Oxford soils. *Am. J. Sci.* **257**, 503–14.

Chorley, R. J. 1969. The elevation of the Lower Greensand ridge, SE England. *Geol. Mag.* **106**, 231–48.

Costa, J. E. and E. T. Cleaves 1984. The piedmont landscape of Maryland: a new look at an old problem. *Earth Surf. Proc. Landforms* **9**, 59–74.

Cox, N. J. 1980. On the relationship between bedrock lowering and regolith thickness. *Earth Surf. Proc.* **5**, 271–4.

Dalrymple, J. B., R. J. Blong and A. J. Conacher 1968. A hypothetical nine-unit land surface model. *Zeitschr. Geomorph.* **12**, 60–76.

Douglas, I. 1982. Editorial. The unfulfilled promise: earth surface processes as a key to landform evolution. *Earth Surf. Proc. Landforms* **7**, 101.

Fink, J. and G. J. Kukla 1977. Pleistocene climates in central Europe: at least 17 interglacials after the Olduvai event. *Quat. Res.* **7**, 363–71.

Finkl, C. W. 1980. Stratigraphic principles and practices as related to soil mantles. *Catena* **7**, 169–94.

Follmer, L. R. 1978. The Sangamon soil in its type area – a review. In *Quaternary soils*, W. C. Mahaney (ed.), 125–65. Norwich: Geo Abstracts.

Furley, P. A. 1971. Relationships between slope form and soil properties developed over chalk parent materials. In *Slopes: form and processes*, D. Brunsden (ed.), 141–63. Inst. Brit. Geogs Spec. Publn No. 3.

Gage, M. 1970. The tempo of geomorphic change. *J. Geol.* **78**, 619–25.

Gerrard, A. J. 1981. *Soils and landforms: an integration of geomorphology and pedology*. London: George Allen & Unwin.

Gunn, J. and S. T. Trudgill 1982. Carbon dioxide production and concentrations in the soil atmosphere: a case study from New Zealand volcanic ash soils. *Catena* **9**, 81−94.

Hack, J. T. 1960. Interpretation of erosional topography in humid temperate regions. *Am. J. Sci.* **258**, 80−97.

Huggett, R. J. 1975. Soil landscape systems: a model of soil genesis. *Geoderma* **13**, 1−22.

Jahn, A. 1968. Denudational balance of slopes. *Geog. Polonica* **13**, 9−29.

Jenny, H. 1941. *Factors of soil formation, a system of quantitative pedology*. London: McGraw-Hill.

Knox, J. C. 1975. Concept of the graded stream. In *Theories of landform development*, W. N. Melhorn and R. C. Flemal (eds), 169−98. Binghamton: Publns in Geomorphology, State University of New York.

Kukla, G. J. 1977. Pleistocene land−sea correlations. I Europe. *Earth Sci. Rev.* **13**, 307−24.

Nikiforoff, C. C. 1942. Fundamental formula of soil formation. *Am. J. Sci.* **240**, 847−66.

Parakshin, Y. P. 1982. Some patterns in the formation and development of Solonetz Soils of Kokchetav Upland in Kazakstan. *Soviet Soil Sci.* **14**, 1−9.

Roy, A., R. S. Jarvis and R. R. Arnett 1980. Soil−slope relationships within a drainage basin. *Ann. Assoc. Am. Geog.* **70**, 397−412.

Ruhe, R. V. and P. H. Walker 1968. Hillslope models and soil formation. I Open systems. In *Transactions of the 9th International Congress on Soil Sci.* Vol. IV, **4**, 551−60.

Trudgill, S. T. 1983. Soil geography: spatial techniques and geomorphic relationships. *Prog. Phys. Geog.* **7**, 345−60.

Vreeken, W. J. 1984. (Re)deposition of loess in Southern Limbourg, the Netherlands. 3. Field evidence for conditions of deposition of the Middle and Upper Silt Loam complexes, and landscape evolution at Nagelbeek. *Earth Surf. Proc. Landforms* **9**, 1−18.

Yaalon, D. H. 1975. Conceptual models in pedogenesis: can soil-forming functions be solved? *Geoderma* **14**, 189−205.

Part I

SOILS AND LANDFORMS

1

Soil properties, slope hydrology and spatial patterns of chemical denudation

T. P. Burt and S. T. Trudgill

Introduction

For more than two decades, a major focus of attention for fluvial geomorphologists has been the hillslope runoff system. Following Hewlett (1961), who first emphasised subsurface flow as the mechanism providing 'partial' source areas for stream runoff, numerous studies have investigated the types and source areas of runoff processes on hillslopes. This preoccupation with process has, to some extent, led hillslope geomorphologists to neglect the erosional development of landforms. Similarly, links between geomorphology and soil science have tended to emphasise soil-water physics rather than pedogenesis. Although the study of runoff processes is a necessary prerequisite, processes of solute uptake must also be investigated to improve understanding of the pattern of chemical denudation on a hillslope and the associated pattern of soil profile development. Given a knowledge of solute processes, two timescales exist for investigation.

In the short term, spatial variations of solute sources and rates of removal may be studied. Current process activity may be defined by examining stream solute levels, nutrient cycling and leaching, and solute yields, with much potential for interdisciplinary work by geomorphologists, soil scientists, plant ecologists, hydrologists and geochemists. Over longer timescales, chemical denudation rates and their relationship to soil and slope profile evolution are the primary concern of the geomorphologist, but such studies of landform evolution must clearly be firmly based in process theory if rigorous explanation is to be achieved. Thus, detailed exposition of solute removal processes must precede consideration of landform response, whether that response is to current process activity or longer-term evolution. It is therefore logical to begin with consideration of hillslope runoff processes, which nevertheless represent the first rather than the final goal of the hillslope geomorphologist.

Subsurface flow processes on the hillslope

It is normal to assume that infiltration is vertical flow into the soil. However, Zaslavsky (1970) has shown that any initiation of a soil profile will impart some degree of anisotropy such that lateral hydraulic conductivity (K_s) will exceed vertical hydraulic conductivity (K_y), and there will be a tendency for water to flow downhill. The horizontal flow component will be proportional to slope angle and the degree of anisotropy. Figure 1.1 shows the components of flow for infiltration into a sloping anisotropic soil; flow normal to the soil layers is q_y and flow parallel to the soil layers is q_s. Zaslavsky (1970) showed that for steady state infiltration, when there is no change in pore water pressure with depth,

$$q_s / q_y = -U \tan \alpha \qquad (1.1)$$

where U is the degree of anisotropy ($U = K_s /K_y$). It can be shown that q_s only equals zero where the land surface is horizontal; elsewhere it must be positive and infiltration is not normal to the surface. The flux vector will be vertical where the soil is isotropic ($U = 1$). The angle β between the soil surface and the flux vector can be expressed as

$$\tan \beta = q_y / q_s = 1/(U \tan \alpha) \qquad (1.2)$$

It is clear that β decreases as $\tan \alpha$ and U increase, so that even for low slope angles and limited anisotropy, the horizontal flow distance will greatly exceed the vertical flow depth of infiltrating water. Figure 1.2 shows the angle of infiltration streamlines for an anisotropy of $U = 5$. Flowlines tend to converge on concave slope profile elements, and whilst divergence occurs on convex elements, this may reinforce the tendency for convergence lower down the slope. When three-dimensional hillslope topography is considered, the effect of flow convergence is accentuated, since it has been shown that soil water movement is parallel to the line of maximum slope gradient (Anderson & Burt 1978). Thus, in a hillslope hollow where both plan and profile are concave, there will be marked convergence of flow into the hollow, particularly where soil anisotropy is also marked.

Many field investigations have defined the factors controlling development of subsurface runoff (reviewed, for example, by Dunne 1978). Soil anisotropy, caused by a less permeable soil horizon at depth or by impermeable bedrock, is, however, crucial to the generation of lateral subsurface flow. A reduction in hydraulic conductivity of several orders of magnitude represents a very large value of U, and thus a flux vector essentially parallel to the hillslope. Particularly important is the build-up of a saturated zone (or 'wedge', since it is usually thickest at the slope base) above the impeding layer (Weyman 1973). Since hydraulic conductivity is maximal when the soil is saturated, such a condition maximises the rate of subsurface runoff. In addition, the saturated layer has important implications for solutional processes in the soil. Simple storage

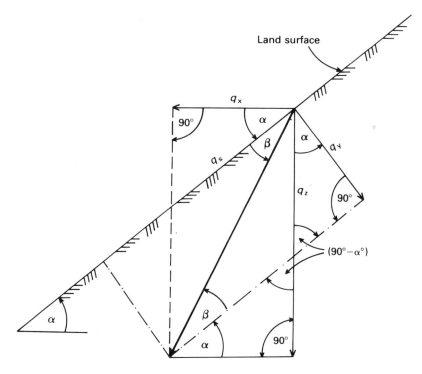

Figure 1.1 Components of flow for infiltration into a sloping anisotropic soil (after Zaslavsky 1970).

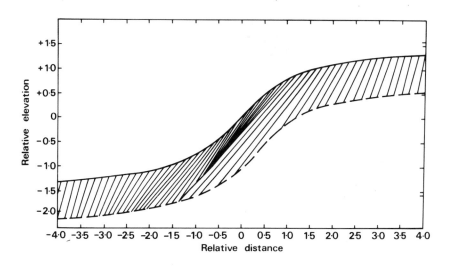

Figure 1.2 Flowlines on a convexo-concave slope for an anisotropy of $U = 5$ (after Zaslavsky 1970).

models demonstrate that soil saturation will build upwards from the impeding layer when input exceeds output (Kirkby 1978). Inputs may be from vertical infiltration (Knapp 1978), lateral flow down slope (Burt & Butcher 1984), or both. Kirkby and Chorley (1967) identified three zones where maximum sub-surface flow will occur: (a) areas of streamline convergence, (b) zones at the slope base, and (c) areas of reduced soil moisture storage. In certain cases, the zone of saturation may reach the soil surface, causing saturation-excess overland flow. Where the excess is due to lateral flow, it is termed 'return flow', and is important since it should be rich in solutes, being 'old' soil water that has been displaced down slope (Anderson & Burt 1982). The input of solute-rich soil water into a stream during a period of storm runoff may be especially important in controlling the overall solute load of the stream. Clearly, therefore, there is a direct link between the source of storm runoff and its chemical composition.

The presence of macropores in the soil (i.e. pores above capillary size) influences the timing, rate and solute load of hillslope runoff. Macropores effectively increase the bulk hydraulic conductivity of a soil layer at a scale relevant to hillslope runoff (Beven & Germann 1981). However, flow rates through macropores are much greater than through the soil matrix, and the implications of such rapid flow for solute uptake are examined in the next section.

Clearly, without lateral soil-water flow, all solutional processes within the soil would be controlled by vertical infiltration. Lateral subsurface runoff, being itself spatially variable, is therefore directly responsible for providing the major source of spatial variability in solutional denudation on hillslopes of uniform parent material.

Solute uptake and transport in the soil profile

Water draining from soil profiles into streams generally has a higher solute concentration than the infiltrating rain water. The uptake of solutes by perco-lating soil water is not, however, a simple process to model. A range of solute losses and gains interact to determine final output concentration; adsorption, exchange processes, plant root uptake and chemical precipitation act to de-crease soil solute concentration while dissolution, desorption and organic matter decay act to increase it. Furthermore, percolating soil water does not necessarily move uniformly through the soil; some moves preferentially through structural voids, giving differential opportunity for solute uptake, and a variety of equilibrium solute concentrations. Transient flows in particular move through structural voids; thus kinetic considerations will control solute concentrations since there is insufficient time to establish solid−solvent equilibrium. The complex amorphous and impure mineral phases in soils further increase the problems of modelling.

Notwithstanding these complications, predictive models for solute uptake have been proposed, initially based on simplifying assumptions that ignore soil structural influences on preferential flow, and kinetics. They consider uniform flow and equilibrium with pure mineral phases, in the contexts of solute uptake and both uptake and losses. Although providing useful first approximations to prediction, such models are not necessarily representative of the complexity of natural processes.

The simplest model involves solute uptake down the soil profile until an equilibrium concentration is reached which can be theoretically defined for pure minerals at given temperatures and pressures (Garrels & Christ 1965, Bohn et al. 1979, Lindsay 1979). The rate of mass transfer of solute over time from solid to solvent (dm/dt) decreases as the difference between the maximum (equilibrium) concentration (C_s) and the actual concentration (C) decreases, according to the model

$$dm/dt = K(C_s - C) \tag{1.3}$$

where K is the transport rate constant appropriate to the mineral phase in question. Thus, solute concentration should increase to a maximum during percolation of water through the soil. Such a model is appropriate for chemical processes dominated by hydrolysis, and when acidity is derived from the upper organic soil layers. It is less relevant if uptake or adsorption occur, since these may decrease concentrations below the theoretical maximum. If equilibrium models were universally applicable, concentrations of solutes draining from soils would be comparable in similar situations. This may be true for base flow conditions, but is certainly not the case when transient flow occurs.

Limestone bedrock, comprising one dominant mineral (calcite), is the simplest to study. Here, equilibrium can be defined relative to carbon dioxide levels in soils, or to soil acidity. Theoretically (Atkinson & Smith 1976), it is unlikely that concentrations of calcium in waters draining from temperate limestone areas will exceed 120 mg l^{-1} Ca^{2+} (300 mg l^{-1} $CaCO_3$) since this is the value equilibrated with the normal maximum of c. 2% by volume carbon dioxide for most moderately or well-drained soils. Observation shows this to be the case (Fig. 1.3). However, this approach is limited to the specification of maximum solute concentration; much of the flow will have concentrations of less than this. Thus, predictions of solutional denudation based on equilibrium concentrations may over-predict when dilution occurs at high flows to give less than maximum concentrations, or under-predict in that, although dilution occurs at high flows, load considerations may mean that more is being removed at high than at low flow.

A further unresolved problem is that of preferential flow in the larger structural voids within soils, often loosely referred to as 'macropore flow'. The terms 'preferential' or 'differential' flow might be preferable, since they only imply that one portion of the soil water is flowing faster than another, without prejudging the precise mode or location of flow. Water may flow differentially through soils even in the absence of macropores. 'Differential' could define any form of differential movement, particularly the subtle movements that may

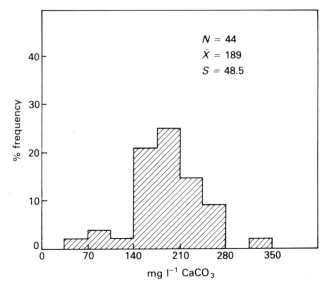

Figure 1.3 Distribution of concentration data for $CaCO_3$ (mg l^{-1}) in drainage water from temperate limestone regions (modified from Atkinson & Smith 1976). For Ca^{2+} divide concentration by 2.497.

occur in more uniform, weakly structured soils. 'Preferential' could refer to more obvious cases where preferential flow follows old root channels or inter-pedal voids, such as commonly occur in cracking clay soils.

The occurrence of preferential flow in strongly pedal soils is relatively easy to predict, since it will occur when rainfall intensity exceeds pedal infiltration capacity (Bouma *et al.* 1981, Smettem *et al.* 1983, Trudgill *et al.* 1983a, 1983b). In weakly pedal soils, peds are more difficult to demarcate, and although differential flow undoubtedly occurs, it is more difficult to define. A common approach distinguishes mobile and immobile water in the soil (Addiscott 1977, Gaudet *et al.* 1977, Skopp *et al.* 1981), with the boundary between the two set at, for example, two bars (Addiscott 1977). Such two-region models are useful since it can be argued that equilibrium calculations apply to the immobile water and kinetic ones to the mobile phase (Travis & Etnier 1981). Such work has used theoretical calculation of soil-water movement and empirical observation of solute concentrations. However, there are in reality three process domains rather than a simple mobile–immobile partition, as follows.

(a)   An immobile phase which is tightly held in soil pores and in which equilibrium concentrations can be reached. Solutes may diffuse from this domain to the more mobile water, but since this water does not appear in drainage water it has no other relevance to prediction of drainage water concentrations.

(b)   Displaced water, which is often equivalent to capillary or matric water, and is held at intermediate tensions which are not, however, universally specified for all soils. Here, solute concentrations may build up in the water as much as the frequency of any displacements allows.

(c)   Rapid, transient flow, which is the most dilute water, associated with high rainfall intensity and transmitted rapidly through the more open, porous pathways in the soil.

The proportions of these water phases will vary with soil texture and structure. Strongly pedal soils such as cracking clay soils will be dominated by domains (a) and (c), while weakly structured soils of a loose, friable nature may be dominated by domain (b).

Thus far, some of the problems of modelling have been highlighted. Attempts at soil-water solute modelling are now reviewed, with emphasis on present' limitations and future prospects and possibilities.

Considerations of models of solute movement often begin with Fick's first law which refers to solute flux in relation to solute gradient (Nye & Tinker 1977). Solute moves in a down-gradient direction through soil from high to low concentrations along the gradient dC/dx:

$$F = -D\,(dC/dx) \qquad (1.4)$$

where F is the flux of solute, D is the coefficient of diffusion, C is the amount of solute per unit volume of soil solution and x is the distance. This equation applies to self-diffusion of solute in static water. If the water itself is moving, convection also occurs; that is, the bodily movement of water from one location to another together with any dissolved solute. In convection, solute flux is equal to the water velocity multiplied by its solute concentration. However, the solute will move down any concentration gradient simultaneously with the convection process, so the two components of solute movement (diffusion and convection) must be added. A further complication in soils is the range of pore-water velocities, which make diffusion and convection difficult to separate as some portions of solute will travel faster than others (in the larger pores or by the more direct routes). The combined movement of diffusion and convection plus the range of pore-water velocities is referred to as 'dispersion', and the dispersion coefficient (D^*) is used for solute flux in soils, thus:

$$F = D^*\,(dC/dx) + vC_1 \qquad (1.5)$$

where v is the water flux in the direction x, and C_1 is the concentration of solute in the soil solution.

Diffusion in water-saturated soil samples can be measured in terms of solute changes in a test chamber (Fig. 1.4). At the start of observation, the left hand chamber has the solute present and the right has no solute. Sampling over time allows measurement of the diffusion from left to right along the gradient dC/dx, until concentrations are equal in the two chambers. Diffusion is naturally retarded if pore connectivity is low, and is rapid if connected macropores

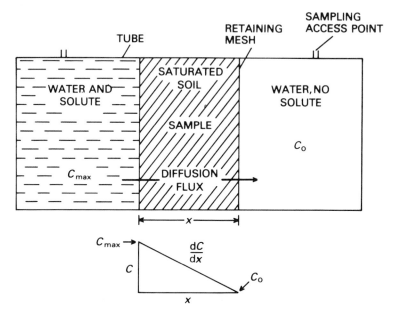

Figure 1.4 Diffusion test equipment (see text for explanation of symbols).

occur. Unsaturated soil moisture conditions in the field result in lower connectivity than in the saturated test chamber, since air spaces act as blocks and decrease the diffusivity of a solute in drier soil. Dispersion can be measured by displacement of water in soil columns. Taking a saturated soil column (Fig. 1.5), tracer-free water is at first present; if tracer is then added, piston flow would produce a breakthrough curve as shown in Figure 1.5a, with appearance when one pore volume of soil water has been displaced. Dispersion (Fig. 1.5b) produces departures from this ideal pattern, with more rapid initial appearance in the faster flow paths and some tailing in the slower, more tortuous, paths. This pattern can be greatly exaggerated by the presence of rapidly flowing macropores, moving the first time of tracer arrival considerably in advance of one pore volume or giving initial peaks of rapid preferential tracer flow (Figs 1.5d & e).

The concept that uniform displacement models may not adequately describe movement of water and solutes in structured soils has become important. For example, Thomas and Philips (1979) assert that water may flow preferentially in structured soils and penetrate more deeply and more rapidly than uniform displacement models predict. Bouma et al. (1981) present a field method for the measurement of preferential flow ('short-circuiting') in soils where flow occurs in the larger soil pores and bypasses much of the soil matrix. A sprinkler device with input intensities of 18.6 and 22.8 mm h^{-1} was used on heavy clay soils; 36% and 47% of the surface-applied water passed through the soil column at each of these intensities, when uniform displacement models predicted no outflow should have occurred. Sprinkler intensities exceeded those in

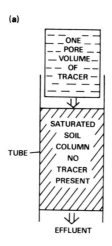

(a)

ONE
PORE
VOLUME
OF
TRACER

TUBE

SATURATED
SOIL
COLUMN
NO
TRACER
PRESENT

EFFLUENT

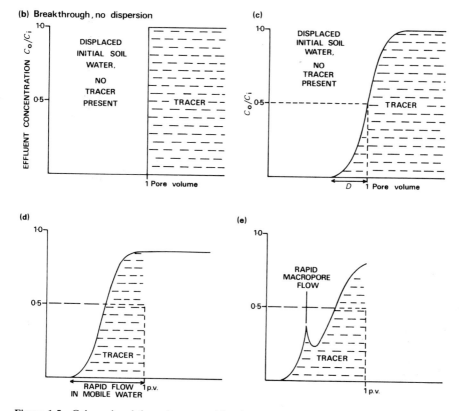

(b) Breakthrough, no dispersion

EFFLUENT CONCENTRATION C_o/C_i

DISPLACED INITIAL SOIL WATER, NO TRACER PRESENT

TRACER

1 Pore volume

(c)

C_o/C_i

DISPLACED INITIAL SOIL WATER, NO TRACER PRESENT

TRACER

D 1 Pore volume

(d)

TRACER

RAPID FLOW IN MOBILE WATER 1 p.v.

(e)

RAPID MACROPORE FLOW

TRACER

1 p.v.

Figure 1.5 Column breakthrough curves: (a) column; (b) no dispersion; (c) dispersion (D = dispersion); (d) rapid flow; (e) macropore flow.

natural rain events, but the results imply that bypassing of soil peds and flow in larger soil pores may increase with increasing rainfall intensity.

Germann and Beven (1981a, 1981b) and Beven and Germann (1981) use a capillary potential (ψ) boundary of -1.0 cm (in units of length, equivalent to energy per unit weight) to indicate the boundary condition between macropore and micropore water. A soil moisture characteristics curve showing the relationship between water potential (ψ) and volumetric water content (θ) can be prepared for a soil, and values of θ greater than the intercept point of -1 cm are assigned to the macropore domain. The volume of this domain in a saturated soil core can be assessed by lowering the head applied to $\psi = -1$ cm and measuring the outflow volume. These authors suggest that transition from micropore to macropore flow occurs when the micropores are saturated or when their infiltration capacity is exceeded. They also suggest that the division between the micropore and macropore domains is arbitrary, and that there may be a third domain in the soil matrix providing a pathway of soil-water flow distinct from the low conductivity micropores.

The actual incidence of macropore flow appears to be related to rainfall intensity (Bouma *et al.* 1981, Bouma & De Laat 1981). However, these observations and those of Beven and Germann were for cracking clay soil where the peds are clearly defined and their infiltration capacity is low. Preferential flow has also been studied in soils where the peds and macropores are less well defined.

Smettem and Trudgill (1983) present evaluations of fluorescent dyes for the identification of water transmission routes in soils. The fluorescent green dye, Lissamine FF, facilitated the observation of cases where dye concentration in microsamples (C_m) differed from that in bulk samples (C_b). If $C_m = C_b$, then uniform flow has occurred; otherwise, differential flow has occurred, with some routes being stained more than other parts of the soil. Preferential flow is identified using the ratio

$$R = C_m / C_b \qquad (1.6)$$

where R is the dye recovery ratio. In a soil with a clay content of $22-24\%$, R values equal to or less than 1 were found most commonly in microsamples from the interior of peds or in apedal soils, whereas values greater than 1 were found in macropores, fissures, cracks and root channels (Smettem & Trudgill 1983). For the same soil type, Trudgill *et al.* (1983a) noted that outflow of surface-applied dyes from *in situ* soils was often in advance of predictions based on uniform displacement. Also, most outflow events could be predicted using 3.6 mm h^{-1} as the maximum pedal acceptance rate of water; rainfall intensities greater than this led to outflow in all but one case. Thus, it was proposed (Trudgill *et al.* 1983b) that the incidence of preferential flow can be predicted when

$$PI/S_p > 1 \qquad (1.7)$$

where PI is the precipitation intensity (mm h^{-1}) and S_p is the sorption (infiltration) rate of the soil peds (mm h^{-1}). The excess of precipitation over pedal sorption capacity was calculated by

$$P_c = P - C_p \qquad (1.8)$$

where P_c is the precipitation excess over C_p (mm), P is the precipitation (mm) and C_p is the pedal sorption capacity (mm). These may all be evaluated on an hourly or per storm basis. Depth of surface-applied solute penetration is predicted by

$$d_f = P_c / \theta_f \qquad (1.9)$$

where d_f is the penetration depth of preferential flow (cm), P_c is as above, but measured in cm, and θ_f is the operational volume of preferential flow (cm^3 cm^{-3}). θ_f can be predicted either using estimates of macropore volume (Beven & Germann 1981) or from tracer breakthrough curves, or it may be calculated from observed tracer penetration (L), where

$$\theta_f = P_c / L \qquad (1.10)$$

Smettem *et al.* (1983) used this approach to predict the movement of surface-applied nitrate fertiliser through soils to streams under conditions of intense rainfall.

The geomorphological application of this work is suggested in Figure 1.6. Here, a bimodal flow model is used where, for an initial rainfall, mobile water passes rapidly through the soil preferentially and contains a low solute level. Retained water may tend to proceed to solid−solvent equilibrium and be displaced in a subsequent rainfall. It is possible to predict (a) the incidence of the rapid flow and (b) the volume of rapid flow, from the treatments described above. Prediction of solute concentrations remains more problematical, as noted in the earlier discussion in this section.

The task is to quantify all the parameters shown in Figure 1.6, and then to aggregate them over time to present a quantitative denudation model for solutional processes in soils. Given intense rainfall and well-structured soils, leaching will be less efficient than uniform flow models would predict because of structural bypassing. Leaching losses will probably originate under these conditions from only a proportion of the soil mass, rather than the whole mass, with solute supplies also derived from diffusion through static water. Preliminary estimates of macropore volumes are around $0.01 - 0.1$ cm^3 cm^{-3} of the soil, and of the mobile soil-water fraction, approximately $0.2 - 0.3$ cm^3 cm^{-3}. Thus the soil mass participating in active leaching could be a fraction of the whole mass, and of these orders of magnitude. This would suggest, in turn, that surface lowering rates would be lower than those predicted by uniform flow models. Such implications are presently untested, but preferential flow in soils

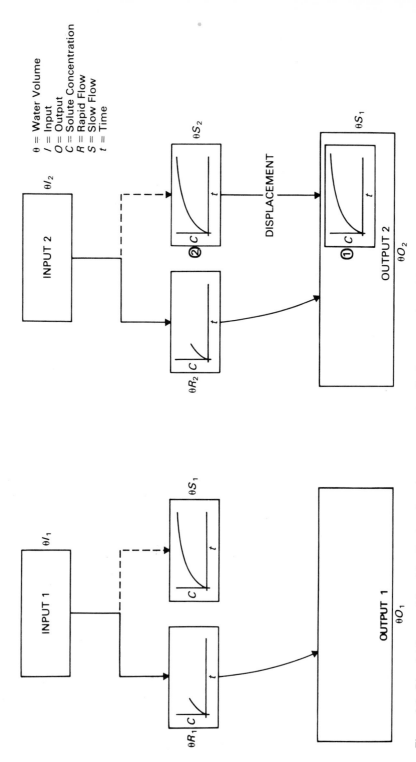

Figure 1.6 Bimodal flow model and implications for solute movement in soils. Input water is partitioned into rapid flow (*R*) and slow flow (*S*).

is clearly an important and neglected aspect of solute movement in soils, should not be ignored in geomorphological studies, and may need to be incorporated in predictive models of landform development in the future.

Patterns of solute removal from hillslopes

Solute output from the hillslope base
Young (1971, p. 46) notes that 'it is remarkable that so little attention has been given to solution loss as a denudational process'. Despite a later observation that solution loss probably accounts for 10−50% of the total removal of rock material from a slope (Young 1971, p. 88), his review of solutional denudation on hilislopes lasts for only one page. Gerrard (1981) provides a similarly brief account. Statham's (1977) slightly longer consideration of solute transport through soils relates particularly to weathering products found in single soil profiles, and to solute transport in rivers. Carson and Kirkby (1972) provide a more extensive survey of solutional denudation on hillslopes, although once again there is a lack of *direct* field evidence obtained on hillslopes, except from single soil profiles. It is clear, therefore, that there is little field evidence available on the *spatial* distribution of solutional erosion on hillslopes.

Numerous studies have sampled subsurface runoff at the base of hillslopes, commonly with the hydrological objective of relating temporal variations in the dissolved load of slope runoff to solute patterns observable in the local stream. Sometimes, hillslope runoff processes have been inferred from such evidence (Walling & Foster 1975, Pilgrim *et al.* 1979), although Anderson and Burt (1982) note the difficulty of using chemical mixing models to identify precise runoff mechanisms. Burt *et al.* (1983) relate the generation of a delayed discharge peak to enhanced nitrate leaching from the Slapton Wood catchment, Devon, UK (Fig. 1.7). Using maps of soil moisture for a major hillslope hollow, the convergence of soil water into its lower part can be associated both with the secondary rise of stream discharge several days after rainfall, and with a significant rise in nitrate concentration of stream water. However, such evidence is circumstantial in the absence of direct observation of soil water over the hillslope during the runoff event. Similarly, Burt (1979) noted the corres-pondence between soil-water convergence into a hillslope hollow at Bicknoller Combe, Somerset, UK, and the concurrent rise in total dissolved load of both stream and throughflow water. Additional leaching from the hollow associated with major runoff events (volumetrically), coupled with the continued presence of a saturated wedge in the hollow, may account for the erosional development of the hollow relative to its adjacent spurs, where production of saturated subsurface runoff is short lived (Burt 1979). Again, however, confirmation of such ideas and further information on the distribution of solutional erosion up the slope profile, demand supportive evidence from the hillslope. Although chemical denudation rates can be estimated from observations at the hillslope base or catchment outlet (e.g. Foster 1979), and distinctive subcatchment units

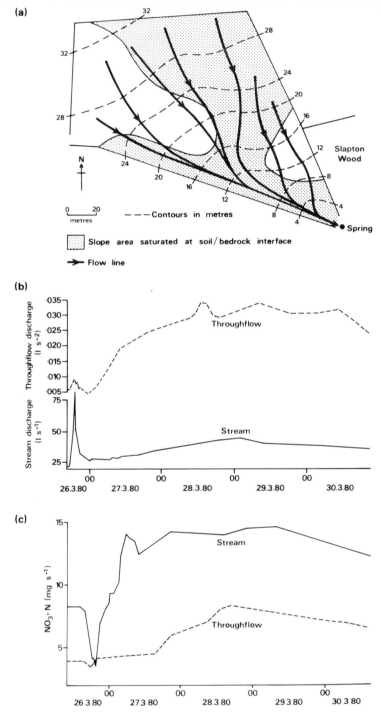

Figure 1.7 Patterns of (a) soil saturation and water movement on a hillslope; (b) stream and throughflow discharge; and (c) nitrate concentration in stream and throughflow water, for the experimental catchment at Slapton Wood, Devon, UK (from Burt *et al.* 1983).

(often defined by land use) can be differentiated, it has not been possible to provide information on the solute yield of different hillslope sections.

Patterns of solute uptake on the hillslope

Direct measurements of solutional denudation may involve two timescales. Short-term variations in the solute concentration of soil water may be investigated using suction samplers (sometimes termed suction lysimeters). The use, efficiency and zone of influence within the soil of such samplers has been reviewed by Cryer and Trudgill (1981) and by Crabtree (1981). Longer-term estimates of the *relative* pattern of solutional denudation may be achieved using weight loss techniques (Trudgill 1975, Crabtree & Trudgill 1984). Rock tablets of known weight are emplaced in the soil in pits, preferably at the soil–bedrock interface. After a period of at least one year, the tablets are exhumed and reweighed. By using a large number of tablets in each soil pit, an average percentage weight loss per tablet may be calculated, and comparing mean values for each pit gives the spatial pattern of relative solutional denudation. Absolute erosion rates are not provided, because of preferential solution of fresh-cut rock faces, and site disturbance during tablet emplacement. Crabtree (1981) discusses a study employing both methods, on a wooded hillslope on Magnesian Limestone at Whitwell Wood, South Yorkshire, UK (Trudgill *et al.* 1981). The dominant direction of water flow in the soil was vertical, except at the slope base where solute-rich groundwater moved laterally towards the stream. The experiment did not unequivocably test the hillslope hydrology control of spatial patterns in solutional denudation, because of soil changes upslope. At the top, the parent material of the Elmton soil series (acid clay glacial drift) differed from that of the midslope Aberford soils (fluvioglacial and solifluction material). Thus the availability of solutes may be controlled as much by initial conditions as by the distribution of hydrological processes. Analysis of soil water showed that solute uptake decreased upslope, implying a consequent upslope increase in solutional denudation at the soil–bedrock interface. Micro-weight loss tablets confirmed that relative rates of solutional denudation increased upslope. These results accord with the Carson and Kirkby (1972) model of slope evolution on humid–temperate limestone slopes (see next section), by implying slope decline. However, the parent materials combine with the prevailing hydrological conditions to control the pattern of solutional denudation of the hillslope in this example.

Modelling solutional denudation on the hillslope

Carson and Kirkby (1972) proposed a conceptual model of solutional denudation on hillslopes in which it is assumed that infiltrating rainwater comes rapidly to equilibrium with the soil solutes. The soil water is also subject to evaporation so that if the concentration of any solute reaches saturation, evaporation loss may cause precipitation of excess solute within the soil. The model requires soil moisture content to increase down slope. Evaporation is assumed to be directly proportional to soil moisture, and so also increases

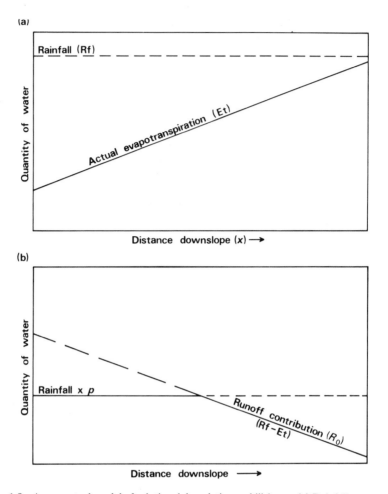

Figure 1.8 A conceptual model of solutional denudation on hillslopes: (a) Rainfall, evaporation and unit runoff; (b) patterns of solutional denudation in relation to rainfall and runoff (from Carson & Kirkby (1972).

down slope (Fig. 1.8a). The quantity of each solute removed at any point on the slope is taken to be equal to the rainfall input (assumed constant over the slope) multiplied by the proportion of the oxide, p, present in the soil, or to the product of the net unit runoff times 1.0 (i.e. the solute reaches its saturated concentration following evaporation), whichever is less (Fig. 1.8b). The model requires that lateral subsurface runoff occurs, in order to produce a downslope increase in soil moisture, and any (solute) precipitation must therefore involve a downslope transfer of solutes. Without lateral flow, all solute transport would be vertical and only parallel slope retreat would be possible. Assuming an initially straight slope with solution acting alone on the slope, Carson and Kirkby show that three types of variation in solutional denudation down slope can be distinguished:

(a) Rainfall .p < runoff for the whole slope; solute removal is constant over the hillslope, and parallel retreat will occur.
(b) Rainfall .p > runoff for the whole slope; solute removal decreases down slope, and slope decline will occur.
(c) An intermediate case with solute removal constant at the top of the slope, but then decreasing down slope in the lower part of the slope. In this case parallel retreat occurs in the upper slope, with slope decline below.

For a limestone, the proportion, p, of calcium carbonate present is almost 1.0, so that case (b) applies whatever the climate (Fig. 1.9a). For igneous rocks, the pattern of solutional denudation progresses from case (b), through (c) to (a) as the ratio of runoff to rainfall increases in more humid climates (Figs 1.9b & c).

Despite the simple attraction of Carson and Kirkby's model, few field experiments have deliberately attempted to verify its predictions. Crabtree's (1981) experiment (see above) is ambiguous because of upslope changes in parent material. Crabtree and Burt (1983) used rock tablets at Bicknoller Combe to investigate solutional denudation on the instrumented hillslope studied by

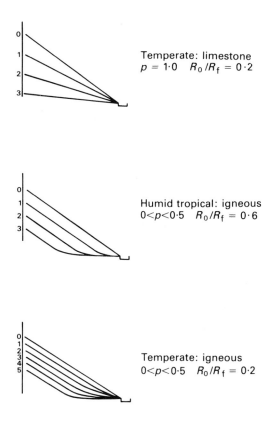

Temperate: limestone
$p = 1{\cdot}0$ $R_0/R_f = 0{\cdot}2$

Humid tropical: igneous
$0 < p < 0{\cdot}5$ $R_0/R_f = 0{\cdot}6$

Temperate: igneous
$0 < p < 0{\cdot}5$ $R_0/R_f = 0{\cdot}2$

Figure 1.9 Generalised patterns of hillslope evolution under solutional denudation for selected rock types and climatic regimes (from Carson & Kirkby 1972).

Anderson and Burt (1978) and Burt (1979). Tablets prepared from Old Red Sandstone weighed about 12.5 g, with a diameter of 31 mm and a thickness of 7 mm. Weight loss of 0.2 − 0.6% occurred over 15 months. Ten tablets were placed in each of 20 pits dug 4 to a slope profile on 5 profile lines; 1 up the centre of the hollow and 2 on each side of this spaced 10 m apart. The pits were 15, 43, 75 and 115 m up slope from the stream. Table 1.1 shows the results obtained. Solutional denudation increased up slope, perhaps related to the general upslope increase in soil acidity. Weight loss was greatest in the upper hollow, though least in the lower hollow. This reflects the continued presence of solute-rich soil water in the lower hollow, whilst in the upper hollow enhanced solutional denudation may have been caused by the rapid convergence of large quantities of short residence time acid throughflow into the hollow from the adjacent spurs, and from the interfluve area. The results suggest that the hydrological contrast between spur and hollow will continue to develop within the general pattern of hillslope decline.

Table 1.1   Changes in mean rock tablet weight loss for the instrumented hillslope at Bicknoller Combe (modified from Crabtree & Burt 1983).

	Distance upslope from stream (m)	Profile	Mean percentage tablet weight loss
upslope changes	115		0.42±0.17
	75		0.38±0.13
	43		0.35±0.08
	15		0.30±0.08
slope profile variations		downstream spur	0.34±0.07
		downstream spur flank	0.35±0.07
		hollow	0.42±0.09
		upstream spur flank	0.33±0.07
		upstream spur	0.31±0.15

Table 1.2   Solute budgets and erosion rates for slope units in the lower basin of East Twin Brook (modified from Finlayson 1977).

	Slope category	Total loss (kg yr^{-1})	Erosion rate (mm 1000 yr^{-1})
upslope	5°	18.39	1.002
upslope	5°−10°	38.20	1.002
upslope	10°−15°	62.24	1.002
upslope	15°−20°	23.22	1.026
downslope	20°	93.34	3.537

Finlayson (1977) calculated a solute budget for different slope categories in the lower basin of the East Twin Brook, Somerset, UK (see Weyman 1973). Distinguishing slow and rapid subsurface runoff, and knowing the source areas and solute concentrations of these two runoff types, he was able to show that the convex slope profile studied would be accentuated by continued solutional denudation (Table 1.2). A similar pattern and rate of erosion was indicated by the use of weight loss tablets. The results are consistent with the Carson and Kirkby model. Using a simple hillslope runoff model, it can be shown that soil moisture will decrease down slope on a convex slope and unit runoff will increase, assuming an evaporation loss proportional to soil moisture storage, and the existence of lateral flow (Burt & Butcher 1984). If solutional denudation is controlled by runoff rather than by rainfall input (Fig. 1.8b), the erosion rate will also increase down slope, as observed by Finlayson (1977) on a convex slope. Such patterns of runoff and solutional erosion will have important implications for the distribution of soil types on the hillslope (see next section).

The Carson and Kirkby model represents one simple case where the solute concentration of infiltrating water is assumed to come rapidly to equilibrium. As indicated in the hillslope hollow at Bicknoller, soil water may still be 'aggressive' even when it has begun to flow laterally downslope. Burt (in press) has discussed the possible patterns of solutional erosion resulting from incorporation of other assumptions about soil-water residence time and solute uptake in a simple hillslope erosion model. However, research in this area remains 'exploratory' (Church 1984), and few data exist which test the Carson and Kirkby model rigorously, although results obtained from the hollow at Bicknoller suggest the need for an alternative model which allows soil water to remain 'aggressive' during its passage down slope. Carefully designed field experiments are needed to test and develop existing models in the light of current concepts of soil hydrology.

Only Huggett (1975) has modelled three-dimensional aspects of solutional denudation, using a fully distributed model. His results demonstrate a solute peak moving as a wave down slope into a hillslope hollow and eventually out into the stream. Whilst not fully developed in the context of hillslope evolution, this model is worthy of further investigation, particularly for its 'event-based' structure. It could be usefully linked with recent hillslope runoff models (e.g. Burt & Butcher 1984) and with soil profile leaching models (e.g. Burns 1974, Addiscott 1977). These models also have the 'isolated' event as their main point of interest, although they could easily accommodate longer timescales.

Relationships with soil properties

Soil properties, solutional denudation, water flow and biological uptake and release of chemical elements are all related to a greater or lesser extent. Thus, the spatial distribution of solutional denudation may be understood in relation to soil properties, given qualifications about biological involvement and hydrological regime. The downslope relationship between varying soil type and solutional denudation patterns (the latter determining the slope profile evolution) is of particular interest. Soil characteristics may be studied to assess whether solutional erosion is greater at the slope foot, mid-slope or slope crest locations. The interaction between soil chemistry and soil hydrological regime will act to constrain or encourage removal of weathering products, and is of considerable importance. In this section, the theoretical patterns of soil—solutional denudation relationships are discussed, and examples of soil distributions on hillslopes are considered.

Soils are often seen as a function of topography, with more leached soils on the slope crest in better-drained conditions, and more solute-rich soils at the base of the slope where mechanical and chemical processes combine to deposit weathering products. However, in the long term, topography can be seen as a function of the distribution of soil types. Two distinct cases can be defined: one in which non-local material such as glacial drift has been deposited over the slope, and one in which the soil has developed *in situ*. In the former case, soil characteristics may become a dominant influence on slope development subsequent to glaciation, with subsoil modification of bedrock form. In the latter case, mutual adjustment of soil characteristics and bedrock form may occur. The various permutations of factors and combinations of slope development models are considered below.

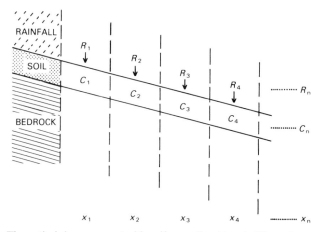

Figure 1.10   Theoretical slope segment with uniform soil and inputs (C = solute concentration).

Consider a soil uniform in chemistry and thickness on a straight initial slope divided into segments x_1, x_2...x_n of equal length (Fig. 1.10). Assuming permeable bedrock and vertically downwards water flux, mineral dissolution rate (dm/dt) will be a function of the concentration in the soil solution (C) and the maximum concentration at saturation (C_s):

$$dm/dt = K(C_s - C) \qquad (1.3)$$

C will tend to approach C_s down the soil profile, resulting in greater dissolution at the top of the profile than at the base, with surface lowering at a maximum near the surface. This pattern will be equal for all segments, leading to parallel retreat of the slope.

For impermeable bedrock, water flow is assumed to be entirely lateral through the soil, and for any one segment (x_n), the value of C will be inherited from the upslope segment (x_{n-1}). Thus solute flux (i.e. mineral weathering, not across-segment transfer) in segment x_n will be:

$$dm_{x(n)} / dt_{x(n)} = K(C_s - C_{x(n-1)}) \qquad (1.11)$$

$C_{x(n-1)}$ will increase down slope until $C_s - C_{x(n-1)}$ approaches zero and dissolution will be negligible. The position on the slope of this zero point will depend on the type of solute, since both K and C_s depend on the specific solute–solvent interaction, and on the nature (volume, velocity) of water flow. In all cases, however, solutional denudation will decrease down slope as $C_s - C_{x(n-1)}$ and acidity decrease, leading to slope decline (i.e. the upper part of the slope is lowered more than the lower).

If C_s is not approached by C during dissolution, because the solid is only slowly soluble and/or of low solubility, dissolution may increase down slope because water flux will increase down slope:

$$Q_{x(n)} = R_{x(n)} + Q_{x(n-1)} \qquad (1.12)$$

where R is the rainfall input and Q is the discharge for a slope segment. Whether or not the solute flux (i.e. mineral dissolution, dm/dt) increases or decreases down slope will depend on the dissolution ratio, R_d:

$$R_d = Q_n / (dm_n / dt_n) \qquad (1.13)$$

If R_d is high, C will not approach C_s and solute removal per segment will tend to increase down slope in proportion to increased water flow. If R_d is low, C approaches C_s and solute removal per segment will tend to decrease down slope, irrespective of water flow, because the soil solution will already be saturated with respect to the solute in question.

The pattern of downslope solutional denudation may be reinforced or counteracted by downslope patterns of inherited soil deposition. This could be because of drift deposition, periglacial colluvial accumulation, or surface wash processes, for example. Surface-acting processes, such as slope wash, may bring more weathered topsoil to the slope foot, whilst other processes may bring less weathered bedrock material to the slope foot (e.g. scree accumulation). Thus the pattern of downslope increase in solutional denudation described above may be reinforced if more solute-rich material is present at the slope foot, or counteracted if the more solute-rich material is present near the slope crest.

In practice, most soil sequences show a downslope increase in solutes or weatherable mineral content from, for example, a slope crest podzol, through a brown podzolic soil to a slope-foot brown earth (e.g. Finlayson 1977, Gerrard 1981). This ignores slope-foot fluvial action, but tends to suggest that slope decline would be the dominant pattern of slope evolution for most soluble material. This would not be the case, however, for the less soluble material where C_s is not reached in hillslope drainage waters and the greater volumes of water down slope lead to increased downslope solutional denudation.

The role of structural bypassing by soil water is not known here, but is liable to reduce solid−solvent contact and minimise the tendency for C to approach C_s. Thus downslope solutional denudation may be more marked and upslope leaching less efficient than a uniform flow model might predict. So where decreasing solutional denudation occurs down slope, the pattern may be less marked than is predicted by uniform flow models, with a less sharp contrast between upslope and downslope positions. This is because upslope water is less equilibrated with the solid phase, allowing solutional potential $(C_s - C)$ to penetrate further down slope.

These models have not been widely tested. Some simulations using a segmented hillslope runoff model (Burt & Butcher 1984) with a leaching component based on a layered soil profile have been attempted. However, the model structure remains simplistic and has not been calibrated or tested in the field. Some model combinations are actually difficult to locate in the field. Many soluble rocks are also permeable (e.g. chalk, limestone) simply because their solubility gives rise to permeability. Thus it may only be possible to consider permeable, soluble rocks and impermeable, or less permeable, less soluble rocks. In addition, many slope forms show strong inheritance from previous erosional regimes, most of which involve mechanical processes such as glacial and periglacial action. This means that testing of such models will be confined to present and future modification of 'fossil' slopes, or to stable forms in equilibrium with a stable climate − if such forms exist.

References

Addiscott, T. M. 1977. A simple computer model for leaching in structured soils. *J. Soil Sci.* **28**, 554−63.

Anderson, M. G. and T. P. Burt 1978. The role of topography in controlling throughflow generation. *Earth Surf. Proc.* **3**, 331−4.

Anderson, M. G. and T. P. Burt 1982. The contribution of throughflow to storm runoff: an evaluation of a chemical mixing model. *Earth Surf. Proc. Landforms* **7**, 565−74.

Atkinson, T. C. and D. I. Smith 1976. The erosion of limestones. In *The science of speleology*, T. D. Ford and R. H. Cullingford (eds), 151−77. London: Academic Press.

Beven, K. and P. Germann 1981. Water flow in soil macropores. II A combined flow model. *J. Soil Sci.* **32**, 15−29.

Bohn, H., B. McNeal and G. O'Connor 1979. *Soil Chemistry* New York: Wiley.

Bouma, J. and P. J. M. De Laat 1981. Estimation of the moisture supply capacity of some swelling clay soils in the Netherlands. *J. Hydrol.* **49**, 247−57.

Bouma, J., L. W. Dekker and C. J. Muilwijk 1981. A field method for measuring short-circuiting in clay soils. *J. Hydrol.* **52**, 347−54.

Burns, I. G. 1974. A model for predicting the redistribution of salts applied to fallow soils after excess rainfall or evaporation. *J. Soil Sci.* **25**, 165−78.

Burt, T. P. 1979. The relationship between throughflow generation and the solute concentration of soil and stream water. *Earth Surf. Proc.* **4**, 257−66.

Burt, T. P. in press. Runoff processes and solutional denudation on humid temperate hillslopes. In *Solutional processes and landforms*, S. T. Trudgill (ed.). London: Wiley.

Burt, T. P. and D. P. Butcher 1984. *A basic hillslope runoff model for use in undergraduate teaching.* Huddersfield Polytechnic, Department of Geography, Occasional Paper No. 11.

Burt, T. P., D. P. Butcher, N. Coles and A. D. Thomas 1983. The natural history of Slapton Ley Nature Reserve, XV: Hydrological processes in the Slapton Wood catchment. *Field Studies* **5**, 731−52.

Carson, M. A. and M. J. Kirkby 1972. *Hillslope form and process.* Cambridge: Cambridge University Press.

Church, M. J. 1984. On experimental method in geomorphology. In *Catchment experiments in fluvial geomorphology*, T. P. Burt and D. E. Walling (eds) 563−80. Norwich: GeoBooks.

Crabtree, R. W. 1981. *Hillslope solute sources and solutional denudation on Magnesian Limestone.* Unpublished Ph.D. thesis, University of Sheffield.

Crabtree, R. W. and T. P. Burt 1983. Spatial variation in solutional dedudation and soil moisture over a hillslope hollow. *Earth Surf. Proc. Landforms* **8**, 151−60.

Crabtree, R. W. and S. T. Trudgill 1984. Two microweight loss techniques for use in hillslope solute studies. *BGRG Tech. Bull.* No. 32.

Cryer, R. and S. T. Trudgill 1981. Solutes. In *Geomorphological techniques*, A. S. Goudie (ed.), 181−95. London: George Allen & Unwin.

Dunne, T. 1978. Field studies of hillslope flow processes. In *Hillslope hydrology*, M. J. Kirkby (ed.), 227−94. London: Wiley.

Finlayson, B. L. 1977. *Runoff contributing areas and erosion.* University of Oxford, School of Geography, Research Paper No. 18.

Foster, I. D. L. 1979. Intra-catchment variability in solute response: an East Devon example. *Earth Surf. Proc.* **4**, 381−94.

Garrels, R. M. and C. L. Christ 1965. *Solutions, minerals and equilibria.* New York: Harper & Row.

Gaudet, J. P., H. Jegat, G. Vechaud and P. J. Wierenga 1977. Solute transfer, with exchange between mobile and stagnant water, through unsaturated sand. *J. Soil Sci. Soc. Am.* **4**, 665−71.

Germann, P. and K. Beven 1981a. Water flow in soil macropores. I An experimental approach. *J. Soil Sci.* **32**, 1−13.

Germann, P. and K. Beven 1981b. Water flow in soil macropores. III A statistical approach. *J. Soil Sci.* **32**, 31−9.

Gerrard, A. J. 1981. *Soils and Landforms.* London: George Allen & Unwin.

Hewlett, J. D. 1961. *Soil moisture as a source of base flow from steep mountain watersheds.* Southeastern Forest Experiment Station, Research Paper No. 132.

Huggett, R. J. 1975. Soil−landscape systems: a model of soil genesis. *Geoderma* **13**, 1−22.

Kirkby, M. J. 1978. Implications for sediment transport. In *Hillslope hydrology*, M. J. Kirkby (ed.), 325−63. London: Wiley.

Kirkby, M. J. and R. J. Chorley 1967. Throughflow, overland flow and erosion. *Bull. Int. Assoc. Sci. Hydrol.* **12**, 5−21.

Knapp, B. J. 1978. Infiltration and storage of soil water. In *Hillslope hydrology*, M. J. Kirkby (ed.), 43−72. London: Wiley.

Lindsay, W. L. 1979. *Chemical equilibria in soils.* New York: Wiley.

Nye, P. H. and P. B. Tinker 1977. *Solute movement in the soil−root system.* Oxford: Blackwell.

Pilgrim, D. H., D. D. Huff and T. D. Steele 1979. Use of specific conductance and contact time relations for separating storm flow components in storm runoff. *Water Res. Res.* **15**, 329−39.

Skopp, J., W. R. Gardner and E. J. Tyler 1981. Solute movement in structured soils: two-region model with small interaction. *J. Soil Sci. Soc. Am.* **45**, 837−42.

Smetten, K. R. J. and S. T. Trudgill 1983. An evaluation of some fluorescent and non-fluorescent dyes for use in the identification of water transmission routes in soils. *J. Soil Sci.* **34**, 45–56.

Smettem, K. R. J., S. T. Trudgill and A. M. Pickles 1983. Nitrate loss in soil drainage waters in relation to by-passing flow and discharge on an arable site. *J. Soil Sci.* **34**, 499–509.

Statham, I. 1977. *Earth surface sediment transport.* Oxford: Oxford University Press.

Thomas, G. W. and R. E. Philips 1979. Consequences of water movement in macropores. *J. Environ. Quality* **8**, 149–52.

Travis, C. C. and E. E. Etnier 1981. A survey of sorption relationships for reactive solutes in soil. *J. Environ. Quality* **10**, 8–17.

Trudgill, S. T. 1975. Measurement of erosional weight-loss of rock tablets. *BGRG Tech. Bull.* No. 17, 13–20.

Trudgill, S. T., A. M. Pickles, T. P. Burt and R. W. Crabtree 1981. Nitrate losses in soil drainage water in relation to water flow rate on a deciduous woodland soil. *J. Soil Sci.* **32**, 433–41.

Trudgill, S. T., A. M. Pickles, K. R. J. Smettem and R. W. Crabtree 1983a. Soil water residence time and solute uptake. 1. Dye tracing and rainfall events. *J. Hydrol.* **60**, 257–79.

Trudgill, S. T., A. M. Pickles and K. R. J. Smettem 1983b. Soil water residence time and solute uptake. 2. Dye tracing and preferential flow predictions. *J. Hydrol.* **62**, 279–85.

Walling, D. E. and I. D. L. Foster 1975. Variations in the natural chemical concentration of river water during flood flows and the lag effect: some further comments. *J. Hydrol.* **26**, 237–44.

Weyman, D. R. 1973. Measurements of the downslope flow of water in a soil. *J. Hydrol.* **20**, 267–88.

Young, A. 1971. *Slopes.* London: Longman.

Zaslavsky, D. 1970. *Some aspects of watershed hydrology.* USDA Research Paper, ARS 41–157.

2

Duricrusts and landforms

A. S. Goudie

Introduction and definitions

The word 'duricrust' was introduced by an Australian geologist (Woolnough 1927) who subsequently defined the term thus (Woolnough 1930, pp. 124–5):

> 'The widespread chemically formed capping in Australia, resting on a thoroughly leached sub-stratum ....The nature of the deposit varies from a mere infiltration of pre-existing surface rock, to a thick mass of relatively pure chemical precipitate ....The mineral matter deposited from solution falls into three main groups: –
> (a) Aluminous and ferriginous
> (b) Siliceous
> (c) Calcareous and magnesian.'

He believed that bedrock control was an important influence on the distribution of these three types, which in effect are broadly equivalent to (a) laterites, bauxites, ferricretes, (b) silcretes and (c) calcretes.

Because of subsequent work on the individual duricrust types, the *crete*-based terminology of which had been laid down by Lamplugh (1907), Goudie (1973, p. 5) proposed a modified definition which resulted from a synthesis of the various definitions that had already been developed for the individual types, and stressed their essentially sub-aerial and near-surface origin and nature:

> 'A product of terrestrial processes within the zone of weathering in which either iron and aluminium sesquioxides (in the case of ferricretes and alcretes) or silica (in the case of silcrete) or calcium carbonate (in the case of calcrete) or other compounds in the case of magnesicrete and the like have dominantly accumulated in and/or replaced a pre-existing soil, rock, or weathered material, to give a substance which may ultimately develop into an indurated mass'

This definition is not greatly at variance with recent definitions of the individual duricrust types. For example, McFarlane (1983a, p. 7) has defined laterites and bauxites as 'Surficial accumulations of the products of vigorous chemical selection developing where conditions favour greater mobility of alkalines, alkali earths and Si than of Fe and Al.' Likewise, Summerfield (1983a, p. 59) has defined silcrete as 'an indurated product of surficial and penesurficial (near-surface) silicification, formed by the cementation and/or replacement of bedrock, weathering deposits, unconsolidated sediments, soil or other materials and produced by low temperature physico-chemical processes and not by metamorphic, volcanic, plutonic or moderate to deep burial diagenetic processes.' Similarly, Watts (1980, p. 663) has suggested the following definition for calcrete:

> '... terrestrial materials composed dominantly, but not exclusively of $CaCO_3$, which occur in states ranging from nodular and powdery to highly indurated and result mainly from the displacive and/or replacive introduction of vadose carbonate into greater or lesser quantities of soil, rock or sediment within a soil profile.'

The purpose of this chapter is not to describe in detail the nature and distribution of duricrusts, information about which can be gleaned from a number of important recent sources (e.g. McFarlane 1976, Summerfield 1978, Netterberg 1980, Bardossy 1981, Watson 1982, Wilson 1983). Its purpose is to examine the geomorphological relationships, palaeoclimatic significance and mechanisms of duricrust formation. Nonetheless, Table 2.1 provides a brief regional guide to some of the most recent literature that describes duricrusts in specific areas.

Duricrusts and landforms

As Table 2.2 indicates, the various duricrust types have a series of important geomorphological effects. These effects result from the thickness of the profiles, the properties of the different components of the profiles (e.g. their occasional ability to harden on exposure) and the topographic situation in which the duricrusts develop.

Ferricrete profiles may frequently be as much as 60 m thick. Calcrete profiles in parts of southern Africa, western Australia and the Texan High Plains may exceed 40 m, while in Zaire and Namibia maximum depths of silicification may also be of the order of 50 m (Veatch 1935).

The hard upper crusts of duricrust profiles form only a limited proportion of the total profile thickness. Typical values for alcrete and ferricrete hardpans are 1 − 10 m (Goudie 1973, Table 18), for calcrete 0.1 − 10 m (with around 0.3 − 0.5 m being most common), while for silcrete values of between 1 and 5 m appear normal.

Beneath the hardpan layer duricrusts display a variety of material types. Ferricretes, for example, often have rather erodible pallid and mottled horizons grading down into more or less coherent bedrock, while calcretes may be

Table 2.1   Recent publications on duricrusts.

Calcrete	
South Africa	Netterberg (1980)
Morocco	Couvreur and Vogt (1977)
France	Vogt (1981)
Namibia	Blümel and Vogt (1981)
Lebanon	Noureddine (1981)
Algeria	Horta (1979)
Spain	Blümel (1982)
Western Australia	Carlisle (1983)
Tunisia	Bonvallot and Delhourie (1978)
Egypt	Hassouba and Shaw (1980)
Puerto Rico	Ireland (1979)
South Australia	Hutton and Dixon (1981)
Silcrete	
Algeria	Smith and Whalley (1982)
Botswana	Summerfield (1982)
South Africa	Summerfield (1983b)
England	Summerfield and Whalley (1980)
Australia	Langford-Smith (1978)
Germany	Wopfner (1983)
France	Rondeau (1975)
Ferricrete and Alcrete	
Uganda	McFarlane (1983a)
Brazil	Esson (1983)
Sierra Leone	Bowden (1980)
India	Subramanian (1978)
Surinam	Aleva (1979)

underlain by friable nodule horizons, and silcretes by kaolinitic clays.

Related to the important geomorphological role of the differences between the properties of hardpans, subhardpan zones, weathered bedrock and bedrock in relatively simple profiles, is the role of alternations of different layers in complex profiles. Not only may there be multiple hardpans of one duricrust type, but in some profiles there may be outcrops of several different types of crust (Fig. 2.1). Summerfield (1978, p. 56), for example, has written:

'In spite of some of the earlier ideas about a "typical" silcrete profile, observations in southern Africa and evidence presented in the more recent literature, in the form of detailed exposure descriptions and profile records, have unequivocably demonstrated that silcrete may occur in a confusing variety of profile relations. Most of the complexity is provided by the often intimate association of silcrete with both calcrete and ferricrete. Furthermore in the Cape coastal zone multiple silcrete horizons occur within clay weathering mantles.'

Table 2.2 Miscellaneous geomorphological effects of duricrusts.

Duricrust type	Effect	Reference
Calcrete	nari detachment lines cap rock	Dan (1962)
calcrete	patterned ground and pseudo-anticlines	Watts (1977)
calcrete	channel geometry	Van Arsdale (1982)
calcrete	karst	Price (1933)
calcrete	suspenparallel drainage	Reeves (1983)
calcrete	rock fracture	Rothrock (1925)
calcrete	flat-irons	Everard (1964)
calcrete	water falls	Vita-Finzi (1969)
silcrete	cuesta, homoclinal or hogback slope forms mesas, buttes	Hutton et al. (1978)
silcrete	resistant carapace of gibbers (stone pavements)	Hutton et al. (1978)
silcrete	sculptured piping forms	Watts (1978)
silcrete	solutional grikes	Twidale and Milnes (1983)
silcrete	sarsen blockstreams	Clark et al. (1967)
ferricrete	slope cambering	Pallister (1956)
ferricrete	mesas with soup-plate form	McFarlane (1969)
ferricrete	reef-like forms	Clare (1960)
ferricrete	bowal pavement	
ferricrete	caves and rock shelters collapse gorges	Bowden (1980)
ferricrete	boxed depressions (baixas)	Vann (1963)
ferricrete	straight pediments	Pallister (1956)
ferricrete	dissolution lakes	Poulouse (1976)

Another general aspect of duricrusts which is relevant to their geomorphological impact is the speed at which they form, and the rapidity with which they may harden on exposure (Table 2.3). Rapid formation helps to preserve otherwise relatively ephemeral landforms (e.g. dunes or alluvial terraces). Quick rates of formation tend to be associated with duricrusts that originate through absolute accumulation rather than relative accumulation (see below). Formation of duricrusts by the weathering *in situ* of bedrock can inevitably be no faster than the time it takes to reach a high degree of decomposition, whereas in unconsolidated weathered material subject to enrichment by outside sources (e.g. dust inputs or lateral water seepage) the rate of development may be considerable. Among observations of rapid formation are those of Alexander (1959) who found bottle tops well cemented into a laterite breccia in Singapore, Pouquet (1966) who noted that in Algeria calcrete crusts had developed over Second World War graves, and Boulaine (1958) who recorded the formation of calcrete in only 75 years beneath a pine forest in central Algeria.

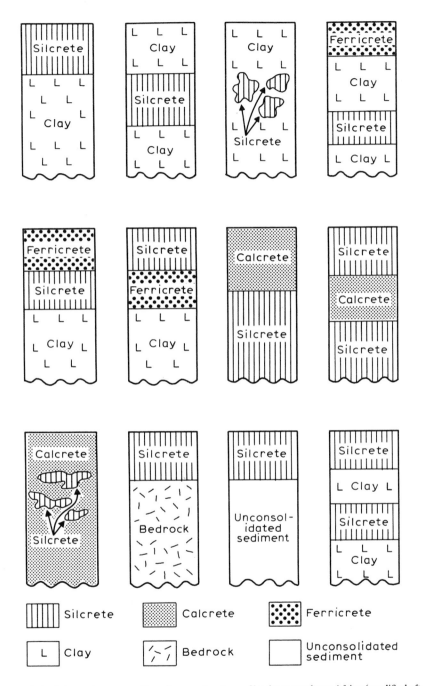

Figure 2.1 Schematic presentation of some silcrete profiles from southern Africa (modified after Summerfield 1978, Fig. 2.10).

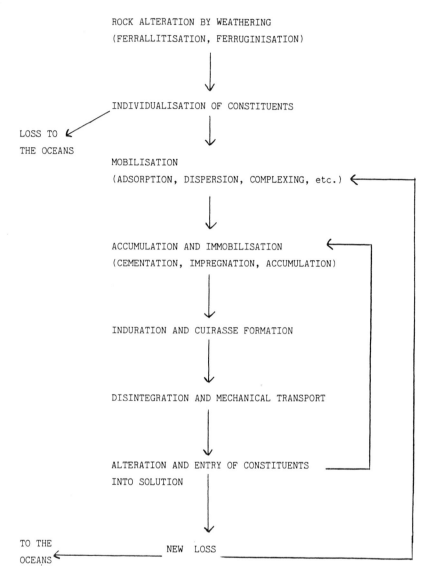

ROCK ALTERATION BY WEATHERING
(FERRALLITISATION, FERRUGINISATION)

INDIVIDUALISATION OF CONSTITUENTS

LOSS TO
THE OCEANS

MOBILISATION
(ADSORPTION, DISPERSION, COMPLEXING, etc.)

ACCUMULATION AND IMMOBILISATION
(CEMENTATION, IMPREGNATION, ACCUMULATION)

INDURATION AND CUIRASSE FORMATION

DISINTEGRATION AND MECHANICAL TRANSPORT

ALTERATION AND ENTRY OF CONSTITUENTS
INTO SOLUTION

TO THE
OCEANS   NEW LOSS

Figure 2.2 Stages in the development of laterites in Guinea (after Maignien 1958).

In spite of these examples of rapid formation it is nonetheless apparent that for some of the great thicknesses of dominant profiles to have developed, a considerable span of time ($10^5 - 10^7$ years) is required, together with a degree of land-surface stability. The Pleistocene was too short and too variable in climate for many of the great duricrust surfaces to have formed, and it may be for these reasons that so many of the world's duricrust formations are of Tertiary age, or even earlier.

Table 2.3 Speeds of duricrust formation.

Type	Method	Approximate rate (mm 1000 yr^{-1})	Source
calcrete (pedogenic)	water balance considerations	0.5 − 350	Netterberg (1981)
calcrete (groundwater)	water balance considerations	10 − 3000	Netterberg (1981)
calcrete (*per descensum*)	from studies of dust inputs	3.3	Goudie (1973)
calcrete	from calculations of deposition from lime-saturated H$_2$O	1.4 − 2.8	Goudie (1973)
ferricrete	from studies of water and rock chemistry	0.4	Trendall (1962)
ferricrete	estimation of silica removal by erosion of plant ash	17	Mikhaylov (1964)
alcrete	estimates of rainfall totals, spring water chemistry and silica content of bedrock	2.9	Cooper (1936)

Thus the silcretes of the Cape coastal zone in South Africa are generally thought to be of Mio-Pliocene age (e.g. Spies *et al.* 1963) while those of Australia are generally placed into one of three periods: the Late Jurassic, Early Cenozoic and Late Cenozoic (Langford-Smith 1978). Likewise much bauxite, on a world-wide basis, is considered to have formed in the Upper Cretaceous, Palaeocene−Eocene and Miocene−Pliocene, with the Eocene deposits being regarded as the most important (Valeton 1972, p. 13). Laterites in Nigeria, Uganda and Australia also tend to have Miocene and Pliocene dates (e.g. de Swardt 1964, Mabbutt 1967, Sombroek 1971).

It is also important to realise that the geomorphological influence of duri-crusts will depend to a considerable degree on the stage of evolution which the feature has reached. This affects both the overall thickness, the nature of the constituents, and the degree of induration. This is brought out in both Maignien's (1958) view of laterite development in Guinea (Fig. 2.2) and in Goudie's model of calcrete development (Fig. 2.3).

Duricrusts and peneplains
In his pioneer study of duricrusts, Woolnough (1927) recognised an association between deeply weathered profiles capped by crusts and widespread planation surfaces. Following Davis (1920, p. 430) he interpreted these surfaces as pene-plains in which chemical weathering and removal in solution are the prime pro-cesses of denudation in the later stages of the cycle of erosion. As Senior and Mabbutt (1979) have pointed out in the context of Australia, some workers

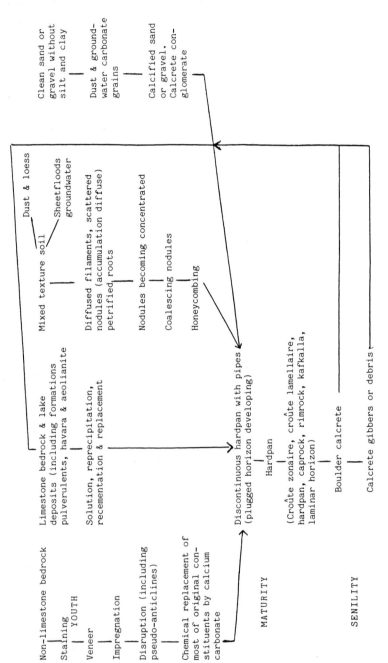

Figure 2.3 Stages in the evolution of calcretes (from Goudie 1975).

accept that the duricrust and the underlying weathering profile are essentially contemporary and that occurrences of this combination over large areas of Australia are also broadly equivalent in age, initially on the basis of an associated planation surface. Regional differences in duricrust character were explained by such secondary factors as lithological control, palaeoclimatic gradient and relief conditions.

However, various doubts about this simple model have arisen, either because there is sometimes an unconformity between the weathered profile and the overlying younger duricrust, or because there may be more than one duricrust in the same area, or because it is possible that some duricrusts may have formed not at the surface, but at depth within a profile of weathering or by selective subsurface induration within unweathered rock sequences.

One consequence of the belief in the association of duricrusts and peneplains was the contention that upland or high-level laterites are essentially 'dead' and being eroded (see, for example, Woolnough 1918). There was also discussion, which had significance in terms of denudation chronology, as to whether the duricrust formed during the later stages of the cycle or whether it formed only after the planation had occurred. McFarlane (1983a, p. 38) has discussed these ideas and has proposed a third possibility:

'... since relative accumulation in the vadose zone is not terminated by lowering of the regional base level of erosion and lowering of the water-table, those pedogenetic laterites which *are* formed in association with erosion surfaces may continue to develop after incision and progressively destroy the topographic evidence for cyclic development of the land surface.'

In her opinion, incision of a laterite surface neither terminates chemical selection nor prevents further movement of the laterite-bearing interface where it survives. In other words a high level laterite need not be 'dead'. Indeed, McFarlane (1983b) has suggested that bauxitisation may result from land-surface rejuvenation:

'The fact that bauxitisation is an extreme stage of lateritisation is difficult to reconcile with the progressively less favourable leaching conditions towards the end of a land-surface reduction cycle and there is growing evidence to indicate that in many cases bauxitisation is initiated by incision of the protore.'

Likewise, bauxitisation appears to be characteristic of areas where drainage is particularly free, such as scarp edges, interfluves and heads of drainage (Fig. 2.4).

Nonetheless, there is little doubt that many duricrusts are associated with low-angle surfaces, whatever their origin. Valeton (1983), for example, has demonstrated that most lateritic bauxites occurred on Cretaceous and Tertiary coastal plains, forming elongated belts, sometimes hundreds of kilometres long, parallel to Lower Tertiary shorelines in India and South America.

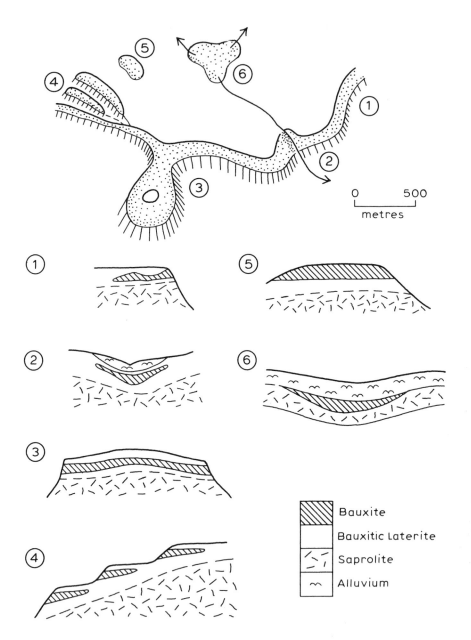

Figure 2.4 Preferred sites for bauxite development in Madhya Pradesh, India (modified after Butt, in McFarlane 1983b, Fig. 21.6):
(1) escarpment; (2) drainage over escarpment;
(3) scarped spur; (4) terraced escarpments;
(5) interfluve; (6) head of drainage and interfluve.

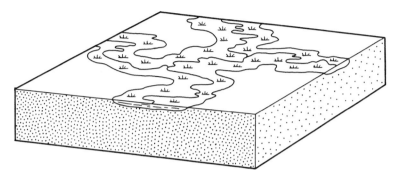

① Silcrete formation in areas subject to inundation
– thickest development near rivers?

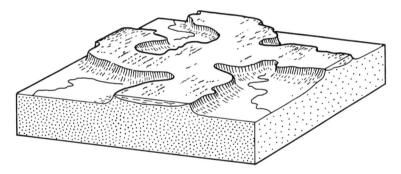

② Rejuvenation - erosion - drainage inversion

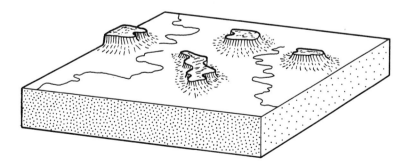

③ Back - wearing to form silcrete - capped residuals

Figure 2.5 Partial area model of formation of silcrete-capped residuals following rejuvenation and scarp retreat (modified after Summerfield 1978, Fig. 7.3).

Relief inversion

Numerous workers have identified the role that duricrusts may play in relief inversion. In the case of laterite, laterite-covered valleys may become ridges or strings of mesas flanking lower, younger valleys, and pediments may become mesas (McFarlane 1976). The relief of laterite surfaces may be modified by pseudo-karstic processes so that the central areas of laterite-capped mesas may become gradually lowered. Thus the periphery stands relatively higher, giving a soup-plate-like form (McFarlane 1969). Likewise the tendency for some calcretes to form preferentially in valleys and depressions sometimes leads to inversion of relief in times of greater erosion, whether by water or wind. Examples of such inverted calcrete relief are provided by Miller (1937) and McLeod (1966).

Summerfield (1978) has also indicated that silcrete can cause relief inversion (Fig. 2.5). In stage 1 of his model silcrete forms in areas subject to inundation and possibly reaches its thickest development in proximity to rivers. In stage 2 rejuvenation of drainage occurs leading to erosion and drainage inversion. Subsequent back-wearing (stage 3) creates silcrete-capped residuals. These may be highly resistant to further destruction by weathering since on a world basis silcretes have a mean silica content of around 96% and may on occasion exceed 99%.

Pseudo-karstic development in duricrust terrains

The presence of duricrust profiles in which there are marked differences in properties between hardpans and some of the more friable and fine-grained materials beneath, creates conditions that favour the formation of pseudo-karstic phenomena produced by subsurface flushing, and in the case of calcrete, solutional effects. Cave formation and roof collapse produce karst-like forms in laterites (Dixey 1920, de Chetelat 1938, Chevalier 1949, McBeath & Barron 1954, Aubert 1963, Thomas 1968, Bowden 1980). Calcretes, because of their high lime contents (on average c. 80%) and relative solubility, frequently show sinkhole development and pipe formation (Price 1933, Gile & Hawley 1966), though many of the closed depressions on calcrete surfaces in the High Plains of Texas may be deflational rather than solutional in origin (Wells 1983).

Palaeoclimatic significance

Many workers have used duricrusts as indicators of past climates, and in broad terms this may be acceptable. Calcretes, for example, are for the most part, though not exclusively, currently forming in semi-arid areas where annual rainfall is around 200 − 500 mm, so that their presence in various Tertiary sediments in western and central Europe (see, for example, Nägele 1962, Freytet 1971) may be used with a fair degree of certainty to infer formerly more desert conditions with an annual water deficit.

Much more controversy surrounds silcrete, however, as indicated by Summerfield (1983b), with a range of inferred climatic conditions ranging from extreme arid (Kaiser 1926) to humid tropical (Whitehouse 1940). Summerfield maintains that silcretes may form under two distinct climatic regimes. He draws a distinction between 'the non-weathering profile' silcretes, which result from localised silica mobility and concentration in high pH environments under a pre-dominantly arid to semi-arid climate, and 'the weathering profile' silcretes whose geochemical and petrographic characteristics are indicative of silicification under a much more humid climate in highly acidic, poorly drained weathering environments. He rejects the notion that silcrete is inevitably diag-nostic of an arid or semi-arid climate.

It is normally accepted that ferricretes and alcretes form under relatively humid conditions. Alternating wet and dry seasons were widely considered to be favourable if not essential to laterite genesis. In particular it was believed that alternating conditions were necessary for sesquioxide precipitation. However, as McFarlane (1976, p. 45) has pointed out, there is some evidence for its for-mation under permanently moist atmospheric conditions.

Origins: some general considerations

To understand the origin and development of these geomorphologically im-portant materials some general considerations need to be borne in mind (Fig. 2.6). Firstly, there is the question of the sources of the materials which contrib-ute to the make-up of duricrusts. The primary elements can be derived from at least four main sources: the weathering of bedrock and sediment, inputs from dust and precipitation, plant residues and the dissolved solids in groundwater. These sources then have to be translocated and concentrated in the geochemical system, either by lateral transfers, or by vertical movements, whether upward (*per ascensum*) or downward (*per descensum*). Thirdly, the transferred ma-terials need to be precipitated, and here a very wide range of processes come into play according to the type of duricrust with which one is concerned. Among the most important of these are changes in chemical equilibria caused by evap-oration, by temperature changes, by pressure changes in the soil air and water systems, by the actions of organisms and by miscellaneous changes caused by interactions of different solution types (the common ion effect, neutralis-ation of strongly alkaline solutions, reactions with cations, etc.).

This three-stage process can be exemplified by Ollier's (1978) model of sil-crete formation:

(a)   weathering releases silica;
(b)   silica is transported either in surface water or groundwater;
(c)   under favourable circumstances silica solidifies and forms silcrete, e.g.
    (i)   in the presence of clean quartz to form nuclei for silica deposition,
    (ii)   where there is a zone of internal drainage where silica from else-where must accumulate (e.g. a playa),
    (iii)   in areas where incoming waters evaporate and so deposit their silica.

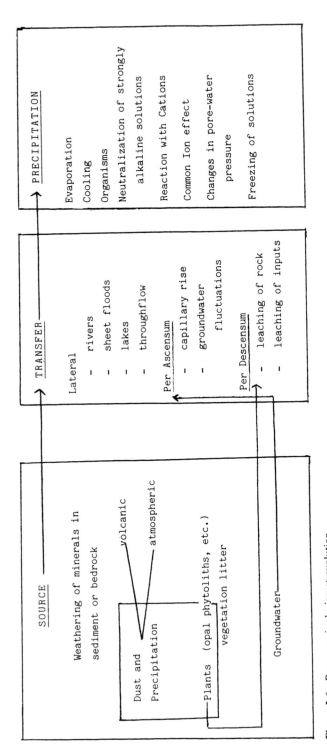

Figure 2.6 Processes in duricrust evolution.

Relative and absolute accumulation

Models for the origin of duricrusts normally fall into one of two categories: those involving relative accumulation and those involving absolute accumulation. Relative accumulations owe their concentrations to the removal of more mobile components, while absolute accumulations owe their concentration to the addition of materials to a profile. However, as McFarlane (1983a, p. 20) has pointed out, the utility of this subdivision depends on important scale considerations. At one extreme the accumulation is entirely relative since laterites would not exist at all were not Fe and Al less readily mobilised than other components during rigorous chemical weathering. At the other extreme, in hand specimens even the residual laterites on interfluves show much addition of Fe, since samples are enriched absolutely in materials which originated above them in the formerly existing column of rock, consumed to provide the residuum.

Furthermore, as Ollier (1978) has pointed out, laterites and silcretes differ in that while ferricretes can result from either relative or absolute accumulation of iron, silcrete can only form by absolute accumulation. Weathering provides the silica and in some cases the material (a weathering profile, for example) in which the silica is deposited.

Lateral and vertical transfer movements

Many of the early models of duricrust formation involved a belief in the power of vertical processes, and especially in the role of capillary rise of solutions from groundwater (e.g. Blake 1902, Woolnough 1918, Fox 1927). Vertical process models of this *per ascensum* type were complemented by *per descensum* models, in which it was believed that material leached from the upper part of a profile would accumulate lower down. Some of the material to be leached downward might be added to the top of the profile in the form of inputs of dust, etc. The dominance of such vertical models resulted from the concern of geologists with stratigraphy and of soil scientists with profiles.

However, in recent decades appreciation of the importance of catenas and toposequences, and of throughflow soil-water movements, has resulted in there being an increasing concern with lateral transfer models. For example, Stephens (1971) has consistently argued that the silcretes of inland Australia formed from silica that was leached during lateritisation in the humid upland areas of the east and then transported by river to low relief areas lying to the west. Similarly the detrital model of calcrete formation (Goudie 1983, p. 115) involves the lateral transportation and redeposition of weathered fragments of calcrete, moving from plateaux surfaces to footslopes.

Furthermore, numerous objections have arisen about the significance of *per ascensum* processes brought about by capillary rise from a water table:

(a)   some duricrusts occur in areas with very low groundwater levels;
(b)   the height over which capillary rise can take place is limited, especially in sandy materials;

(c)   some duricrusts overlie impermeable rocks or very coarse sediments through which capillary rise would be limited;

(d)   the growth of an indurated layer of duricrust, however thin, and also of multiple layers (which are often observed in section), would greatly reduce the rate of capillary rise;

(e)   duricrusts often mantle the topographic surface rather than corresponding to the disposition of the water table.

One slightly unusual explanation for duricrust formation is that proposed for the silcretes of parts of Australia, where, it has been suggested, overlying or adjacent basalt sheets have played a role. Even amongst those who have proposed this association there is little agreement as to whether the supposed basaltic effect has been hydrothermal alteration (Harper 1924), contact metamorphism (Brown 1926), á release of silica from weathered basalt (Gunn & Galloway 1978), or a reduction in the migration of pore waters caused by the presence of a basaltic caprock (Taylor & Smith 1975). Some doubt, however, whether such a special mechanism is justified (e.g. Young & McDougall 1982) for what is such a widespread phenomenon.

The role of organic processes

Another general feature of models of duricrust formation over recent years has been the appreciation of the importance of organic processes. For example, in the case of calcrete, laboratory simulations with microflora (Krumbein 1968), studies of petrography which have revealed calcified organic filaments of soil fungi, algae, actinomycetes and root hairs of vascular land plants (Esteban 1974, Klappa 1978, 1979), and consideration of the role of organically produced hydrogen sulphide in causing precipitation of calcite by gypsum reduction (Lattman & Lauffenburger 1974) have caused the role of organisms to be given the attention they deserve.

In the case of laterite various organic agencies have also been mooted. Termites, for example (see Lee & Wood 1971), may play a role. They do, for instance, transport large quantities of material from depth to the surface, contributing to surface enrichment of the upper horizons of pedogenetic laterites. They may also create tubulo-aveolar cavities (Machado 1982). Micro-organisms could contribute to both mobilisation and precipitation of materials. Thus Chukrov (1981) considers the transition from goethite to haematite in laterite profiles to be the result of iron bacteria activity, and Groudev and Genchev (1978) have reported that desilicifying bacteria could be used to remove combined silica (kaolin) from bauxite. Some experimental verification of such views has recently become available. For example, Heydemann *et al.* (1982) have successfully isolated micro-organisms, *Bacillus*, from samples taken aseptically out of laterites and bauxites from Australia, India and Africa. They have also demonstrated the role they play in the rapid mobilisation of aluminium from aluminosilicates.

Likewise as Lovering (1959) demonstrated, many plants accumulate certain elements in their leaves, thereby potentially contributing to both weathering and concentration of silica.

Conclusion

Duricrusts are widespread features, especially in low latitudes, though relict forms occur in more temperate areas. They have a multitude of geomorphological effects, but their association with peneplains is a subject of some debate. Also controversial is the question of their palaeoclimatic significance. They result from a complex interplay of different source materials, transfer processes and precipitation mechanisms in the surface and near-surface environment. In the past the roles of lateral translocations and organic processes have tended to be neglected.

Acknowledgements

The author is greatly indebted to various colleagues for their frequent inspiration during the preparation of this paper, including Dr Mike Summerfield, Dr Andrew Watson, Dr Marty McFarlane and Professor C. C. Reeves.

References

Aleva, G. J. J. 1979. Bauxitic and other duricrusts in Suriname: a review. *Geol. Mijn.* **58**, 321–36.

Alexander, F. E. S. 1959. Observations on tropical weathering: a study of the movement of iron, aluminium and silicon in weathering rocks at Singapore. *Q. J. Geol Soc. Lond.* **115**, 123–44.

Aubert, G. 1963. Soils with ferruginous or ferralitic crusts of tropical regions. *Soil Sci.* **95**, 235–42.

Bardossy, G. 1981. *Karst bauxites.* Budapest: Akadémiai Kiadó.

Blake, W. P. 1902. The caliche of southern Arizona: an example of deposition by the vadose circulation. *Trans. Am. Inst. Mining Metal. Engrs.* **31**, 220–6.

Blümel, W. D. 1982. Calcretes in Namibia and SE Spain. Relations to substratum, soil formation and geomorphic factors. *Catena* Suppl. 1, 67–82.

Blümel, W. D. and T. Vogt 1981. Croûtes calcaires de Namibie — problemes geomorphologiques et étude micromorphologique. *Récherches Géogr. Strasbourg* **12**, 55–68.

Bonvallot, J. and J.-P. Delhourie 1978. Étude de différentes accumulations carbonatées d'une toposéquence due centre Tunisien (Djebel Semmama). *103 Congrès national des sociétés savantes, Nancy, Sciences*, Vol. IV, 281–9.

Boulaine, J. 1958. Sur la formation des carapaces calcaires. *Trav. Collab. Bull. Serv. Carte Géol. l'Algérie* New Ser. **20**, 7–19.

Bowden, S. J. 1980. Sub-laterite cave systems and other pseudo-karst phenomena in the humid tropics: the example of the Kasewe Hills, Sierra Leone. *Zeitschr. Geomorph.* **NF 24**, 77–90.

Brown, I. A. 1926. Some tertiary formations on the south coast of New South Wales, with special reference to the age and origin of the so-called 'silica' rocks. *J. Proc. R. Soc. New South Wales* **59**, 387–99.

Carlisle, D. 1983. Concentration of uranium and vanadium in calcretes and gypcretes. In *Residual deposits: surface related weathering processes and materials*, R. C. L. Wilson (ed.) 185–95. Oxford: Blackwell.

Chevalier, A. 1949. Points de vue nouveaux sur les sols d'Afrique tropicale sur le degradation et leur conservation. *Bull. Agric. Congo Belge* **40**, 1057–92.

Chukrov, F. V. 1981. On transformation of iron oxides by chemogenic eluvium in tropical and subtropical regions. In *Proceedings of the International Seminar on Laterisation Processes*, Brasil, 11–14.

Clare, K. E. 1960. Roadmaking gravels and soils in Central Africa. *Road Research Laboratory Overseas Bull.* No.120

Clark, M. J., J. Lewin and R. J. Small 1967. The sarsen stones of the Marlborough Downs and their geomorphological implications. *Southampton Res. Ser. Geog.* **4**, 3–40.

Cooper, W. G. G. 1936. The bauxite deposits of Gold Coast. *Bull. Gold Coast Geol Surv.* **7**.

Couvreur, G. and T. Vogt 1977. Sur quelques croûtes calcaires du Haut-Atlas Central (Maroc) et leur signification Geomorphologique. *Récherches Géogr. Strasbourg* **3**, 5–28.

Dan, J. 1962. Disintegration of nari lime crust in relation to relief, soil and vegetation. *Archives Int. Photogrammetine* **14**, 189–94.

Davis, W. M. 1920. Physiographic relations of laterite. *Geol. Mag.* **57**, 429–31.

de Chetelat, E. 1938. Le modèle lateritique de l'ouest de la Guinée francaise. *Rev. Géog. Phys. Géol. Dyn.* **11**, 5–120.

de Swardt, A. M. 1964. Lateritisation and landscape development in parts of equatorial Africa. *Zeitschr. Geomorph.* **NF 8**, 313–33.

Dixey, F. 1920. Notes on lateritization in Sierra Leone. *Geol. Mag.* **57**, 211–20.

Esson, J. 1983. Geochemistry of a nickeliferous laterite profile, Liberdade, Brazil. In *Residual deposits: surface related weathering processes and materials*, R. C. L. Wilson (ed.), 91–9. Oxford: Blackwell.

Esteban, M. 1974. Caliche textures and 'microdium'. *Bol. Soc. Geol Ital.* **92**, 105–15.

Everard, C. E. 1964. Climatic change and man as factors in evolution of slopes. *Geog. J.* **130**, 65–9.

Fox, C. S. 1927. *Bauxite*. London: Crosby, Lockwood.

Freytet, P. 1971. Paléosols résiduels et paléosols alluviaux hydromorphes associés aux dépôts fluviatiles dans le Crétacé Superieur el l'Eocene basal du Languedoc. *Rév. Geogr. Phys. Géol. Dyn.* **13**, 245–68.

Gile, L. H. and J. W. Hawley 1966. Periodic sedimentation and soil formation on an alluvial fan piedmont in Southern New Mexico. *Proc. Soil Sci. Soc. Am.* **30**, 261–8.

Goudie, A. S. 1973. *Duricrusts in tropical and sub-tropical landscapes*. Oxford: Clarendon Press.

Goudie, A. S. 1975. The geomorphic and resource significance of calcrete. *Prog. Geog.* **5**, 79–118.

Goudie, A. S. 1983. Calcrete. In *Chemical sediments and geomorphology*, A. S. Goudie and K. Pye (eds), 93–131. London: Academic Press.

Groudev, S. and F. Genchev 1978. Bioleaching of bauxites by wild and laboratory-bred microbial strains. In *4th International Congress for the Study of Bauxites, Alumina and Aluminium*, Vol. 1, 271–8.

Gunn, R. H. and R. W. Galloway. 1978. Silcretes in south-central Queensland. In *Silcrete in Australia*, T. Langford-Smith (ed.), 51–72. Armidale: University of New England.

Harper, L. F. 1924. Comment in *Bull. New South Wales Geol Soc.* **10**, 12.

Hassouba, H. and H. F. Shaw 1980. The occurrence of palygorskite in Quaternary sediments of the coastal plain of north-west Egypt. *Clay Minerals* **15**, 77–83.

Heydemann, M. T., A. M. Button and H. D. Williams 1982. *Preliminary investigation of microorganisms occurring in some open blanket lateritic silicate bauxites*. Paper presented at IInd International Seminar on Laterisation Processes, São Paulo.

Horta, O. S. 1979. Les encroûtements calcaires et les encroûtements gypseux en géotechnique routière. *B.E.T. Lab. Mechanique Sols, Alger, mémoire Technique* **1**, 105.

Hutton, J. T. and J. C. Dixon. 1981. The chemistry and mineralogy of some South Australian calcretes and associated soft carbonates and their dolomitisation. *J. Geol Soc. Australia* **28**, 71–9.

Hutton, J. T., C. R. Twidale and A. R. Milnes 1978. Characteristics and origin of some Australian silcretes. In *Silcrete in Australia*, T. Langford-Smith (ed.), 19–39. Armidale: University of New England.

Ireland, P. 1979. Geological variation of 'case-hardening' in Puerto Rico. *Zeitschr. Geomorph.* **NP 32**, 9–20.

Kaiser, E. 1926. *Die diamanterwüste Sudwest-Afrikas*. Berlin: Reimer.

Klappa, C. F. 1978. Biolithogenesis of microdium: elucidation. *Sedimentology* **25**, 489–522.

Klappa, C. F. 1979. Calcified filaments in Quaternary calcretes: organo-mineral interactions in the subaerial vadose environment. *J. Sedimen. Petrol.* **49**, 955–68.

Krumbein, W. E. 1968. Geomicrobiology and geochemistry of the Nari lime-crust (Israel). In *Recent developments in carbonate sedimentology in Central Europe*, G. Müller and G. M. Friedman (eds), 138–47. Heidelberg: Springer Verlag.

Lamplugh, G. W. 1907. Geology of the Zambezi basin around Batoka Gorge. *Q. J. Geol Soc. Lond.* **63**, 162–216.

Langford-Smith, T. (ed.) 1978. *Silcrete in Australia*. Armidale: University of New England.

Lattman, L. H. and X. Lauffenburger 1974. Proposed role of gypsum in the formation of caliche. *Zeitschr. Geomorph. Suppl.* **20**, 140–9.

Lee, K. E. and T. G. Wood 1971. *Termites and soils*. London: Academic Press.

Lovering, T. S. 1959. The significance of accumulator plants in rock weathering. *Geol Soc. Am. Bull.* **70**, 781.

Mabbutt, J. A. 1967. Denudation chronology in Central Australia. In *Landform studies from Australia and New Guinea*, J. N. Jennings and J. A. Mabbutt (eds), 144–81. Cambridge: Cambridge University Press.

McBeath, D. M. and C. N. Barron 1954. Report on the lateritic ore deposit at Iron and Warmara Mountains, Berbice. In *Report of the Geological Survey Dept. British Guiana for the Year 1953*, 87–100.

McFarlane, M. J. 1969. *Lateritisation and landscape development in parts of Uganda*. Unpublished Ph.D. thesis, University College London.

McFarlane, M. J. 1976. *Laterite and landscape*. London: Academic Press.

McFarlane, M. J. 1983a. Laterites. In *Chemical sediments and geomorphology*, A. S. Goudie and K. Pye (eds), 7–58. London: Academic Press.

McFarlane, M. J. 1983b. Laterites. In *AGID guide to mineral resource development 1983*, 441–88.

Machado, A. de B. 1982. *The contribution of termites to the formation of laterites*. Paper presented at IInd International Seminar on Laterisation Processes, São Paulo.

McLeod, W. N. 1966. The geology and iron deposits of the Hammersley Range area, western Australia. *Bull. Geol Surv W. Australia*, No. 117.

Maignien, R. 1958. *Le cuirassement des sols en Guinée, Afrique occidentale*. Memoires du Service de la Carte Geologique D'Alsace et de Lorraine, No. 16, 239.

Mikhaylov, B. M. 1964. Part played by vegetation in the lateritisation of mountainous areas of the Liberian shield. *Doklady Akad. Nauk. SSSR* **157**, 23–4.

Miller, R. P. 1937. Drainage lines in bas relief. *J. Geol.* **45**, 432–38.

Nägele, E. 1962. Zur petrographie und Entslehung des Albsteins. *Neues Jahrbuch Geol. Palaeontol. Abh.* **115**, 44–120.

Netterberg, F. 1980. Geology of southern African calcretes: 1. Terminology, description, macro-features and classification. *Trans Geol Soc. South Africa* **83**, 255–83.

Netterberg, F. 1981. Rates of calcification estimated from water balance considerations. In *Abstracts, International Conference on Aridic Soils*, Jerusalem.

Noureddine, L. 1981. Remarques sur les croûtes calcaires dans la Beqa'a meridionale (Liban). *Récherches Géog. Strasbourg.* **12**, 5–12.

Ollier, C. D. 1978. Silcrete and weathering. In *Silcrete in Australia*, T. Langford-Smith (ed.), 13–7. Armidale: University of New England.

Pallister, J. W. 1956. Slope development in Buganda. *Geogr. J.* **122**, 80–7.

Poulouse, K. V. 1976. Dissolution lakes in lateritised landscape and their genetic significance. *Records Geol Surv. India* **107**, 126–33.

Pouquet, J. 1966. *Initiation geopedologique.* Paris: CDU.

Price, W. A. 1933. The Reynosa problem of South Texas and the origin of caliche. *Bull. Assoc. Am. Petrolm. Geol.* **17**, 488–52.

Reeves, C. C. Jr 1983. Pliocene channel calcrete and suspenparallel drainage in West Texas and New Mexico. In *Residual deposits: surface related weathering processes and materials,* R. C. L. Wilson (ed.), 179–83. Oxford: Blackwell.

Rondeau, A. 1975. |Cuirasses|de fer et cuirasses de silice en France occidentale. *Bull. Assoc. Géograph. Francais,* No. 424–5, 161–64.

Rothrock, E. P. 1925. On the force of crystallisation of calcite. *J.Geol.* **33**, 80–2.

Senior, B. A. and J. A. Mabbutt 1979. A proposed method for defining deeply weathered rock units based on regional geological mapping in southwest Queensland. *J. Geol Soc. Australia* **26**, 237–54.

Smith, B. J. and W. B. Whalley 1982. Observations on the composition and mineralogy of an Algerian duricrust complex. *Geoderma* **28**, 285–311.

Sombroek, W. G. 1971. Ancient levels of plinthisation in NW Nigeria. In *Paleopedology: origin, nature and dating of paleosols,* D. H. Yaalon (ed.), 329–36. Jerusalem: Israel University Press.

Spies, J. J., L. N. J. Engelbrecht, S. J. Mahlerbe and J. J. Viljoen 1963. *Die geologie van die gebied tussen Bredasdorp en Gansbaai.* Pretoria: Geological Survey.

Stephens, C. G. 1971. Laterite and silcrete in Australia: a study of the genetic relationships of laterite and silcrete and their companion materials, and their collective significance in the weathered mantle, soil, relief and drainage of the Australian continent. *Geoderma* **5**, 5–52.

Subramaniam, K. S. 1978. How old are the laterites in the Indian peninsula? *J. Geol Soc. India* **17**, 353–58.

Summerfield, M. A. 1978. *The nature and origin of silcrete with particular reference to southern Africa.* Unpublished D.Phil. thesis, University of Oxford.

Summerfield, M. A. 1982. Distribution, nature and genesis of silcrete in arid and semi-arid southern Africa. *Catena,* Suppl. 1, 37–65.

Summerfield, M. A. 1983a. Silcrete. In *Chemical sediments in geomorphology,* A. S. Goudie and K. Pye (eds), 59–91. London: Academic Press.

Summerfield, M. A. 1983b. Silcrete as a palaeoclimatic indicator: evidence from southern Africa. *Palaeogeog. Palaeoclim. Palaeoecol.* **41**, 65–79.

Summerfield, M. A. and W. B. Whalley 1980. Petrographic investigation of sarsens (Cenozoic silcretes) from southern England. *Geol. Mijn.* **59**, 145–53.

Taylor, G. and I. E. Smith 1975. The genesis of sub-basaltic silcretes from the Monaro, New South Wales. *J. Geol Soc. Australia* **22**, 377–85.

Thomas, M. F. 1968. *Some outstanding problems in the interpretation of the geomorphology of tropical shields.* British Geomorphological Research Group Occasional Paper, No. 5, 41–9.

Trendall, A. F. 1962. The formation of 'apparent peneplains' by a process of combined lateritisation and surface wash. *Zeitschr. Geomorph.* NF6, 183–97.

Twidale, C. R. and A. R. Milnes 1983. Slope processes active late in arid scarp retreat. *Zeitschr. Geomorph.* **27**, 343–6l.

Van Arsdale, R. 1982. Influence of calcrete on the geometry of arroyos near Bucheye, Arizona. *Geol Soc. Am. Bull.* **93**, 20–6.

Valeton, I. 1972. *Bauxites.* Amsterdam: Elsevier.

Valeton, I. 1983. Palaeoenvironment of lateritic bauxites with vertical and lateral differentiation. In *Residual deposits: surface related weathering processes and materials,* R. C. L. Wilson (ed.), 77–90. Oxford: Blackwell.

Vann, J. H. 1963. Development processes in laterite terrain in Amapa. *Geog. Rev.* **53**, 406–17.

Veatch, A. C. 1935. Evolution of the Congo Basin. *Mem. Geol Soc. America,* No. 3.

Vita-Finzi, C. 1969. *The Mediterranean valleys.* Cambridge: Cambridge University Press.

Vogt, T. 1981. Croûtes calcaires et sols rouges en France mediterranéenne: trois examples. *Récherches Géog. Strasbourg.* **12**, 13–22.

Watson, A. 1982. *The origin, nature and distribution of gypsum crusts in deserts.* Unpublished D.Phil. thesis, University of Oxford.

Watts, N. L. 1977. Pseudo-anticlines and other structures in some calcretes of Botswana and South Africa. *Earth Surf. Proc.* **2**, 63–74.

Watts, N. L. 1978. Displacive calcite: evidence from recent and ancient calcretes. *Geology* **6**, 699–703.

Watts, N. L. 1980. Quaternary pedogenic calcretes from the Kalahari (southern Africa): mineralogy, genesis and diagenesis. *Sedimentology* **27**, 661–86.

Wells, G. L. 1983. Late-glacial circulation over central North America revealed by aeolian features. In *Variation in the global water budget*, M. Bevan and R. Ratcliffe (eds), 317–30. Berlin: Reidel.

Whitehouse, K. W. 1940. *Studies in the late geological history of Queensland.* Paper, Dept. of Geology, Univ. Queensland, No. 2, 2–22.

Wilson, R. C. L. (ed.) 1983. *Residual deposits: surface related weathering processes and materials.* Oxford: Blackwell.

Woolnough, W. G. 1918. The physiographic significance of laterite in Western Australia. *Geol. Mag.* (Decade 6), 385–93.

Woolnough, W. G. 1927. The duricrust of Australia. *J. Proc. R. Soc. New South Wales* **61**, 24–53.

Woolnough, W. G. 1930. Influence of climate and topography in the formation and distribution of products of weathering. *Geol. Mag.* **67**, 123–32.

Wopfner, H. 1983. Environment of silcrete formation: a comparison of examples from Australia and the Cologne Embayment, West Germany. In *Residual deposits: surface related weathering processes and materials*, R. C. L. Wilson (ed.), 151–66. Oxford: Blackwell Scientific.

Young, R. W. and I. McDougall 1982. Basalts and silcretes on the coast near Ulladulla, southern New South Wales. *J. Geol Soc. Australia* **29**, 425–30.

3

Pre-Quaternary weathering residues, sediments and landform development: examples from southern Britain

C. P. Green

Introduction

In this account the aim is to trace a consistent scheme of pre-Quaternary land-form development for southern Britain (Fig. 3.1). The discussion is necessarily selective and concentrates on the evidence obtained from weathering residues and sediments. In studying landform development, evidence of weathering or soil formation is valuable in two ways. First, weathered rocks and soil horizons may survive *in situ* and so indicate the position and environmental context of palaeosurfaces. Second, weathering products preserved in sediments yield a record of the environmental conditions in the erosional province from which those sediments are derived.

The pre-Quaternary origin of significant elements in the relief of the British Isles has long been accepted. On the Chalk of south-east England, debate has centred on whether the bulk of denudation took place in the Palaeogene (Pinchemel 1954) or in the Neogene (Wooldridge & Linton 1955). On the Palaeozoic rocks of south-west England there is the added possibility that land-forms may be of Mesozoic age (Guilcher 1949, Linton 1951).

The synthesis first published by Wooldridge and Linton in 1939, *Structure, surface and drainage in south-east England*, is fundamental for understanding pre-Quaternary landform development in south-east England. The two essential beliefs underpinning this work are (a) that no significant structural movement occurred in southern England in the later Tertiary, and (b) that the Neogene was occupied by a major erosional episode in which a large part of the denudation of the Chalk was effected, and the present summit relief of the Chalk was developed as an erosional land surface − the Mio-Pliocene Peneplain.

This synthesis, restated in 1955, profoundly influenced work elsewhere in southern Britain, both on the Mesozoic rocks of Wessex (Waters 1960), and on the older, harder rocks of upland Britain in Wales (Brown 1960) and the south-west peninsula (Brunsden 1963). Already in 1954, however, an alternative explanation of Tertiary landform development in south-east England was provided by Pinchemel (1954), who stressed the importance of Palaeogene denudation of the Chalk of southern England. Although the views of Wooldridge and Linton remained influential in the 1960s (Linton 1964), research has gradually moved the balance of opinion towards the Pinchemel model (Small 1964, Clark *et al.* 1967, Hodgson *et al.* 1967, Green 1969, 1974c, Catt & Hodgson 1976), leading to a new geomorphological synthesis for the Tertiary in south-east England by Jones (1980, 1981).

This new model recognises a sub-Palaeocene surface of marine origin beneath the Tertiary sediments of the Hampshire and London Basins, and widely preserved on the Chalk dip-slopes. It also proposes the existence of a Late Palaeogene sub-aerial surface which lay only a little above the present Chalk summits. Around the Weald, the regional focus of this new model, only remnants of the earlier sub-Palaeocene marine surface can be recognised, and the interpretation of the later sub-aerial surface is largely inferential. Further west, in the west of the Hampshire Basin and towards the south-west peninsula, a more complete record of Palaeogene sub-aerial denudation is preserved. Parts of this record have been the subject of recent reinvestigation (Edwards 1973, 1976, Hamblin 1973a, 1973b, Small 1980, Plint 1982), and this area between Salisbury and Dartmoor is the focus of the present account. Jones (1980, p. 43) notes that his own interpretation 'raises questions concerning the reality and age-range of the various "geomorphological staircases" that have been proposed, both in areas adjacent to the Weald and also the harder rock terrains of Upland Britain'. In the western counties of southern England some of these questions can be addressed.

Pre-Tertiary denudation

The culmination of the Upper Cretaceous marine transgression (Hancock 1969) was a turning point of far-reaching significance in the long-term development of landforms in southern Britain (Table 3.1). The most elevated parts of the south-west peninsula may have escaped submergence (Hancock 1969), but over the rest of southern Britain a continuous cover of chalk was deposited. If landforms of Mesozoic age are present in the relief, they have been re-exposed in post-Cretaceous times from beneath this cover. Attention must be given first, therefore, to possible exhumed relics of pre-Cretaceous denudation.

Pre-Cretaceous denudation episodes
There is good evidence for massive denudation of the rocks of south-west England in the Permian and Triassic. The granites of Dartmoor range in age

Table 3.1 Key to stratigraphic terms used. The absolute timescale prior to the Cenozoic is not proportional. Timescale based on Harland *et al.* (1964).

My	Era	Period		Epoch / Stage	Formations and Beds
7.	CENOZOIC	TERTIARY	NEOGENE	Pliocene	Coralline Crag, Lenham Beds; St Erth Beds
26.				Miocene	
37.			PALAEOGENE	Oligocene	Bembridge Marl; Bovey Formation
53.				Eocene	Poole Formation, London Clay, Reading Beds; Bagshot Beds; Aller Gravel, Wooley Grit, Bullers Hill Gravel, Tower Wood Gravel
				Palaeocene	
65.	MESOZOIC	CRETACEOUS	upper	Danian	
70.				Maastrichtian	
76.				Campanian	
82.				Santonian	
88.				Coniacian	
94.				Turonian	
100.				Cenomanian	Combepyne Soil
106.			lower	Albian	Upper Greensand
112.				Aptian	Lower Greensand – Hythe Beds
136.		JURASSIC			Lulworth Beds
195.		TRIASSIC			New Red Sandstone; Budleigh Salterton Pebble Beds
225.		PERMIAN			
280.		CARBONIFEROUS			

from Late Carboniferous to Early Permian (Hawkes 1982) and must originally have been intruded at depths of between 5 and 9 km from the surface. Nevertheless, debris from the Dartmoor igneous complex occurs in sediments dating from the Early Permian, by which time erosion had cut down to the top of the pluton, probably with granite exposed in the bottom of the canyons (Laming 1982). Material from the Dartmoor igneous complex is also present in Triassic conglomerates, and Laming (1982) further suggests that the granite may never have been completely covered by sediments of New Red Sandstone age. Dangerfield and Hawkes (1969) believe that post-Permian denudation of the Dartmoor granite has been relatively limited (50–200 m) and that the present erosion surface may not be much lower with respect to the rocks than that existing at the close of the Permian.

Both Permian and Triassic rocks provide excellent and familiar evidence of geomorphological conditions at the time of their deposition. Breccias, aeolian and fluvial sands, conglomerates and mudstones indicate deposition under semi-arid conditions around the margins of a dissected upland. The preservation in the sediments of limestone clasts and primary feldspars demonstrates the lack of intense chemical weathering. Specific evidence of surface weathering conditions in the Triassic deserts is preserved in the Budleigh Salterton Pebble Beds where pebbles are coated with iron and manganese oxides, suggestive of Triassic 'desert varnish'.

The sub-Permian unconformity outcrops at quite high levels around the Crediton and Tiverton cuvettes (up to 180 and 250 m respectively) and these depositional trenches undoubtedly had expression in Permian times, as is shown by sediment transport trends which indicate movement into them in both cases from both the north and south (Laming 1982). Despite these traces of the Late Palaeozoic land surface and relief, the sub-Permian surface in south-west England is nowhere recognisable beyond its present outcrop. By the end of Triassic times the rocks of south-west England appear to have been reduced to a surface of very low relief. Progressive denudation is reflected in the passage upward in the thick (more than 1000 m) Permian succession in Devon from basal breccias through sandstones to mudstones. Renewed erosion in the Triassic led to the deposition of a second, and similar, sequence of continental red beds.

Hart (1982) adopts the view that south-west England remained a land area for much of the Jurassic. Certainly most of the Jurassic rocks in the Wessex Basin are marine shelf and shallow water sediments. Cosgrove (1975) has described from this area kaolinitic clay mineral assemblages, to which he ascribes a Cornubian origin, in sediments of all ages from Lower Lias to Portlandian. Terrigenous material is, however, scarce in these rocks, so any Jurassic land area in south-west England was evidently of low relief and no land surfaces of Jurassic age can be recognised on the Palaeozoic rocks.

Specific evidence of Jurassic terrestrial weathering in southern England is confined to the Lulworth Beds of the uppermost Portlandian in south Dorset, where carbonaceous shales ('Dirt Beds'), sometimes enclosing fossil tree-stumps,

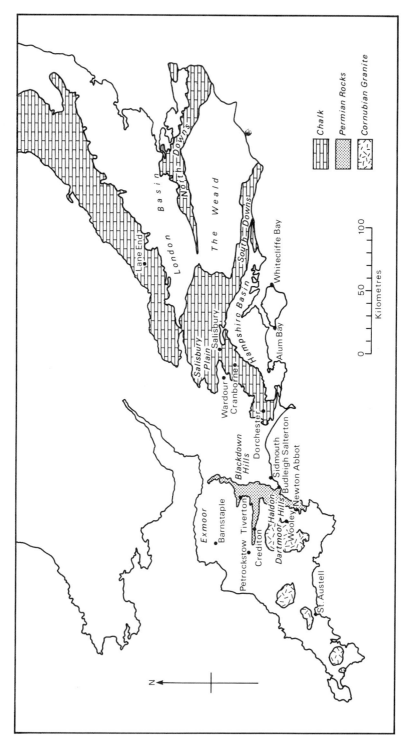

Figure 3.1 Southern England: localities and selected geological outcrops.

represent soils formed during periods of emergence in a coastal lowland of marshes and lagoons very close to sea level. A rich fauna indicates warm climatic conditions, but associated evaporites suggest a tendency towards aridity.

Cretaceous marine transgressions

Very shallow water conditions persisted in southeastern England in the early part of the Lower Cretaceous (Wealden) and the rocks of south-west England almost certainly continued to form part of a substantial but generally low-lying land area. Towards the end of the Lower Cretaceous and in the early part of the Upper Cretaceous, the intensity of erosion appears to have increased in south-west England, probably as a result of uplift. Arenaceous sediments in the Lower and Upper Greensands reflect this intensification. Strongly transgressive conditions are also apparent and pronounced unconformities exist at the base of several formations. For example, the Upper Greensand of Albian–Cenomanian age rests on an eroded surface of earlier rocks down at least to the Devonian limestone of the Newton Abbot area (Brunsden *et al.* 1976). The Chalk is itself transgressive across earlier Cretaceous rocks, but nowhere in southern England is it found resting directly on pre-Albian rocks.

There are several important interpretative points which must be considered in any attempt to decipher the geomorphological significance of the Upper Cretaceous (Chalk) transgression:

(a)   For large parts of south-west England, the preservation of flint in residual deposits or incorporated in later sediments is the only evidence of a former more extensive cover of Chalk.

(b)   The presence of flint suggests that the Chalk was of similar facies to that preserved in south-east England.

(c)   Microfossils preserved in residual flint indicate that Chalk ranging in age from mid-Turonian to mid-Campanian was originally present.

(d)   Whatever the character of the relief across which the transgression extended, details of the sub-Chalk surface must have been shaped by marine processes. Where the Upper Cretaceous rocks overlie earlier Mesozoic sediments, the surfaces of unconformity are indeed effectively plane, but on the harder rocks of the south-west peninsula planation may have been incomplete. Marine transgression does not necessarily mean complete marine planation, and benches on the flanks of the Dartmoor granite have been explained (Simpson 1964) as the product of transgressive Upper Cretaceous seas.

(e)   The evidence does not show whether the submergence of Cornubia was complete.

Based on these interpretative assumptions, several conclusions are possible. First, the survival without modification of terrestrial surfaces and associated pedological features of pre-Chalk age in the present topographical landscape of south-west England is unlikely. Second, despite this conclusion, the gross morphology of the region could well predate the deposition of the Chalk and be the product of Cretaceous or earlier Mesozoic denudation. Finally, if the ancient rocks of south-west England were deeply weathered in pre-Chalk times, weathered rock could have survived the Upper Cretaceous transgressions and a pre-Chalk date for the kaolinised granites of the south-west peninsula might be considered. Indeed, a Cretaceous or pre-Cretaceous age has been suggested for similar kaolinisation elsewhere in Europe (Millot 1970, Lidmar-Bergström 1982, Esteoule-Choux 1983).

Palaeogene denudation

The Chalk of south-west England
Residual flint (Tower Wood Gravel) is known on the summits of the Haldon Hills (Hamblin 1973a), in solution pipes in Devonian limestone near Kingsteignton (Brunsden *et al.* 1976), and at Orleigh Court near Barnstaple in North Devon (Edwards & Freshney 1982). The flints are nodules, or broken pieces of nodules, and being unrolled, represent the most westerly evidence of *in situ* weathering of the Chalk. Further west, flints are an important component of Tertiary and Quaternary gravels at various levels, but their source remains uncertain. Flint appears to be completely absent from the highest ground of the south-west peninsula, but Upper Chalk occurs offshore at no great distance from the coast both in the Western Approaches and in the Bristol Channel (Curry *et al.* 1970, Hamilton 1979). Derivation of flint from offshore Chalk outcrops must therefore be considered as a possible alternative to derivation from a formerly continuous cover of Chalk over the Palaeozoic rocks of Devon and Cornwall, with consequent tectonic implications.

The traces of chalk sedimentation in south-west England throw some light on the problem of dating the denudation of the Chalk:

(a) Chalk sedimentation appears to have terminated in south-west England in the mid-Campanian, and this may indicate the end of marine conditions there at that time. Removal of the Chalk by sub-aerial agencies could therefore have begun in the Upper Cretaceous.

(b) Maastrichtian (Upper Cretaceous) and Danian (Lower Palaeocene) sediments are absent in southern England, but offshore they are all pure limestones, consistent with denudation of the Chalk in contemporary land areas.

(c) The clay matrix of the residual Tower Wood Gravel consists largely of well-ordered kaolinite (Hamblin 1973b), derived, according to Hamblin, from a hydrothermal source in the Dartmoor granite. Others, however, have suggested a weathering origin (Green 1974b, Isaac 1981), and Isaac (1983b) argues that the clay mineralogy reflects a long period of weathering of Upper Cretaceous rocks, and the corresponding advanced maturity of the residues.

(d) Flint is rare in the Tertiary Bovey Formation which occupies several fault-guided basins in south-west England. The Bovey Formation is in part older than the Upper Eocene (Wilkinson *et al.* 1980, Edwards & Freshney 1982), so the denudation of the bulk of the Chalk and the removal of residual flint deposits cannot be later than this.

The Palaeozoic rocks of south-west England

The Bovey Formation provides useful evidence for reconstructing the Palaeogene geomorphological environment. It contains kaolinitic clays derived from the weathering of granite and Palaeozoic sediments, together with alteration products from the granite (Bristow 1969), indicating that these basement rocks were apparently quite extensively exposed in Eocene times.

Further evidence for dating the erosion of the basement rocks is present in Tertiary gravel deposits from south Devon (Table 3.2). Resting on the Tower Wood Gravel on the Haldon summits, and directly on the Upper Greensand in the Kingsteignton pipes and around the margin of the Bovey Basin are flint-rich gravels (Bullers Hill Gravel of Hamblin 1973b) which represent a fluvial reworking of the residual Tower Wood Gravel. Small amounts of Palaeozoic debris may indicate renewed exposure of the basement, or the material may be

Table 3.2 Representative analyses of Palaeogene gravels (6.3–25.4 mm).

	Flint	Upper Greensand Chert	Quartz	Silicified Purbeck Limestone	Other	Number of samples	Average sample size
Palaeocene Reading Beds: transgressive marine – Salisbury Plain	99.7	0.0	0.2	0.0	0.1	2	1092
Eocene Poole Formation: fluvial							
– Dorset	62.2	6.5	26.7	0.0	4.5	2	308
– Salisbury Plain	59.0	10.9	26.4	0.5	3.2	3	453
Plateau Gravels west of Dorchester ?Poole Formation	20.8	2.8	63.4	1.0	12.0	6	447
Aller Gravel: Kingsteignton pipe	9.9	0.0	71.1	0.0	19.0	1	1291
Bullers Hill Gravel							
– Haldon Hills	83.4	·0.1	11.6	0.0	4.8	1	1094
– Kingsteignton pipe	93.5	0.0	3.8	0.0	2.7	1	735

derived from marine pebble beds of Cretaceous age. These flint-rich gravels are the earliest Tertiary sediments in the area. In the Kingsteignton pipes they are overlain by the Aller Gravels (Edwards 1973) in which flint is relatively much less important. The main part of the Bovey Formation appears to overlie these gravel deposits.

Kaolinite in the Bovey Formation, including the Bullers Hill Gravel, is less well ordered than that in the Tower Wood Gravel and is associated with illite and mixed-layer clays. Isaac (1983b) argues that this suite of clay minerals reflects a weathering phase separate from, and of shorter duration and/or lesser intensity than, the phase responsible for the Tower Wood Gravel. *In situ* deep weathering profiles (Bristow 1969) yielding secondary minerals similar to those in the Bovey Formation are present beneath Oligocene Bovey Formation sediments on the margin of the Petrockstow Basin, and confirm the existence of a deeply weathered terrain in south-west England during the Palaeogene.

Isaac (1983b) places the formation of the Tower Wood Gravel and its equivalents in the Lower Palaeocene Danian and regards this as an episode of intense lateritic weathering during which any former cover of chalk was reduced to a mantle of weathering residues. There is in fact no reason why this weathering of the Chalk should not have begun even earlier, in the Upper Cretaceous. Further east, in the Hampshire Basin, there is good evidence for at least partial removal of the Chalk before the deposition of the Upper Palaeocene Reading Beds. The flint-rich Bullers Hill Gravel seems likely, therefore, to be at least as old as the Reading Beds and possibly older. The deposition of the bulk of the Bovey Formation could then have occupied the Eocene and part or all of the Oligocene. Throughout this interval, sediment was derived from the erosion of deep but relatively immature weathering profiles developed on both granite and Upper Palaeozoic metasediments under subtropical or warm temperate climatic conditions.

The Chalk of the western counties

The uplands of east Devon, south Somerset and west Dorset are underlain by weathered Cretaceous rocks, Tertiary residual deposits and Tertiary sediments, all of which provide important insights into the course of weathering and long-term landform development in southern Britain.

Isaac (1981, 1983b) has suggested that in the area north of Sidmouth, residual flint gravels with a matrix of kaolinitic clay (Peak Hill Gravel) are the periglacially disturbed remnants of lateritic Palaeocene weathering profiles developed on Cretaceous rocks. In a few places Isaac identifies the original lateritic profile (Combpyne Soil) in an undisturbed state, displaying characteristic pallid and mottled zones and an overlying red-earth horizon. Elsewhere in the same area, decalcification of the Blackdown facies of the Upper Greensand had previously been explained by Tresise (1960) in terms of weathering. Isaac (1983b) demonstrates that the weathered Greensand here includes similar weathering products to those in the overlying Tertiary residues, and he refers it to the same phase of lateritic weathering.

On the basis of clay mineralogy, Isaac equates these residues with the Tower Wood Gravel. A Danian age is inferred by analogy with lateritic weathering of that age in the Intrabasaltic Formation of Northern Ireland. That the timescale of weathering episodes on the east Devon plateaux was broadly similar to that around the Bovey Basin is demonstrated by the Tertiary sediments found to the north and east of the area examined by Isaac, including those in the main Tertiary outcrop of the Hampshire Basin.

Although flint is the main component in gravels of Upper Palaeocene Reading Beds age, heavy mineralogy (Blondeau & Pomerol 1968), clay mineralogy (Gilkes 1968) and pebble content (Green 1969) indicate that the basement rocks of south-west England were exposed in Reading Beds times. The Reading Beds themselves are succeeded by the London Clay, an early Eocene marine sequence which, even at the western extremity of its outcrop, shows little evidence of close proximity to land. It rests directly on the Chalk at Cranbourne and is shown in the following paragraph to have overstepped the Reading Beds extensively further to the west.

In the Hampshire Basin the London Clay is succeeded by the Poole Formation of fluvial sands and gravels, including the Agglestone Grit which contains, in addition to flint and Upper Greensand chert, significant quantities of quartz and other hard siliceous rocks (Green 1969, Plint 1982). Similar gravels are found (Reid 1896, 1898) on the summits of the Chalk and Upper Greensand as far west as the Blackdown Hills (Table 3.2), and were associated with eastward draining rivers whose catchments extended even further west into areas where the rocks beneath the Chalk were exposed. These Plateau Gravels include, however, well-rounded, chatter-marked flint pebbles which indicate the presence of material of marine origin (Waters 1960). In addition, occasional small pebbles of silicified shelly limestone are usually present, this limestone being similar to that of beds in the Purbeck rocks of south-east Dorset. The widespread distribution of these limestone pebbles further west in association with flint shingle of marine aspect suggests that it was originally carried westward in the beach deposits of the transgressive London Clay sea. A few pebbles of similar silicified limestone have been recovered from gravels in the Reading Beds, while gravels similar to those on the Chalk and Upper Greensand summits to the west of Dorchester are also found on Salisbury Plain and on the summits of the Chalk at the western margin of its outcrop in Wiltshire (Green 1969, 1974c).

The evidence in all these early Tertiary sediments confirms that the basement rocks of the south-west peninsula were exposed before the end of early Eocene times. Further, the Chalk in south-east Dorset had been removed and the exposed Upper Jurassic rocks were already subject to erosion in Late Palaeocene Reading Beds times, and were more widely exposed in Early Eocene (London Clay) times. An extensive erosional surface, refined by Early Eocene marine and fluvial processes, is indicated in the broad region between Salisbury in the east and Dartmoor in the west. Remnants of this surface are still recognisable where the higher summits in this area are underlain by Upper Cretaceous rocks (Small 1980). Areas of Palaeozoic rock further to the west drained towards the Early Eocene seas of southern England, but there is no evidence that the sea, even at its maximum extent in London Clay times, extended further west than the basin of the upper Otter (Green 1974a).

The Chalk of south-east England

The work of the Soil Survey of England and Wales (Catt & Hodgson 1976) has shown fairly conclusively that the Clay-with-flints of south-east England comprises a mixture of the insoluble residue of the Chalk, mainly flint, and the debris of Tertiary sediments, mainly the Reading Beds. This mixing is attributed to processes acting during the Quaternary in the final stages of the exhumation of the Chalk from beneath its cover of Tertiary sediments. It is argued that the Chalk summits and dip-slopes of south-east England represent a dissected sub-Palaeocene surface, and form an up-dip extension of the well-marked unconformity beneath the Reading Beds. Thus the bulk of the denudation of the Chalk in this region occurred before deposition of the Reading Beds. The re-exposure of the present Chalk outcrop to sub-aerial weathering and denudation is placed in the Quaternary. Complete removal of the Chalk up-dip from its present outcrop in south-east England occurred early in the Palaeogene, as is demonstrated by quartzose gravels in the Reading Beds at Lane End (Wooldridge & Ewing 1935), and by the presence of chert pebbles derived from the Hythe Beds of the Weald in the Early Eocene Bagshot Beds of the London Basin (Dewey & Bromehead 1915).

Soils and Silcretes in Palaeogene sediments

The preservation of palaeosols or weathering horizons within the Palaeogene sediments of southern Britain has rarely been recorded. Mottled clays in the Reading Beds are described by Montford (1970) as redistributed ferralitic clays, and by Catt (1983) as containing material (kaolinite) which may well have been formed by tropical weathering. Some features of the Reading Beds are attributed by Ellison (1983) to post-depositional leaching of lagoonal and barrier-complex sediments during phases of temporary emergence in Reading Beds times. Palaeosols have been identified by Buurman (1980) in the Reading Beds at Alum Bay, where the hydromorphic soils with red iron (haematite) compounds are compared to profiles in Surinam and south-east Asia, currently experiencing high annual temperatures and a marked dry season. In the Early Oligocene Bembridge Marls at Whitecliff Bay, Daley (1967) has identified pseudomorphs after gypsum and associated casts of desiccation cracks, both of which he relates to near-surface processes in environmental conditions characterised by high rates of evaporation, such as those now found in semi-arid areas. This evidence suggests the occurrence in Early Oligocene times of weathering environments at least locally drier than those of the Palaeocene and Early Eocene.

Silicified deposits (often termed sarsens) are widespread in southern Britain, and their origin as silcretes resulting from surface weathering under tropical or subtropical climatic conditions has been proposed (e.g. Clark et al. 1967). In south-east England most sarsens are silicified Late Palaeocene (Reading Beds) or Early Eocene sediments, and a Palaeogene date for the silicification process has been preferred by most authors, because the subtropical or tropical climatic conditions generally thought necessary for silica mobilisation are unknown in Europe later in the Tertiary.

Summerfield and Goudie (1980) have shown that although some sarsens in south-east England are similar to pedogenic silcrete in contemporary tropical and subtropical areas, most lack the features associated with formation within the weathering profile itself.

In south-west England siliceous breccias occur in association with the Peak Hill Gravels in East Devon (Isaac 1981, 1983a). Isaac finds pedogenic silcretes to be relatively more common in this area than in the samples from south-east England described by Summerfield and Goudie. Isaac believes that in east Devon, pedogenic and non-pedogenic silcretes are intimately associated with one another, and their further relationship with lateritic weathering profiles indicates a long and complex history of pedogenesis and diagenesis. Evidence for this complexity is present in the Early Eocene (Poole Formation) Plateau Gravels of the same area, in which cobbles of silicified flint breccia are recorded (Waters 1960), and which are themselves silicified to form puddingstone conglomerates (Waters 1960). Elsewhere in Devon, the silicification of the Wooley Grit (Blyth & Shearman 1962), probably equivalent to the Bovey Formation (Edwards & Freshney 1982), may belong to the same complex history.

Overview of Palaeogene geomorphological events

Throughout southern Britain there are consistent indications that the denudation of the Chalk was largely effected under tropical or subtropical conditions before the end of the Palaeocene. In the Late Palaeocene and Early Eocene the Chalk around the western fringe of the Hampshire Basin, and the rocks exposed by the removal of the Chalk both to the west and to the north of the surviving Chalk outcrop in southern England, formed an erosional province in which a surface of low relief was widely developed. During the same interval, the area occupied by the Chalk outcrop in south-east England became an essentially depositional province and was buried beneath the Thanet Sands, the Reading Beds and a substantial thickness of Eocene sediments, derived largely from the northern part of the erosional province previously mentioned (Morton 1982). Towards the end of the Palaeogene (mid-Eocene onwards) the area of marine sedimentation in southern Britain became progressively smaller (Murray & Wright 1974), although there is no evidence for the substantial production of terrigenous sediment. In fact, in the offshore record of Palaeogene sedimentation in the English Channel (Curry et al. 1970), carbonate rocks predominate throughout and form the whole recorded sequence for the Middle and Late Eocene. The volume of Oligocene sediments is small, comprising in addition to the onshore outcrops, only one offshore record – of freshwater limestone. This scarcity of terrigenous sediment when most of southern England formed a land area suggests a Late Palaeogene terrain of low relief close to base level.

Neogene geomorphology in southern England

The sedimentary record

The record of Neogene landform development in southern Britain is tenuous. It is not clear whether this reflects the deficiencies of the record, or a prolonged morphostatic phase. Neogene deposits on land are limited and scattered, and either represent Pliocene near-shore conditions (Coralline Crag, St Erth Beds) or are of uncertain age and environmental significance (Lenham Beds). Neogene sediments are represented offshore around southern Britain by Miocene marine carbonates in the Western Approaches (Curry *et al.* 1970, Evans & Hughes 1984). No deposits of undoubted Pliocene age have been recorded, and there is no indication in this limited depositional record of any significant production of terrigenous sediment by erosion of Neogene land areas.

South-east England

The recognition of a Neogene erosional surface on the Chalk of south-east England (the Mio-Pliocene peneplain) was central to the work of Wooldridge and Linton (1955) on Tertiary landform development. Of more conceptual significance was their belief that the peneplain represented the culmination of a Neogene erosional phase in which a large part of the post-Cretaceous denudation of the Chalk took place. The reality of the Mio-Pliocene peneplain has often been challenged (Pinchemel 1954, Clark *et al.* 1967) and it now seems probable that the denudation of the bulk of the Chalk occurred very early in the Tertiary (Jones 1981).

In the interpretation of the Mio-Pliocene peneplain, Wooldridge and Linton attributed the Clay-with-flints to weathering during its formation. Catt (1983) emphasises that there is no evidence of pre-Quaternary pedogenetic processes in the Clay-with-flints. He concludes that no part of the land surface was inherited without modification from the Tertiary, but notes that the erosion needed to remove evidence of pre-Quaternary soil development was not necessarily very great.

It is implicit in the findings discussed by Catt and Hodgson (1976) and Catt (1983) that in areas now occupied by the Chalk outcrop, Neogene land surfaces were developed across Early Tertiary sediments which protected the Chalk from Neogene weathering and erosion. For the Weald, Jones (1980, Fig. 9) shows the present outcrop of the Chalk in the North and South Downs to have been covered by Palaeogene sediments at the end of the Palaeogene. He indicates negligible Neogene denudation of the Chalk in these areas, but where the Chalk and earlier rocks outcropped on the Late Palaeogene sub-aerial surface in the area lying in front of the present Chalk escarpments, he shows substantial erosion amounting to the complete removal of the Late Palaeogene surface.

In the western counties, by contrast, remnants of the Late Palaeogene surface still survive even in areas where the surface was developed across Chalk and earlier rocks and was not protected by a substantial cover of Palaeogene sediments. In this region the amount of Neogene erosion is either very small, as in east Devon, or is localised in areas such as the Vale of Wardour, in which episodic structural elevation may have continued during the Neogene (Green 1974c). Around the Vale of Wardour, small remnants of a sub-Palaeogene surface mantled with disturbed Palaeogene sediments are preserved on the highest summits. These rise some 40–50 m above the general level of well-marked and more widespread summit conformities which are interpreted (Green 1974c) as Neogene erosional surfaces.

Residual deposits on these surfaces are described by Cope (1976) from a small area near Wilton and appear to be indistinguishable from Clay-with-flints elsewhere in southern England. Heavy mineralogy suggests the presence of Reading Beds material, and the yellowish-red colour of the Clay-with-flints is attributed to weathering under warm, moist climatic conditions in one or more interglacial periods. The Clay-with-flints further west on the same surfaces is less well known but differences may be expected since the Reading Beds, if they were ever present there, were removed early in the Eocene.

The western counties

The problem of identifying evidence for Neogene stages of landform development is also present on the summits underlain by Upper Cretaceous rocks in west Dorset, south Somerset and east Devon. Here, Waters (1960) has argued that a surface occupied by the debris of Early Tertiary sediments (the Poole Formation) has been partly replaced by a surface of Neogene age occupied only by weathering residues. These residues are now in part reinterpreted by Isaac (1981) as being Palaeogene in age (Combpyne Soil, Peak Hill Gravel), although a second phase of soil formation (the Seven Stones Soil) is also recognised by Isaac. This represents a phase of less intense weathering following the breaching and dissection of the Palaeocene weathering profile. Isaac refers the Seven Stones Soil to the Eocene, but a later Tertiary date is not impossible, and might reconcile the views of Isaac and Waters. In either case the modification of the Early Tertiary surface during the Neogene was slight.

Overview of post-Palaeogene geomorphological events

Judging from the evidence from the western counties, burial beneath Palaeogene sediments was only one factor in the survival of Palaeogene landforms during the Neogene. Another key factor appears to have been the relation of the Neogene land surface to base level, especially in so far as this was conditioned by structural activity. It is significant in this respect that in the western counties not only is the development of Neogene landforms localised, but in addition the Chalk there escaped substantial surface lowering by solution during the Neogene. This situation might be explained if rates of surface lowering were much smaller in the Neogene than those indicated by studies of present-day limestone outcrops (Young 1974), but it seems more likely that it arises because the Late Palaeogene land surface remained close to base level during the Neogene except in zones of structural elevation such as the Central Weald and the Vale of Wardour, which experienced the continuing effects of episodic uplift.

If the Neogene was characterised in this way by surfaces of low relief close to base level, the present elevation of the summits of southern England is the result of late epeirogenic movements which seem to be indicated for onshore areas in the offshore record of Cenozoic sedimentation and sea-level behaviour (Evans & Hughes 1984).

Tertiary landform development in south-west England

In the study of long-term landform development on the Palaeozoic rocks of south-west England, the highest summit relief of Dartmoor (Brunsden 1963) and Exmoor (Balchin 1952) is generally interpreted as an Early Tertiary sub-aerial surface surmounted by residual summits rising to an intra-Cretaceous (?sub-Chalk) level (Linton 1951). At lower levels in this landscape, episodes of partial planation are identified and have been referred to the Neogene. In re-examining this interpretation several unresolved problems are apparent:

(a) Although there is evidence of Palaeogene weathering and erosion of the rocks of south-west England in the sediments of the Bovey Formation and in the Tertiary deposits of the Hampshire Basin, there is no sedimentary record of Neogene erosion in either the fault-guided basins of the south-west peninsula itself, in the Hampshire Basin, or in offshore areas.

(b) Relic weathering residues are not consistently related to any of the supposed Tertiary erosion surfaces. On the granites, a superficial layer of weathered rock (growan) a few metres thick is often present, but most of this material on Dartmoor has suffered only partial kaolinisation of the feldspars (Eden & Green 1971). Formation by weathering processes similar to those of today seems likely, and there is no evidence demanding pre-Quaternary weathering or soil formation in the weathered granite.

In several areas more completely kaolinised granite is present. In some cases such kaolinisation is the result of relatively high temperature hydrothermal alteration (Sheppard 1977), although in other cases, the formation of the kaolinite at atmospheric temperatures is indicated and a weathering-profile origin is probable.

Explaining the localisation of the more completely kaolinised weathered granites and establishing the age of the weathering episode(s) is difficult. Localisation may reflect differential susceptibility of the original granites to both hydrothermal and weathering processes. In the St Austell granite advanced decomposition is localised on the lithionite granite forming the western part of the intrusion and is rare or absent on the biotite granite forming the eastern part (Bristow 1969). Bristow (1977) has argued that weathering may be favoured in areas already affected by hydrothermal alteration.

Given the evidence of the nature of successive pre-Quaternary weathering episodes elsewhere in southern Britain, already summarised in this account, it seems most likely that the formation of the intensely kaolinised deep-weathering profiles on the granites of south-west England occurred during the Upper Cretaceous and Palaeocene; a later date seems unlikely, but an earlier date is possible.

(c)   The disposition of Cretaceous and Early Tertiary sediments and weathering residues in relation to the broad pattern of relief on the Palaeozoic rocks in south-west England seems inconsistent with the production of an important part of that relief by erosion in the Neogene. This situation is illustrated in the case of the Dartmoor granite. On the Haldon summits the sub-Albian surface truncates New Red Sandstone rocks at about 230 m. The sub-Chalk unconformity is also present beneath the Tower Wood Gravel, and the level of the summits themselves is determined by an Early Eocene surface associated with the Bullers Hill Gravel. The difference in elevation of nearly 300 m between these surfaces and others thought to be of similar age on the Dartmoor granite requires a gradient of about 10 m km^{-1} between the two localities. Mineralogical evidence (Hancock 1969) showing that Dartmoor contributed little or no detritus to Upper Cretaceous sediments suggests that this difference in elevation did not exist in Upper Cretaceous times.

As a simple regional uplift of western Britain is inadequate to account for the differences in elevation involved, differential post-Cretaceous uplift may be indicated. Vertical movement on the dextral wrench faults of south-west England would provide one suitable structural explanation, and has already been proposed for the Exmoor area by Shearman (1967). Elsewhere in south-west England the low level of surviving Tertiary sediments and residual deposits relative to the higher plateau surfaces suggests that relief elements of broadly similiar Early Tertiary age but widely differing elevation have been separated by faulting, notwithstanding the fault-controlled sites of some deposition.

Pre-Quaternary geomorphological synthesis

Re-evaluation of long-term landform development in southern Britain has become possible in the last 20 years mainly as a result of interest in the associated sediments and weathering residues, rather than in the erosional landforms themselves which were the focus of previous attention. In the present account an attempt has been made to show that evidence preserved in sediments and residual deposits permits a consistent interpretation of landform development for the pre-Quaternary period for the whole region. The sequence of development may be summarised as follows:

(a)   Permo-Triassic erosion effected the primary shaping of the present relief of the Palaeozoic rocks of western Britain.

(b)   Relatively minor modification of Permo-Triassic forms occurred during the Jurassic and Lower Cretaceous.

(c)   The Upper Cretaceous saw the burial of Palaeozoic rocks beneath a cover of Chalk.

(d)   Palaeocene denudation of the Chalk under tropical climatic conditions reduced the Chalk over the Palaeozoic basement to a mantle of weathering residues.

(e)   Late Palaeocene and Eocene erosion then occurred under subtropical climatic conditions. The erosional morphology of the summit relief on the Palaeozoic rocks of south-west England was probably acquired during this interval. Faulting contributed significantly to relief development and may have recurred later in the Tertiary. In south-east England the Chalk acquired the essential features of its present summit morphology before the end of the Palaeocene and was buried quite deeply beneath Tertiary sediments. On the Chalk of the western counties and over the Weald modification of the relief continued in the Eocene and possibly throughout the Palaeogene (Jones 1980). A thin veneer of sediments and weathering residues was repeatedly reworked. The total volume of erosion during this early Tertiary erosional episode was probably nowhere very large.

(f)   A Neogene morphostatic episode then followed. Relief elements that can be confidently referred to the Neogene are virtually absent in southern England. In some areas the Late Palaeogene surface survives without convincing evidence of either erosional encroachment or lowering by solution in the Neogene. In other areas, erosional surfaces have developed at the expense of the Late Palaeogene surface and form the summit relief. They predate the record of Quaternary geomorphological events preserved in the terrace sequences of the main river valleys, and a Neogene age seems to be indicated for them. This differential relief development in the Neogene is most readily explained as the result of localised structural activity.

Conclusion

Research during the last ten years in the western counties of southern England, mainly on Tertiary sediments and weathering residues, confirms the geomorphological model developed for the Tertiary in south-east England during the same period. In particular it confirms that a Late Palaeogene sub-aerial surface, which can only be inferred in the Weald region, survives extensively further west.

It is largely on the basis of evidence in sediments and weathering residues, or through the lack of these materials, that it is possible to distinguish clearly between morphologically similar elements of the summit relief. These include (a) remnants of a sub-Palaeocene marine surface, often occupied by Clay-with-flints incorporating Palaeocene Reading Beds material; (b) remnants of a Late Palaeogene sub-aerial surface occupied either by lateritic weathering profiles or by a thin but complex veneer of Palaeogene fluvial and marine sediments;

and (c) relief of possible Neogene origin from which Palaeogene debris has been completely removed.

The evidence, while showing that the summit relief is not all of the same age and origin, can nevertheless be accommodated in a consistent geomorphological synthesis in which differences in terms of the origin and history of the relief are best explained by differences in the structural context.

References

Balchin, W. G. V. 1952. The erosion surfaces of Exmoor and adjacent areas. *Geogr. J.* **118**, 453–76.

Blondeau, A. and C. Pomerol 1968. A contribution to the sedimentological study of the Palaeogene of England. *Proc. Geol Assoc.* **79**, 441–5.

Blyth, F. G. H. and D. J. Shearman 1962. A conglomeratic grit at Knowle Wood, near Wooley, south Devon. *Geol. Mag.* **99**, 30–2.

Bristow, C. M. 1969. Kaolin deposits of the United Kingdom of Great Britain and Northern Ireland. *23rd Int. Geol. Congr. (1968)* **15**, 275–88.

Bristow, C. M. 1977. A review of the evidence for the origin of the kaolin deposits in SW England. In *Proceedings of the 8th Kaolin Symposium and Meeting on Alunite*, Madrid – Rome, K2, 1–19.

Brown, E. H. 1960. *The relief and drainage of Wales*. Cardiff: University of Wales Press.

Brunsden, D. 1963. The denudation chronology of the River Dart. *Trans. Inst Brit. Geogs* **32**, 49–63.

Brunsden, D., J. C. Doornkamp, C. P. Green and D. K. C. Jones 1976. Tertiary and Cretaceous sediments in solution pipes in the Devonian limestone of South Devon, England. *Geol. Mag.* **113**, 441–7.

Buurman, P. 1980. Palaeosols in the Reading Beds (Palaeocene) of Alum Bay, Isle of Wight. *Sedimentology* **27**, 593–606.

Catt, J. A. 1983. Cenozoic pedogenesis and landform development in southeast England. In *Residual deposits: surface related weathering processes and materials,* R. C. L. Wilson (ed.), 251–8. Oxford: Blackwell.

Catt, J. A. and J. M. Hodgson 1976. Soils and geomorphology of the Chalk in SE England. *Earth Surf. Proc.* **1**, 181–93.

Clark, M. J., J. Lewin and R. J. Small 1967. The sarsen stones of the Marlborough Downs and their geomorphological significance. *Southampton Res. Ser. Geog.* **4**, 3–40.

Cope, D. W. 1976. Soils in Wiltshire 1. Sheet SU03 (Wilton). *Soil Survey Record* No. 32. Harpenden: Soil Survey of England and Wales.

Cosgrove, M. E. 1975. Clay mineral suites in the post-Armorican formations of South West England. *Proc. Ussher Soc.* **3**, 243.

Curry, D., D. Hamilton and A. J. Smith 1970. Geological and shallow subsurface investigations of the Western Approaches to the English Channel. *Instn. Geol Sci. Lond. Rep.* **70/3**, 1–12.

Daley, B. 1967. Pseudomorphs after gypsum in the Bembridge Marls. *Proc. Geol Assoc.* **78**, 319–24.

Dangerfield, J. and J. R. Hawkes 1969. Unroofing of the Dartmoor granite and possible consequences with regard to mineralisation. *Proc. Ussher Soc.* **2**, 122–31.

Dewey, H. and C. E. N. Bromehead 1915. The geology of the country around Windsor and Chertsey. *Mem. Geol. Survey UK.*

Eden, M. J. and C. P. Green 1971. Some aspects of granite weathering and tor formation on Dartmoor, England. *Geogr. Ann.* **53A**, 92–9.

Edwards, R. A. 1973. The Aller Gravels, Lower Tertiary braided river deposits in South Devon. *Proc. Ussher Soc.* **2**, 608–16.

Edwards, R. A. 1976. Tertiary sediments and structures of the Bovey Basin, south Devon. *Proc. Geol Assoc.* **87**, 1–26.

Edwards, R. A. and E. C. Freshney 1982. The Tertiary sedimentary rocks. In *The geology of Devon*, E. M. Durrance and D. J. C. Laming (eds), 204–37. Exeter: Exeter University Press.

Ellison, R. A. 1983. Facies distribution in the Woolwich and Reading Beds of the London Basin, England. *Proc. Geol Assoc.* **94**, 311–20.

Esteoule-Choux, J. 1983. Kaolinitic weathering profiles in Brittany: genesis and economic importance. In *Residual deposits: surface related weathering processes and materials*, R. C. L. Wilson (ed.), 33–8. Oxford: Blackwell.

Evans, C. D. R. and M. R.Hughes 1984. The Neogene succession of the South Western Approaches, Great Britain. *J. Geol. Soc. Lond.* **141**, 315–26.

Gilkes, R. J. 1968. Clay mineral provinces in the Tertiary sediments of the Hampshire Basin. *Clay Min.*7, 351–61.

Green, C. P. 1969. An early Tertiary surface in Wiltshire. *Trans Inst. Brit. Geogs.* **47**, 61–72.

Green, C. P. 1974a. Pleistocene gravels of the River Axe in south western England, and their bearing on the southern limit of glaciation in Britain. *Geol Mag.* **111**, 213–20.

Green, C. P. 1974b. The Haldon Gravels of south Devon. *Proc. Geol Assoc.* **85**, 293–4.

Green, C. P. 1974c. The summit surface on the Wessex Chalk. In *Progress in geomorphology*, E. H. Brown and R. S. Waters (eds), 127–38. Inst. Brit. Geogs. Spec. Publn. 7.

Guilcher, A. 1949. Aspects et problèmes morphologiques du massif de Devon–Cornwall comparés à ceux d'Armorique. *Rev. Geog. Alpine* **37**, 689–717.

Hamblin, R. J. O. 1973a. The clay mineralogy of the Haldon Gravels. *Clay Min.* **10**, 87–97.

Hamblin, R. J. O. 1973b. The Haldon Gravels of south Devon. *Proc. Geol Assoc.* **84**, 459–76.

Hamilton, D. 1979. The geology of the English Channel, south Celtic Sea and continental margin, South Western Approaches. In *The north-west European shelf seas. 1. Geology and sedimentology,* F. T. Banner, M. B. Collins and K. S. Massie (eds), 61–87. Amsterdam: Elsevier.

Hancock, J. M. 1969. Transgression of the Cretaceous sea in south west England. *Proc. Ussher Soc.* **2**, 61–83.

Harland, W. B., A. Gilbert Smith and B. Wilcock (eds) 1964. *The Phanerozoic timescale.* London: Geological Society.

Hart, M. B. 1982. The marine rocks of the Mesozoic. In *The geology of Devon*, E. M. Durrance and D. J. C. Laming (eds), 179–203. Exeter: Exeter University Press.

Hawkes, J. R. 1982. The Dartmoor granite and later volcanic rocks. In *The geology of Devon*, E. M. Durrance and D. J. C. Laming (eds), 85–116. Exeter: Exeter University Press.

Hodgson, J. M., J. A. Catt and A. H. Weir 1967. The origin and development of Clay-with-flints and associated soil horizons on the South Downs. *J. Soil Sci.* **18**, 85–102.

Isaac, K. P. 1981. Tertiary weathering profiles in the plateau deposits of East Devon. *Proc. Geol Assoc.* **92**, 159–68.

Isaac, K. P. 1983a. Silica diagenesis of Palaeogene kaolinitic residual deposits in Devon, England. *Proc. Geol Assoc.* **94**, 181–6.

Isaac, K. P. 1983b. Tertiary lateritic weathering in Devon, England and the Palaeogene continental environment of south west England. *Proc. Geol Assoc.* **94**, 105–15.

Jones, D. K. C. 1980. The Tertiary evolution of south-east England with particular reference to the Weald. In *The shaping of southern England*, D. K. C. Jones (ed.), 13–47. London: Academic Press.

Jones, D. K. C. 1981. *The geomorphology of the British Isles: south east and southern Britain.* London: Methuen.

Laming, D. J. C. 1982. In *The geology of Devon*, E. M. Durrance and D. J. C. Laming (eds), 148–78. Exeter: Exeter University Press.

Lidmar-Bergström, K. 1982. Pre-Quaternary geomorphological evolution in southern Scandinavia. *Sveriges geologiska undersökning, Ser. C*, No. 785.

Linton, D. L. 1951. Problems of Scottish scenery. *Scot. Geog. J.* **67**, 65–85.

Linton, D. L. 1964. Tertiary landscape evolution. In *The British Isles: a systematic geography*, J. Wreford-Watson and J. B. Sissons (eds), 110–30. London: Nelson.

Millot, G. 1970. *Geology of clays*. New York: Springer-Verlag.

Montford, H. M. 1970. The terrestrial environment during Upper Cretaceous and Tertiary times. *Proc. Geol Assoc.* **81**, 181–204.

Morton, A. C. 1982. The provenance and diagenesis of Palaeogene sandstones of southeast England as indicated by heavy mineral analysis. *Proc. Geol Assoc.* **93**, 263–74.

Murray, J. W. and C. A. Wright 1974. Palaeogene Foraminiferida and palaeoecology, Hampshire and Paris basins and the English Channel. *Spec. Pap. Palaeontology,* Special paper, No. 14.

Pinchemel, P. 1954. *Les plaines de Craie du nord-ouest du Bassin Parisien et du sud-est du Bassin de Londres et leur bordures*. Paris: Armand Colin.

Plint, A. G. 1982. Eocene sedimentation and tectonics in the Hampshire Basin. *J. Geol Soc. Lond.* **139**, 249–54.

Reid, C. 1896. The Eocene deposits of Dorset. *Q. J. Geol Soc. Lond.* **52**, 490–6.

Reid, C. 1898. The Eocene deposits of Devon. *Q. J. Geol Soc. Lond.* **54**, 234–8.

Shearman, D. J. 1967. On Tertiary fault movements in north Devonshire. *Proc. Geol Assoc.* **78**, 555–91.

Sheppard, S. M. F. 1977. The Cornubian batholith, SW England: D/H and $^{18}O/^{16}O$ studies of kaolinite and other alteration minerals. *J. Geol Soc. Lond.* **133**, 573–91.

Simpson, S. 1964. The supposed 690ft marine planation in Devon. *Proc. Ussher Soc.* **1**, 89–91.

Small, R. J. 1964. Geomorphology. In *A survey of Southampton and its region*, F. J. Monkhouse (ed.), 37–50. Southampton: British Association.

Small, R. J. 1980. The Tertiary geomorphological evolution of south-east England: an alternative interpretation. In *The shaping of southern England*, D. K. C. Jones (ed.), 49–70. London: Academic Press.

Summerfield, M. A. and A. S. Goudie 1980. The sarsens of southern England: their palaeoenvironmental interpretation with reference to other silcretes. In *The shaping of southern England*, D. K. C. Jones (ed.), 71–100. London: Academic Press.

Tresise, G. R. 1960. Aspects of the lithology of the Wessex Upper Greensand. *Proc. Geol Assoc.* **71**, 316–39.

Waters, R. S. 1960. The bearing of superficial deposits on the age and origin of the upland plain of east Devon, west Dorset and south Somerset. *Trans. Inst. Brit. Geogs.* **28**, 89–97.

Wilkinson, G. C., R. A. B. Bazley and M. C. Boulter 1980. The geology and palynology of the Oligocene Lough Neagh Clays, Northern Ireland. *J. Geol Soc. Lond.* **137**, 65–76.

Wooldridge, S. W. and C. J. C. Ewing 1935. The Eocene and Pliocene deposits of Lane End, Bucks. *Q. J. Geol Soc. Lond.* **91**, 293.

Wooldridge, S. W. and D. L. Linton 1939. *Structure, surface and drainage in south-east England*. Inst. Brit. Geogs, 10.

Wooldridge, S. W. and D. L. Linton 1955. *Structure, surface and drainage in south-east England*. London: Philip.

Young, A. 1974. The rate of slope retreat. In *Progress in geomorphology*, E. H. Brown and R. S. Waters (eds), 65–78. Inst. Brit. Geogs, Special Publn. 7.

4

Soil erosion and landscape stability in southern Iceland: a tephrochronological approach

A. J. Gerrard

Introduction

The extremely dynamic nature of the geological and geomorphological processes operating in Iceland provides an unrivalled natural laboratory within which to study a variety of features. It is not surprising that the volcanic, glacial and periglacial features have attracted most geomorphological attention but Iceland also offers unique possibilities for the study of general extraglacial landscape change. Mass movement and gullying are widespread on most slopes and wind erosion is important over much of the landscape, and is of increasing concern to many of the inhabitants (Runólfsson 1976). The frequent volcanic eruptions, followed by the deposition of tephra, provide a valuable chronological sequence within which to examine landscape change. An additional advantage of working in Iceland is that the written record since the time of first settlement around AD 890 is excellent, and most of the historic volcanic eruptions can be dated precisely. Many of the tephra falls associated with these eruptions are distinctive and they trap and preserve the soils, sediments and landforms on which they fall. This chapter examines a number of such buried sequences in Eyjafjallasveit, southern Iceland; the sites examined in detail are Seljaland, Thórsmörk and Laufatungur (Fig. 4.1). The area is sufficiently close to the most frequently active volcanoes Hekla and Katla to allow the principles of tephrochronology to be used to interpret landscape change. Some of these principles are now examined.

Tephrochronology

Bishop Gisli Oddson of Skalholt, south Iceland, is thought to be the first author to mention the volcanic ash layers that occur in Icelandic soils (Thórarinsson

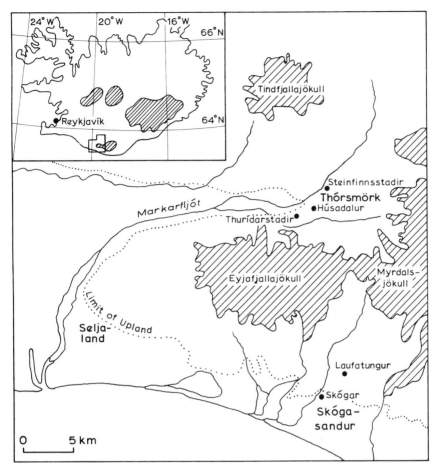

Figure 4.1 General location of the study area.

1980). In his *De mirabilibus Islandiae* (On the wonders of Iceland) written in 1638, he states that in Icelandic humus soils there are thick ash layers separated by humus layers and that portions of trees are embedded in the lower layers. Unwittingly, the Bishop was describing the characteristics of ash falls which later were to develop into the procedure known as tephrochronology. Sigurdur Thórarinsson, in his doctoral thesis, was the first to promote tephra as an important tool in volcanological, glaciological and geomorphological research (Thórarinsson 1944). Thórarinsson chose the name tephra as a collective term for all pyroclasts because he was unhappy with the inconsistent use of the term ash. Furthermore, ash was also often used to denote a certain grain-size fraction of the volcanic ejection. Therefore, he proposed that the vague classification with regard to grain size into dust, ash, sand, lapilli, etc. should be replaced by classification into groups using the standard grain-size scale. From this beginning, tephrochronology has now developed into a valuable and consistently used tool in all landscapes that have experienced tephra falls (Westgate & Gold 1974).

There have been about 200 volcanic eruptions in Iceland in Postglacial time, with between 30 and 40 having occurred since settlement began (Thórarinsson 1979). Tephra layers are more extensive and frequent than would be expected in a mainly basalt-producing area because many of the volcanic sites are under the ice caps and are more explosive than they would otherwise be (Thórarinsson 1980). The basaltic tephras are generally dark coloured, often black, whereas the more acidic eruptions produce a lighter coloured tephra. There are 12 widespread light coloured tephras and they are easiest to trace across the landscape. However, great care must be taken in using colour alone as a means of identification because of the effects of subsequent weathering changes and because the composition and colour of the tephra can change within the course of a single major eruption.

The oldest tephra accurately dated through written sources is that produced by the Hekla eruption of AD 1104. The tephra falls since then can be dated reasonably accurately and the major eruptions to have affected southern Iceland in the historic period are listed in Table 4.1. One of the most important stratigraphic horizons is provided by the dark olive-green tephra called the Landnám ash because the eruption occurred approximately at the time of settlement (Landnám). Thórarinsson (1970) has also called it the 'settlement niveau' and it is thought to have originated from a site in the vicinity of Eldgja. It was dated originally at AD 934 (Hammer et al. 1980) but it is now thought to be nearer AD 890.

The complete sequence of tephra falls is very rarely encountered in the deposits of a particular region. Some will have been removed by subsequent erosion and the most complete sequences are found in sediment sinks such as peat bogs, but the main determinant of tephra availability is the wind direction and strength at the time of the eruption. Thus, one of the prehistoric eruptions of Hekla (H₄, c. 4000 BP) was carried mostly to the north. The tephra is about

Table 4.1 Historic volcanic eruptions that affected Eyjafjallasveit, south Iceland.

Volcano	Date of eruption					
Hekla	1104	1158	1206	1222	1300	1341
	1389	1510	1597	1634	1639	1766
	1845	1947	1970	1980	1982	
Katla	c.900	c.1000	1179	1245	1262	1311
	c.1357	1416	c.1490	1580	1612	1625
	1660	1721	1755	1823	1860	1918
Eyjafjallajökull	1612?	1821				
Landnám (Eldgja?)	c.890					

200 cm thick in Thórsárdalur, to the north-west of the volcano, but it is only 0.5 cm to the south near the Markar river. Also, the AD 1357 eruption of Katla deposited 14 cm of tephra on Skeidararsandur but only 1−2 cm on Skógasandur, 15 km to the west.

Tephras are identified in the field by their colour, thickness, grain size and stratigraphic relationships and can be analysed in the laboratory for their mineralogical, physical and chemical characteristics. The establishment of stratigraphic relationships is made easier because the tephra layers are widely separated by wind-blown material. Also, if the landscape is stable and there is sufficient time between eruptions, well-developed soils will form and may be buried by the next tephra fall. However, the identification of such buried soils or palaeosols can be extremely difficult (Valentine & Dalrymple 1976). It is the nature of the soils themselves that produces the greatest problems, namely the ephemeral nature and susceptibility to change of many soil features and the similarity of many soils to sediments (Gerrard 1981). This is especially true of Iceland where climatic conditions slow down the rate of soil formation and where surface erosion is so prevalent. This problem was encountered by Bruce (1973) in New Zealand where A and B soil horizons had been removed prior to the deposition of succeeding loess. Also, many tephra falls are not sufficient to kill the surface vegetation and are incorporated into the developing soil as was certainly the case with the tephra associated with the 1947 eruption of Hekla. The methodology adopted by Butler (1959) and Walker (1962) in identifying sequences of ground surfaces and alternating stable and unstable phases on hillslopes can be used with great success in Iceland. The precise dating of many of the former surfaces by tephrochronology gives this approach even greater scope and has been used by many workers to describe, in general terms, the extent of soil erosion in Iceland.

Soil erosion in Iceland

Soil erosion is a major problem in Iceland. Thoroddsen (1914) has described the phenomenon of *mistur*, which is the yellowish-brown cloud of dust that is carried to the remote quarters of the island. It has been estimated that more than half of the vegetation cover present at the time of settlement has been destroyed and the fertility of the remaining soil greatly reduced (Bjarnasson 1978). The natural climax vegetation of birch forest has been reduced to about 1000 km^2 from an original estimated cover of 40 000 km^2, and about half the area below 400 m (20 000 km^2) is now practically devoid of soil (Thórarinsson 1970). The evidence for soil erosion in historic times, and its effects, have been described in detail by Thórarinsson (1962).

His extensive work, based on tephrochronology, has consistently produced soil thickening curves as represented in Figure 4.2. Much of the evidence has been obtained from sections in eroded remnants of vegetation and soil known as 'rofbards'. However, rofbards are much higher (thicker) than the soil cover

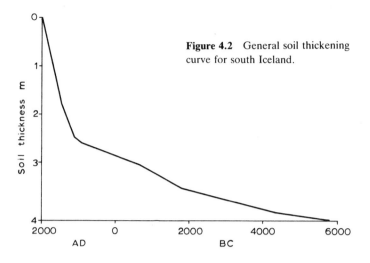

Figure 4.2 General soil thickening curve for south Iceland.

was before deflation started and profiles measured in them exaggerate average soil thickening. Notwithstanding these problems, a consistent pattern of soil erosion has emerged partially substantiated by examination of the inorganic content of peat layers (Einarsson 1963). During the greater part of the Post-glacial warm period the rate of soil thickening was slow, with Iceland covered in much more vegetation than is the case now. From about 2500 to 600 BC birch woods grew on many of the present bogs and there appeared to be a reasonable equilibrium between soil erosion and soil formation. Climatic deterioration at about 600 BC lowered the vegetation limit and seems to have led to increased wind erosion, but the greatest increase followed the period of settlement at about AD 890. The rate of thickening of loessic material was on average 4–5 times greater after AD 1104. Climatic deterioration during the Little Ice Age and volcanic activity have contributed to the increase in soil erosion, but Thórarinsson (1962) is convinced that the main cause of the erosion is man and his grazing animals, depleting the birch woods and breaking the vegetation cover.

Pollen analysis has produced similar results (Einarsson 1963). Four major pollen zones can be identified. In the Late glacial and pre-Boreal period there was a small birch maximum in northern Iceland whereas southern Iceland appears to have been birch-free. The first birch maximum occurred in the Boreal period followed by a decline and an increase in sphagnum caused by the wetter conditions of the Atlantic period. During the sub-Boreal period there was a second peak of birch followed by a gradual decline to c. AD 890 when birch forests declined very rapidly and there was a sharp increase of grasses and cultural indicators. A similar sequence was found at Thórsárdalur by Thórarinsson (1944), in which, at the time of settlement, the ratio of grass to birch pollen increased dramatically together with a sudden increase in herbaceous pollen types associated with farmsteads.

Soil erosion by wind is more prevalent in south and central Iceland. Research into the effects of katabatic winds from the glaciers and ice caps of central Iceland has shown that wave formation and the descent of dry air coming predominantly from Arctic air masses is associated with the belt of severe soil erosion in southern and central Iceland (Ashwell 1966, 1972, Ashwell & Hannell 1958). These results do not contest the role of man in creating the conditions for modern soil erosion but they suggest why severe erosion is restricted to specific zones.

The synthesis that is emerging is of increasing soil erosion since the time of settlement. Although wind erosion may be the most widespread process, gullying and various forms of mass movement are important in creating the initial eroded areas and there have also been considerable river and channel changes (Sveinbjarnardóttir et al. 1982). It is within this context that three specific sites in south Iceland are examined.

Seljaland

The plateau areas above Seljaland suffer extensive soil erosion, and stone pavements and patterned ground are becoming exposed below the 3–4 m of superficial material. Wind deflation is rapidly removing the poorly consolidated silts and sands that comprise the majority of the slope materials. The initial breaking of the vegetation cover to expose the materials is a combination of overgrazing by sheep and natural processes such as spring sapping, gullying and shallow landsliding. As soon as the vegetation cover is breached, the soil is immediately susceptible to the strong winds that blow unhindered from the sea a few kilometres to the south. The slopes still vegetated possess a dense network of earth hummocks or thúfur. Detailed analysis of the sections exposed in the eroding rofbards substantiates the general findings already described but also provides new information (Fig. 4.3). Buried thúfur can be seen in the upper 1.5 m of the sequence because of the way in which the tephra layers 'drape' the pre-existing micro-relief. There is no indication of buried thúfur below the Katla 1500 tephra and it is interesting to speculate that the influx of silt, a material susceptible to frost heave, has meant a greater development of thúfur.

The increase in the amount of loessic material above the Landnám ash is very clear with an even greater increase in the rate of deposition after 1728. The rate of thickening of material below the Landnám ash shows an average annual rate of 0.14 mm yr^{-1}, whereas that above it shows an increase of 1.2 mm yr^{-1}. After 1728, the annual rate of deposition is c. 3 mm, but these later figures are probably exaggerated for the reasons discussed earlier. The sequence below the Landnám ash is more condensed with only relatively thin loessic layers. Also, many of the tephra layers, especially those at 230, 247, 256 and 320 cm contain abundant plant roots and indicate, at the very least, a rudimentary soil and vegetation cover. Many of the loessic layers contain pipes which are important in the rapid transmission of water through the slope materials.

Analysis of samples taken from various parts of the succession show little

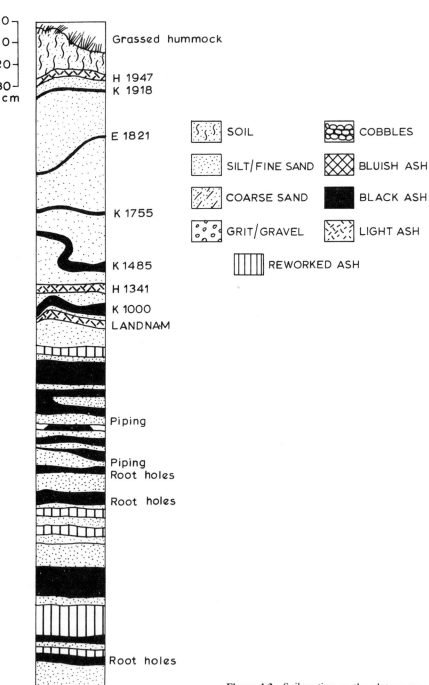

Figure 4.3 Soil section on the plateau area above Seljaland.

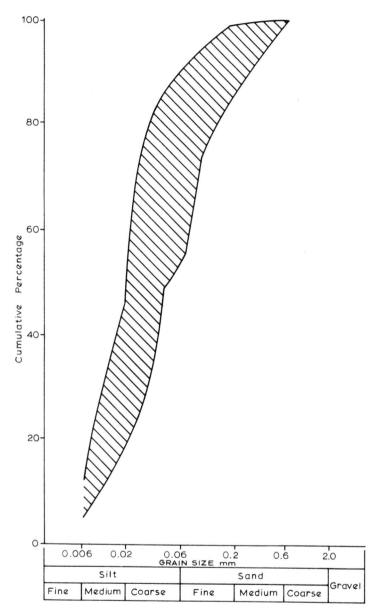

Figure 4.4 Grain-size envelope curves for loessic material in the Seljaland section.

variability in grain-size characteristics (Fig. 4.4). The material is well sorted and predominantly in the medium to coarse silt range. Microscopic analysis has revealed that the samples contain a variety of volcanic material including Pele's hair, regular and irregular glass shards and aggregates. Generally, the larger particles are the most angular although many subhedral and euhedral crystal fragments occur. Comparison with a sample of Katla 1357 tephra extracted from nearby peat deposits shows that the tephra is composed of angular to sub-angular fragments of crystalline minerals. This tephra sample also contained much clear glass, ranging in appearance from transparent to black and there was little evidence of the weathered material present in the soil samples.

The origin of the loessic soil is difficult to isolate. Much of it must have come from the eroded interior deserts but it is likely to have been reworked many times. Some has undoubtedly come from wind deflation on the great expanses of valley and coastal sandur and a smaller, but presumably increasing, amount of material is coming from local eroded areas. The sequence just described is typical of the open plateau areas above the former sea cliff that runs almost continuously along the southern coast of Iceland. It demonstrates that move-ment of material by wind has been the dominant process and that it has increased considerably since the settlement of this part of Iceland. The other sites suggest the same general increase in erosion but with different processes involved.

Thórsmörk

Thórsmörk is located between the ice caps Tindfjallojökull, Eyjafjallajökull and Myrdalsjökull, at the inner end of the valley sandur of the Markar River (Fig. 4.5). It is a heavily dissected hilly area bounded by the glacier-fed rivers Markar and Krossá. The sites examined in detail are adjacent to the abandoned farmsteads of Thurídarstadir and Steinfinnsstadir and in the secluded valley of Húsadalur. Identification of diagnostic tephra allows local variations in pro-cesses at these sites to be identified.

Húsadalur is now part of a conservation area and a National Park as it con-tains some of Iceland's little remaining birch woodland. Some areas have been fenced off to exclude sheep and the contrast between the grazed and ungrazed areas is considerable. The higher slopes are unvegetated and wind erosion, only partially hindered by more resistant tephra layers, has removed much of the loessic soil (Gerrard 1983). Within the protected area there is strong regen-eration of birch woodland and the relatively secluded and mostly wooded lower parts of Húsadalur possess relatively thin loessic layers between the recent tephra units. The most distinctive tephra, which is very useful as a marker horizon, is the very pale ash associated with the 1821 eruption of Eyjafjallajökull. On the lower slopes of Húsadalur this tephra is usually followed by a simple sequence of loess layers and thin black Katla tephras (Fig. 4.6). Occasional process changes are represented by slope wash layers of coarse grit. Activity on the steeper neighbouring slopes during the same period has been somewhat different. The material down the sequence to the 1821 layer cannot be differentiated into separate units and is composed of a mixture of sands and grits up to 50 cm thick. The Hekla 1947 tephra, which is usually very distinctive near the top of the sequences, is not identifiable, nor are the various recent Katla tephras. Lack of stratification and the considerable mixing of material implies a movement more akin to a fluid mass movement than a sequence of slope washes. It also appears

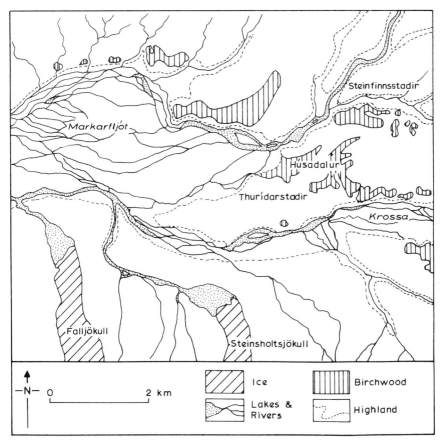

Figure 4.5   The Thórsmörk area, south Iceland.

to infill a small slope depression. The 1821 tephra seals 24 cm of reworked material which includes particles up to 3 cm in length. This is preceded by a coarse, iron-stained sand resting on a 2 cm loess layer which in turn seals a further 20 cm of reworked material. Loessic material of 1 cm rests on 0.5 cm of very compressed small birch twigs. The mechanism of formation of this twig layer is unclear but it could represent a natural concentration on the slope as similar concentrations of twigs occur in other exposures on the far side of the Markar River. The base of the sequence is occupied by a mixture of very coarse, frost-shattered pebbles. The whole sequence indicates successive periods of slope readjustment separated by periods when wind-blown material could accumulate undisturbed. Some of the sedimentary units thicken and thin down slope indicating movement up and down slope of the junction between erosion and deposition.

Most of the exposures on the gentler upper slopes indicate considerable aeolian deposition and are similar to the sequence described above Seljaland (Fig. 4.6). The various thicknesses of loessic material are largely a function of position,

VALLEY BOTTOM   LOWER SLOPE   PLATEAU AREAS

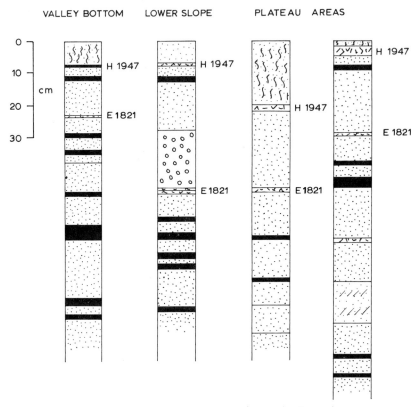

Figure 4.6 Soil sections in Húsadalur, Thórsmörk (key as for Fig. 4.3).

local topography and vegetation cover. Rofbards at Thurídarstadir and Steinfinnsstadir illustrate this variability and also the difficulty of determining absolute depositional rates. At Thurídarstadir, 47 cm of loessic material separates the 1821 and 1947 levels whilst there is 83 cm of similar material at Steinfinnsstadir. Also, two black tephras, the upper one probably associated with the 1918 Katla eruption, are present at Steinfinnsstadir but missing from the Thurídarstadir site, indicating that the latter sequence is probably truncated. However, both sequences emphasise the increasing amounts of material deposited since 1821.

A remarkable sequence of deposits exposed in a gully on the River Krossá side of Thurídarstadir illustrates the tremendous activity that has taken place in the recent past. Material 2.6 m thick exists above the 1821 layer and consists of 74 different sedimentary units, mostly alternations of grit or sand and silts. Some of the grit layers possess particles up to 7 mm in length and when traced up slope, gradually merge with coarse material and pebbles up to 10 cm in length. The sequence continues for a further 2 m below the 1821 level with the material becoming much coarser. These pebble layers, when traced laterally, possess many cut-outs and finer lenses and appear to represent a series of infilled channels on a series of fans. The alternating coarse and fine layers higher up the sequence may represent an annual layer of slope wash during later spring or early summer followed by aeolian deposition in the summer months. This is a sequence of events which occurs at the present time on the severely eroded areas. The larger gullies become infilled with wind-blown material during the summer months, and this is eventually flushed out to mix with the glacio-fluvial sediments of the Markar valley sandur.

The considerable activity in the past, represented by the slope materials just described, is still continuing. The vegetated slopes are evolving through a combination of accelerated soil creep and shallow earth slumping whilst the bare slopes are being eroded by a combination of processes. Wind erosion in this area is widespread and moves particles loosened by frost action, desiccation or trampling by sheep. Movement of material by wind in the Thórsmörk area is complicated by the effects of the nearby ice caps, as winds from many directions occur. All eroded sites are characterised by a series of major gullies, some up to 6 m deep, with numerous smaller subparallel rills. Water flow within the slope materials is high and a complex network of pipes occurs, with some up to 25 cm in diameter. The rills appear to be developing by the collapse of the larger pipes, some of the collapse being initiated by sheep breaking through the surface. Eluviation, resulting in the formation of pipes, tends to occur in soils possessing initial weaknesses caused by low bulk density, an unusual particle size distribution or a structure which has been altered by chemical effects (Gilman & Newson 1980). Many of the Icelandic slope materials possess some of these characteristics. The considerable landscape changes that these deposits indicate and the alternation in dominant erosional process have also occurred in Laufatungur.

Laufatungur

The landscape above the steep former marine cliff rises in a series of gently sloping plateaux to the ice cap of Myrdalsjökull. The successive plateaux are separated by steeper scarp sections through which the rivers have cut gorges. The area described here is dissected by two streams, one of which has cut a deep gorge. The slopes of the steeper sections are mantled by infilled gullies and a series of fossil gravel fans. The major stream appears to have downcut very recently as upper benches in the gorge are graded to the same level as the fossil fans. The stream is currently depositing a very coarse-grained fan at a lower level. These slopes are now partially vegetated but wind erosion has enlarged rofbards on the flatter areas and exposed rock, till and patterned ground of polygons and stripes (Fig. 4.7). The area is exposed to winds from the coast and to katabatic winds from the ice cap.

A fairly complete tephra stratigraphy is present in the exposures and allows a detailed interpretation of landscape change within the historic period. Most of the tephra are from nearby Katla beneath the ice of Myrdalsjökull, but a

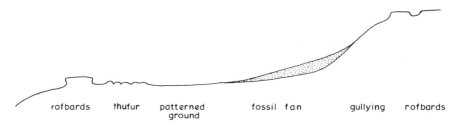

Figure 4.7 Generalised sequence of landforms on the plateau area above Skógar.

number of Hekla ashes are also present. The Landnám ash is present and is fortunately very distinctive, occurring low down in the succession. Four sections
are described in detail from contrasting positions to examine the absolute and
relative landscape changes that have taken place. Two of these exposures are in
fossil fans leading off the higher slopes and two are on the plateau surface. It is
anticipated from the previous results that the time of settlement is a crucial
threshold in the evolution of the Icelandic landscape, therefore the sequences
are described from the marker horizon of the Landnám ash.

The thickest sequence of deposits, up to 4.5 m, is exposed in the side of the
former gravel fan (Fig. 4.8). The material immediately below the Landnám ash
is a reddish sandy silt resting on 16 cm of dark brown sandy silt possessing features such as organic staining and root holes, suggesting that it represents the
remnants of a fossil soil. This is followed down the sequence by 4.5 cm of black
tephra, probably from Katla, resting on a more distinctive dark brown, iron-
stained fossil soil. This fossil soil contains prominent root holes which penetrate
some way into an underlying thick black ash. The material immediately above
the Landnám ash is a yellowish-brown sandy silt which also possesses the features characteristic of a fossil soil noted above. This is sealed by a thin loessic
layer followed by the Katla 1000 tephra. The sandy silt above the Katla 1000
tephra is also iron-stained and exhibits some rudimentary soil features.

The entire character of the sediments then changes with 110 cm of reworked
and partially stratified silt and sand with lenses of coarser particles. This unit is
sealed by the Hekla 1300 tephra which is followed by a similar sequence of reworked sediments and tephra until the level of the Hekla 1597 tephra. There is
a sudden change in the character of the material at this level. Whereas the sediments from before 1597 are stratified into coarse and fine layers up to 2 cm
thick, the material deposited since 1597 is almost entirely a silt with the well-
rounded and sorted grains indicative of wind action. The succession then becomes one of loess layers separated by the various tephras. A similar but thinner
sequence of sediments occurs a little way down slope at the outer limit of the
fossil fan.

The succession in a rapidly eroding rofbard on the plateau below the fan
shows a complete reversal of the fan sequence (Fig. 4.9). The traces of fossil
soils around the Landnám ash are still present but the material up to the Katla
1485 layer is mainly loessic, whilst the sediments for the rest of the succession

LAUFATUNGUR

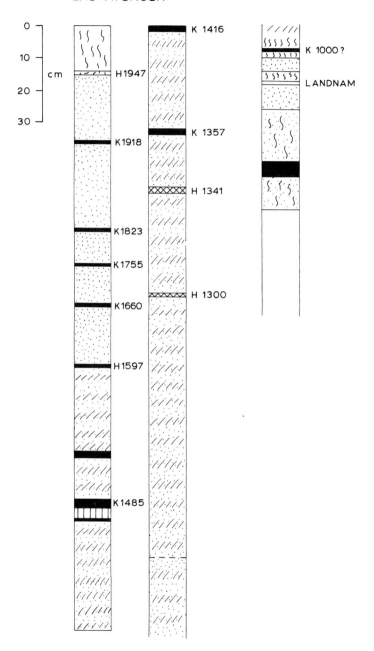

Figure 4.8 Section exposed in the side of a fossil fan, Laufatungur (key as for Fig. 4.3).

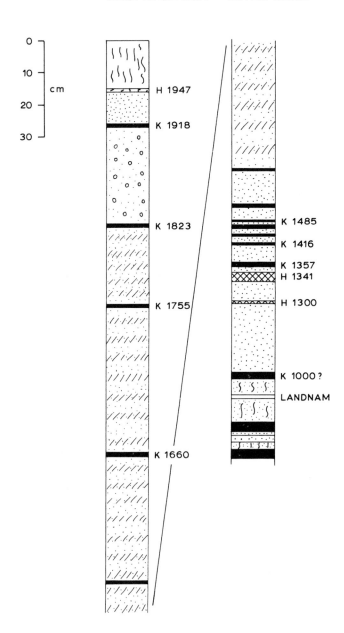

Figure 4.9 Soil section in an eroding rofbard on the plateau area at Laufatungur (key as for Fig. 4.3).

Table 4.2   Rates of soil thickening (cm yr^{-1}) at Laufatungur, south Iceland.

Date	Fossil fan	Process	Rofbard	Process
1000–1300	0.37	slope wash	0.07	aeolian
1300–1341	0.73	slope wash	0.15	aeolian
1341–1357	1.00	slope wash	0.13	aeolian
1357–1416	0.53	slope wash	0.10	aeolian
1416–1485	0.48	slope wash	0.16	aeolian
1485–1597	0.35	slope wash	0.57	slope wash
1597–1660	0.29	aeolian	0.57	slope wash
1660–1755	0.13	aeolian	0.48	slope wash
1755–1823	0.16	aeolian	0.35	slope wash
1823–1918	0.28	aeolian	0.33	slope wash
1918–1947	0.72	aeolian	0.34	aeolian
post 1947	0.39	aeolian	0.42	aeolian

are laminated coarse and fine sands. The contrast between the two sites is shown clearly in Table 4.2. The net rates of accumulation also change with high rates on the fan until 1485 and high rates on the plateau after 1485. This clearly indicates a marked change in landscape stability over quite short distances. Evidence in other exposures suggests that the change in character of the sediments on the fan reflects incision into the fan by the major stream, leaving little potential for water flow over the surface. The surface of the fan then began to stabilise and received only aeolian input. The lower plateau areas then seem to have come under the influence of water flow, possibly an overbank flood effect. This changes after 1823 to predominantly aeolian deposition which implies much of the erosion of the plateau has occurred since 1823. The section from higher up the slope towards the upper plateau shows an entire sequence of loessic material from the Landnám ash upwards and is very similar to the sequence described at Seljaland. The depositional units also increase in thickness in the more recent past attesting to the relative increase in wind-blown soil.

These results demonstrate very clearly the immense contribution tephrochronology can make to an understanding of the considerable recent changes that have occurred in the landscape of southern Iceland. Not only do the tephra layers enable the surfaces to be dated very precisely, but they also enable those surfaces to be traced from one exposure to the next.

Conclusions

The sites in south Iceland examined in detail have confirmed the general increase in soil erosion since the time of settlement that previous researchers have suggested. This has been clearly visible in the increasing build-up of wind-blown silt on the gentler plateau surfaces. The relative stability of the landscape at the time of settlement is evident in the nature of the materials immediately below and above the Landnám (settlement) ash. Above the Landnám ash there is a relatively sudden increase in the non-tephra components.

However, within this trend of gradually increasing erosion a number of specific local variations have been identified. Many of the steeper slopes have experienced phases of both erosion and deposition with evidence of cut and fill in fossil alluvial fans and the build-up of lower slope colluvial deposits. Some sites, such as Laufatungur, have also indicated marked local variations in the ratio of wind-blown to water-lain deposits, while all three sites indicate an increasing rate of deposition after about 1800.

The amount of current erosion at all the sites is considerable, with the most severely eroded areas being the bare slopes above the wooded and protected valley of Húsadalur. This area can be seen as an extension of the interior 'desert' area of Iceland where almost the entire soil cover has been removed. The plateau areas of Seljaland and Laufatungur are, in comparison, well vegetated but the eroded areas are continually increasing in extent and many Icelandic scientists are concerned lest the erosion reaches the level of that in Thórsmörk. The historical evidence suggests that the landscape will stabilise but probably only if the amount of sheep grazing is reduced.

Acknowledgements

The research was made possible by a grant from the University of Birmingham Field and Expedition Fund and by the assistance of Paul Buckland and Gudrún Sveinbjarnardóttir who were financed by an award from the Leverhulme Trust Fund. Fieldwork was carried out in Iceland under National Research Council permits 1/80 and 48/80. The tephrochronological studies of Gudrún Larsen have provided a framework into which landscape change can be fitted and her assistance in the field is gratefully acknowledged. The help in the field of Pete Foster and Cynthia Greig is also acknowledged.

References

Ashwell, I. Y. 1966. Glacial control of wind and soil erosion in Iceland. *Ann. Assoc. Am. Geog.* **56**, 529–40.
Ashwell, I. Y. 1972. Dust storms in an ice desert. *Geogr. Mag.* **44**, 322–7.
Ashwell, I. Y. and F. G. Hannell 1958. Notes on a föhn wind in Iceland. *Weather* **13**, 295–7.
Bjarnasson, H. 1978. Erosion, tree growth and land regeneration in Iceland. In *The breakdown and restoration of ecosystems*, M. W. Holdgate and M. J. Woodman (eds), 241–8. New York: Plenum Press.
Bruce J. G. 1973. Loessial deposits in southern South Island, with a definition of Stewarts Claim Formation. *N.Z. J. Geol. Geophys.* **16**, 533–48.
Butler, B. E. 1959. *Periodic phenomena in landscape as a basis for soil studies.* CSIRO Aust. Soil Publn. No. 14, Canberra.
Einarsson, T. 1963. Pollen-analytical studies on the vegetational and climatic history of Iceland in

late and postglacial times. In *North Atlantic biota and their history*, A. Lowe and D. Lowe (eds), 355–65. London: Oxford University Press.

Gerrard, A. J. 1981. *Soils and landforms*. London: George Allen & Unwin.

Gerrard, A. J. 1983. Contemporary soil erosion in Pórsmörk, southern Iceland. *Arbok. Hins. islenzka fornleifafelags* **1982**, 57–9.

Gilman, K. and M. D. Newson 1980. *Soil pipes and pipeflow*. Brit. Geomorph. Res. Group Res. Monogr. 1. Norwich: GeoBooks.

Hammer, C. V., H. Clausen and W. Dansgaard 1980. Greenland ice-sheet: evidence of Postglacial volcanicity and its climatic impact. *Nature* **288**, 230–5.

Oddson, G. 1638. *De mirabilibus Islandiae*.

Runólfsson, S. 1976. Soil conservation in Iceland. In *The breakdown and restoration of ecosytems*, M. W. Holgate and M. J. Woodman (eds), 231–40. New York: Plenum Press.

Sveinbjarnardóttir, G., P. C. Buckland and A. J. Gerrard 1982. Landscape change in Eyjafjallasveit, southern Iceland. *Norsk Geogr. Tidsskr.* **36**, 75–88.

Thórarinsson, S. 1944. Tephrokronologiska pä Island. *Geogr. Ann.* **26**, 1–217.

Thórarinsson, S. 1962. L'érosion éolienne en Islande a la lumiére des études tephrochronologiques. *Rev. Geomorph. Dynam.* **13**, 107–24.

Thórarinsson, S. 1970. Tephrochronology and medieval Iceland. In *Scientific techniques in medieval archaeology*, R. Beyer (ed.), 295–328. Los Angeles: University of California Press.

Thórarinsson, S. 1979. Tephrochronology and its application in Iceland. *Jökull* **29**, 33–6.

Thórarinsson, S. 1980. The application of tephrochronology in Iceland. In *Tephra studies*, S. Self and R. S. J. Sparks (eds), 109–34. London: Reidel.

Thoroddsen, T. 1914. An account of the physical geography of Iceland with special reference to the plant life. In *The botany of Iceland*, L. K. Rosenvinge and E. Warning (eds), 191–334. Copenhagen: Ejnar Munksgaard,

Valentine, K. W. G. and J. B. Dalrymple 1976. Quaternary buried paleosols: a critical review. *Quat. Res.* **6**, 209–22.

Walker, P. H. 1962. Soil Layers on hillslopes: a study at Nowra, New South Wales, Australia. *J. Soil Sci.* **13**, 167–77.

Westgate, J. A. and C. M. Gold (eds) 1974. *World bibliography and index of tephrochronology*. Edmonton: University of Alberta.

Part II

SOIL PROPERTIES AND
SLOPE PROCESSES

5

Aggregate stability, runoff generation and interrill erosion

J. De Ploey and J. Poesen

Aggregate stability

Aggregate stability is one of the main factors controlling topsoil hydrology, crustability and erodibility. There is considerable information on the dominant structural and textural characteristics of the topsoils of many soil groups, obtained by extensive soil mapping and the analysis of standardised mechanical and chemical parameters. These data are used for the determination of the K-factor in the Universal Soil Loss Equation, which is supposed to represent the long-term average erodibility of topsoils. More limited, however, is our knowledge concerning the short-term structural dynamics, which depend on the annual impact on slopes of changing weather conditions, tillage, crop cover, etc. For this and other reasons the modelling of runoff generation and interrill erosion is still problematic and the complexity of both these phenomena is expressed by the concept of 'partial area contribution to runoff' and by what may be called the 'partial area differentiation of interrill erosion'. A tentative discussion of these topics may start with considerations on the nature of aggregated topsoils and changing microstructures which have a major impact on ponding, runoff generation and erosion by splash and wash.

According to the USDA Soil Survey Manual (1951) and soil micromorphologists (Brewer 1964) peds, pedological features and S-matrices constitute the basic units of organisation of a soil material. Peds (macro- or micropeds) are referred to as 'compound particles, or clusters of primary particles, which are separated from adjoining aggregates by surfaces of weakness'. The definition of a ped is in contrast to (a) a 'clod', caused by disturbances such as ploughing or digging which mould the soil to a transient mass that slakes with repeated wetting and drying; (b) a 'fragment', caused by rupture of the soil mass across natural surfaces of weakness; and (c) a 'concretion', caused by local concentrations of compounds that irreversibly cement the soil grains together.

The primary shape of aggregates or peds is often angular or subangular, but after erosion they become more rounded and their transportability and possibly their 'detachability' therefore increase. Erosion may also change the aggregate size composition of the topsoil, but little is known about this mechanical aspect of aggregation. According to Gardner (1956) aggregate size distribution often corresponds to a logarithmic-normal distribution which plots as a straight line on log-probability co-ordinates.

For Brewer and other micromorphologists the fundamental structure of aggregates is the S-matrix, 'the material within the simplest (primary) peds, or composing apedal soil materials, in which the pedological features occur; it consists of the plasma, the skeleton grains, and voids that do not occur in pedological features other than plasma separations'. This definition is specified for aggregates by the models of Emerson (1959, 1977), describing the geometrical arrangement of the S-matrix with additional consideration of the relative stability of this arrangement. Quartz and other mineral grains compose the skeleton, while clay domains, organic matter, sesquioxides and concretions constitute the porous plasma. The stability of this matrix depends on different linkages, owing to (a) water menisci, (b) electrostatic linkages within and between the clay domains, and (c) linkages due to humus or sesquioxides, or to micelles or soil conditioners, with the additional intervention of hydrogen bonds or Van der Waals' forces. Consequently soil micromorphology is a fundamental tool in aggregate studies; comprehensive accounts of the results obtained by this discipline are given by Mücher and Morozova (1983).

Both biological agents and physicochemical factors play an important role in aggregation. Consider the mechanical effects of biological agents. Surface crusts can be formed by fungal hyphae and, after colonisation of topsoils by algae, may considerably reduce infiltration rates. Root press contributes to the formation of aggregates, particularly in densely rooted grassland soils. Earthworms are well-known producers of aggregated material within surface horizons, and there is a continuous microbiological production and decomposition of soil-binding organic materials related to the intervention of bacteria, different fungi and streptomycetes (Harris *et al.* 1966). Polysaccharides and fungal mycelia are also responsible for assisting soil aggregation. In many forested areas the soil fauna bring to the surface considerable amounts of aggregated, erodible material.

An extensive literature discusses the positive relationships between aggregate stability and the organic matter content of topsoils. Kemper and Koch (1966) analysed over 500 samples from western USA and Canada, and found a marked increase of aggregate stability up to an organic matter content of about 2%, beyond which aggregate stability increased comparatively little. Stability was also promoted by an increasing nitrogen content of the soils. These conclusions are corroborated by data obtained for the loess loam areas of Belgium by researchers from the Leuven Laboratory of Experimental Geomorphology. Topsoils with less than 2% organic matter are often marked by low aggregate stability and high crustability. Aggregation is thus related more to organic matter content than to clay content. Shrinking and swelling are well-known agents for the production of cracks and surfaces of weakness separating natural peds (Marshall & Holmes

1979), and it is the general experience of farmers that freezing promotes aggregation within cultivated soils.

The impact of clay particles or clay domains on aggregate stability cannot be summarised in a simple manner, although a positive relationship is most frequently mentioned, for example by Kemper and Koch in the USA and Canada. Swelling clays, such as montmorillonites, often reduce the stability of aggregates and the tendency to swelling and dispersion is more readily apparent the lower the valence of the exchangeable cation and, within each valence group, the smaller the ion. In particular, a high exchangeable sodium percentage (ESP) and a high electrolyte concentration appear to promote clay dispersion (Agassi *et al.* 1981, Imeson & Verstraten 1981). In addition, micromorphological clay particle arrangements are of importance. In Luxembourg farmland aggregates, Imeson and Jungerius (1976) found relative instability due to the presence of oriented clay domains. Mücher (1981) reviews many microstructural factors controlling aggregation. However, there is specific evidence that organic matter plays a dominant role in aggregate and topsoil stability, especially when the clay content is relatively low (Baver *et al.* 1972, Hartmann & De Boodt 1974). There is also general agreement about the high stabilising effects of iron and aluminium oxides. Mücher and Morozova (1983) conclude that the most resistant aggregates are probably derived from laterite and plinthite formations, while De Boodt *et al.* (1961) emphasise the positive impact on aggregate stability of calcium carbonate contents.

According to Harris *et al.* (1966) mulches in general will promote aggregation because of their conversion into soil binding agents, although rapid deterioration of the aggregate status may result from a frequent cultivation of annuals with little organic matter supply. These authors claim that cultivation may advance aggregation when the soil moisture content lies within a certain range, and when soils are tilled with suitable implements and without excessive physical pressures. Johnson *et al.* (1979) mention that clods resulting from ploughing wet soil were quite unstable compared with clods resulting from ploughing nearer the liquid limit.

A very important phenomenon with regard to pipeflow, pipe erosion and gully development is the formation of shrinkage cracks, especially on clay loam soils, subject to contrasting seasonal weather conditions. Sometimes cracking may start in the subsoil, particularly in Bt horizons with a prismatic-columnar structure, or in plough soles. There are increasing indications that specific chemical conditions, such as a high exchangeable sodium percentage and a low electrolytic concentration, favour shrinkage, dispersion and swelling (Imeson *et al.* 1982). Swelling and dispersion are closely related to the chemical composition of the soil solution, dispersion generally being preceded by swelling.

One commonly applied aggregate stability test is a wet sieve technique, which defines the content of water stable aggregates (WSA) of varying sizes. Studying the erodibility of a great number of English and Canadian soils, Bryan (1974, 1976) reports a significant negative correlation between erodibility and WSA contents, especially WSA larger than 0.5 mm. Reliable and reproducible measures of raindrop impact are water-drop tests that count the number of drop impacts of known force required to break down a soil aggregate to a certain state of disruption (Low 1954). Ultrasound techniques are used to achieve similar results (North 1976). According to Imeson and Vis (in press) both methods are suitable for assessing soil aggregate stability but the former seems to be more suitable for rather unstable soils.

Runoff generation

Important temporal and spatial variations characterise runoff generation on slopes because of its dependency on a wide range of factors, including soil moisture status, intensity and duration of rainfall, interception and surface storage, slope angle, lateral drainage after infiltration and, of course, textural and structural variations of the topsoils within a toposequence.

From a methodological point of view it is initially useful to investigate runoff generation studies on pure sediments rather than aggregated topsoils. By rainfall simulator experiments, Poesen (1981) compared conditions of runoff generation for a range of sandy, silty and clayey sediments and registered the following effects: (a) raindrop impacts compact the top layer which provokes an increase of cohesion; (b) with time, silt and clay particles migrate into the topsoil, over a depth of several millimetres, and constitute a less permeable 'filtration pavement', as mentioned by Bryan (1973); (c) drop impact forces cause topsoil liquefaction resulting from increasing pore-water pressures (De Ploey 1971). These effects reduce the infiltration rates (f_i), times to ponding (T_p) and times to runoff (T_r), and are proportional to rainfall intensity (I_r) and to the precipitation's kinetic energy (KE). The effective intensity, for a supposed vertical rain, decreases on steeper slopes (S) according to $I_r \cos S$. Therefore, f_i and the infiltration capacity (f_c) can be higher on steeper slopes as Moeyersons (1975) concluded for granite grus slopes in Nigeria.

Simulator tests on sandy sediments, with a saturated hydraulic conductivity (K_s) between 120 and 800 mm h^{-1} show that the runoff thresholds at f_c vary between 30 and 120 mm h^{-1}. These results suggest that f_c is from one-fifth to one-sixth of K_s. This discrepancy between f_c and K_s can be explained by the liquefaction effect, even without considering the effects of any filtration pavement. In addition, one must remember the influence of water temperature on liquid viscosity and its impact on f_i, f_c and K_s values.

The presence of thin, hydrophobic organic coatings on sediments may considerably reduce their infiltrability. This was confirmed by simulator tests on a dune plot at Kalmthout, northern Belgium, where an episodic colonisation by algae permits runoff to start at I_r values of about 30 mm h^{-1} (De Ploey 1977, 1980).

There seems to be an ambivalent effect of stone covers on runoff generation. On the one hand, gravel and stones intercept raindrops and dissipate their impact energy, so that surface sealing is impeded and infiltration relatively increases (Epstein *et al.*1966). Yair and Lavee (1976), however, found that stone covers promote runoff yield on steep talus slopes in the Sinai desert because of the delivery of water by the stones to the interspaces.

For many topsoils, aggregation, disaggregation and crusting are important factors of runoff generation. The process of crusting is directly related to rain-drop impact, aggregate stability, progressive wetting, and splashing. In experiments, Farres (1978) has clearly shown the role of aggregate size and time in the evolution of the crusts which he defines as 'areas with continuous particle cover without any definable aggregate boundaries'. Farres found that crusting progressed in time according to a sigmoidal function. Disaggregation is initially slow with the intervention of air-slaking, but beyond a certain critical amount of rain, crusting advances rapidly by: (a) detachment of micropeds or micro-aggregates by splash erosion and their redistribution by splash transport; (b) compaction and flattening of surface aggregates by the beating action of rain-drops and liquefaction; (c) downward washing of material and the eventual formation of filtration pavements. Micromorphological analyses of loamy, stable clods illustrate how rain provokes a surficial compaction and how clods become more and more protected from raindrop impact by a progressive concentration of large sand grains, rain-stable microaggregates and plant roots and root fragments (De Ploey & Mücher 1981). Unstable clods are flattened by raindrop impact and this process contributes largely to crusting. At the same time there is a marked reduction in the type and number of original pores whereas rounded, closed voids, called vesicles, become prominent. Finally the millimetre-thick crusts on fairly stable topsoils will contain a 'conglomerate' of residual rain-stable soil clods or aggregates, redeposited microaggregates and individual particles, organic material and vesicles, with a grain size composition often quite different from the subsoil (P. Farres, personal communication). Two processes at least may be responsible for this phenomenon. Poesen and Savat (1980) have shown that splash has a selective preference for the modal size classes of sand mixtures, while there may be a selective downwash of particles. McIntyre (1958) and Baver *et al.* (1972) mention the dispersion of fine particles, followed by transport into the soil and clogging of the pores as a possible cause of crust formation, but this phenomenon has not been observed in the thin sections of loess loamy topsoils studied by De Ploey and Mücher (1981).

According to Farres the thickness of the crust can be described by the equation

$$Y = a \log X + b \qquad (5.1)$$

where Y is the mean crust depth, X is the volume of water applied and a and b are regression constants, depending on the nature and the size of the aggregates. The thickness and rough structure of crusts can be evaluated under the microscope, preceding a detailed micromorphological analysis of thin sections.

Crustability is related to consistency and to a consistency index (C_{5-10}) derived from the liquid limit curve (De Ploey 1977, De Ploey & Mücher 1981).

$C_{5-10} = w_5 - w_{10}$, in which w_5 and w_{10} are the water contents, in per cent dry weight, for which the two sections in a Casagrande cup touch each other over a distance of 1 cm, after respectively 5 and 10 blows. This test reveals that relatively consistent topsoils, with a high aggregate stability and a low crustability, absorb more water and energy than unstable topsoils, which lower their consistency and become crusted. From tests and field observations, on Belgian loamy soils, it was found that consistency index values are closely related to the percentage of clay and to 5 times the organic matter content. It was also found, for winter corn fields, that global crusting during the winter characterises topsoils with a value of C_{5-10} less than 2, whereas less than 50% of the total surface of those fields having topsoils with C_{5-10} exceeding 3 were sealed by the end of March. Crusts have greater bulk densities and lower porosities than the underlying subsoil. Bulk densities of between 1.1 and 1.7 are frequent for crusts, whereas in the subsoil the unit weight may be lower than 1.0 g cm^{-3} (Lemos & Lutz 1957, Epstein & Grant 1967). Crust formation in general is less pronounced on structured topsoils with dense rooting systems, such as grassland or cultivated prairie soils (Luk 1979). Different biological factors may impede or reduce crust formation. The activity of soil fauna, the presence of a leaf litter horizon and digging or burrowing by animals all have a marked impact on topsoil structures and soil erosion in many environments (Yair & Rutin 1981, Hazelhoff *et al.* 1981). The effect of pipeflow in abandoned root canals and the development of soil horizons with different lateral and vertical hydraulic conductivities explain why on many forested slopes subsurface flow is so important, even when it is limited to basal slope sections (Nortcliff *et al.* 1979). Subsurface flow may produce saturation overland flow during periods of high intensity rainfall as recently described in tropical NE Queensland by Bonell *et al.* (1983) and as is currently being studied by the Amsterdam Laboratory for Physical Geography in the Luxembourg Ardennes.

Strong crusting may produce Hortonian overland flow or what Kirkby (1978) called 'infiltration excess overland flow', whereby runoff is reached at high rainfall intensities for a small value of V, the volume of water infiltrated prior to runoff. In this case runoff generation is related to the presence of a two-layered soil (Hillel 1971).

Recently Morin *et al.* (1981) investigated the relation

$$K_u / d = f_t / h_b \qquad (5.2)$$

where K_u is the hydraulic conductivity of the crust, d is the crust thickness, f_t is the infiltration rate at time t and h_b is the suction at the interface between the crust and the layer below. Thereby h_b is directly proportional to L, the thickness of the soil layer down to the wetting front, or the free water surface. Morin *et al.* found that for loamy and sandy soils in Israel, K_u / d values are high for shallow layers under conditions of high energy rainfall, but are stable and lower when the L layer is greater than 3−4 cm, or with low intensity precipitation. Therefore, the effect of a crust on infiltration is minimal when high intensity rainfall occurs on a dry, crusted topsoil. Conversely, the crust effect is more pronounced in more humid areas because of higher L values and a higher interface suction (h_b), the latter promoting a sealing effect by fixing clay particles. During heavy rain, drop impacts may detach fine particles from the crust, with a consequent increase of K_u / d values.

Although it is generally assumed that Hortonian overland flow is typical for many semi-arid and arid environments, Scoging and Thornes (1979) showed that for semi-arid Spain 'saturation overland flow' is dominant on soils with a limited storage volume (V) whereby

$$V = I_r B/(I_r - A) \tag{5.3}$$

and I_r is the rainfall intensity, A is the final infiltrability and B is a constant depending on soil conditions. V converges to B when $(I_r - A)$ approaches I_r, i.e. when A is zero or very small or I_r is very large. The time to surface saturation (T_s) is then given by

$$T_s = B/(I_r - A) \tag{5.4}$$

As pointed out by Kirkby (1978), this result differs from the application of the Philip equation (Philip 1957) for Hortonian flow, whereby

$$V = I_r B/(I_r - A)^2 \tag{5.5}$$

Unlike Equation 5.4 this means that V tends to zero when I_r is very large or A is negligible.

In field situations, T_s, the time to ponding (T_p) and the time to runoff (T_r) are difficult to observe. Imeson and Kwaad (1980) and Imeson (1983) admit that, in northern Morocco, ponding is reached when approximately 40% of the surface has attained this state. A value of 100% ponding occurs a relatively long time after ponding and overland flow have commenced in the microtopographic depressions. Scoging and Thornes describe 'a visible flow at the surface', and this stage can be visualised during simulator tests by using a dye such as potassium permanganate. Imeson found in Morocco, according to the work of Smith and Parlange (1978), that the amount of rain required to pond the soil (p_r) could be predicted reasonably well by the equation

$$\int_o^{T_p} p_r dt = (A/K_s) \ln \{ I_r / I_r - K_s)\} \tag{5.6}$$

where $A = s^2/2$ (s is the sorptivity), I_r is the rainfall intensity and K_s is the saturated hydraulic conductivity of the topsoil. Equation 5.6 was applied to calculate the infiltration envelopes after rainfall simulator tests, plotting rainfall to ponding against rainfall intensity. According to Dunin (1976), sorptivity is defined by plotting cumulative infiltration curves against time ($T^{0.5}$). On his plot sites Imeson found that the infiltration envelopes were greatly influenced by the development of a surface crust.

Also on the basis of rainfall simulator tests Bork and Bork (1981) discussed runoff hydrographs obtained in semi-arid south-east Spain. They found that, with a very high rainfall intensity, maximum runoff occurs at the beginning of the rainstorm according to a component A_s, which is additional to the evolution of the runoff coefficient, following the Philip equation (Philip 1957):

$$A_s = R.T^B. \exp(C.T) \qquad (5.7)$$

where R is the maximum runoff coefficient at the beginning of the rainstorm, T is the time from the start of runoff and B and C are time-dependent parameters.

The progressive reduction of the runoff coefficient, from its maximum value (R), might well be explained by the initiation of lateral subsurface flow and, therefore, increased drainage of the topsoil. Lateral subsurface flow is known to be an important component of hillslope hydrology as was discussed in detail by Chorley (1978) and Whipkey and Kirkby (1978), and on a theoretical basis by Zaslavsky and Sinai (1981). Under many circumstances it would promote runoff generation, especially on the basal slope sections.

The uneven distribution of precipitation (Yair et al. 1978), crusting and surface runoff roughness, pipeflow and lateral subsurface flow, and the differentiation of soils and vegetation, are factors which explain why in toposequences T_p and T_r values vary so much in space and time. The same is true for areas where snow-melt and glacier-melt are the main sources of runoff (Zeman & Slaymaker 1975).

Further supportive evidence is the often erratic response of runoff to rainfall in 'badlands' environments (Bryan et al. 1978). In Alberta these authors have observed runoff starting earlier on shale plots with low slope angles than on steep slopes. In Spain, Scoging and Thornes (1979) discussed T_r measurements and concluded that there was also a poor relationship with the slope gradient except on very low slope angles, where the relationship between vegetation and T_r gave higher values of T_r. Similarly, in the Rif Mountains of Morocco, Imeson (1983) found no clear relationship between slope angle (S) and T_p. Experiments proceeding in the Leuven laboratory, in collaboration with Toronto University (Bryan), confirm that for the relationship of S and T_r the simple rule that 'the steeper the slope the shorter the T_r' does not hold true. There are also indications that the influence of antecedent rainfall on infiltration rates can be shortlived. There are good reasons for T_r being smaller on steep slopes because of the gravity component, the oblique impact of rain, the relatively reduced depression storage preceding runoff, etc. Opposing these effects, however, is the role of $I_r \cos S$ (I_r is rainfall intensity) and its effect on crusting, for supposed vertical rain. Crust formation on steeper slopes may be further opposed by topsoil erosion, and in addition there will be more lateral, subsurface drainage on these sections. These are some factors which increase the time to runoff, and therefore one may expect different models of runoff generation to be representative of specific, regional geographical environments.

In Figure 5.1 data from the literature concerning the relationship T_r (min) to

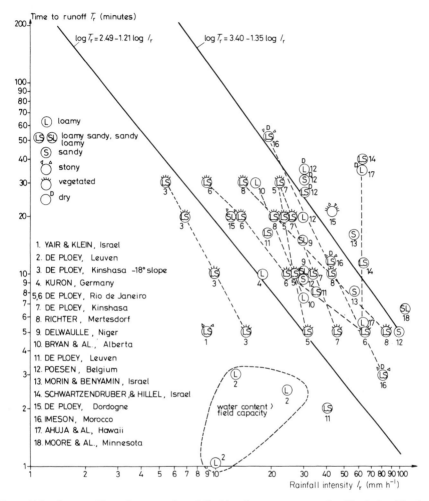

Figure 5.1   Proposed boundary equations delimiting the most common $I_r - T_r$ relationships for runoff starting on drained, loamy to sandy topsoils. References: see De Ploey (1983).

I_r (mm h^{-1}) are plotted. These infiltrometer data, resulting from both field and laboratory measurements, relate to bare loam and sandy loam soils, tilled fields, pure sediments and plots under a steppic vegetation. The slope angle of the plots was between 1° and 10° and most topsoils were drained before the rain started. A plot of the T_r data on log−log paper suggests the following boundary conditions, whereby T_r was calculated as a safe, maximum value (De Ploey 1983):

$$\log T_r(\min) = 2.49 - 1.21 \log I_r \qquad (5.8)$$

$$\log T_r(\max) = 3.40 - 1.35 \log I_r \qquad (5.9)$$

T_r values were lower than indicated by Equation 5.8 when the moisture content of the topsoil was definitely above field capacity. They were also lower on a stony topsoil in Israel and on fairly steep slopes near Kinshasa, Zaire, with a tufted, steppic grass vegetation. Equation 5.9 excludes very dry sediments, uncrusted soils or sandy topsoils with a high infiltration capacity.

Figure 5.2 shows that the $T_r(\text{min})$ values correspond to an amount of precipitation ($P_r(\text{min})$) of no more than 2–3 mm and a kinetic energy (KE) of about 60 J m^{-2}. One could call this a case of 'saturation overland flow', characterised by a constant volume P_r. It implies that there is a minimal infiltration and ponding before runoff commences, a situation existing when the topsoil is fairly wet and crusted, and/or because runoff generation is promoted by such factors as liquefaction, hydrophobic coatings, stemflow and the concentration of rainwater by stone covers. $T_r(\text{max})$ and $P_r(\text{max})$ values may correspond to either Hortonian overland flow or saturation overland flow, and there is a negative

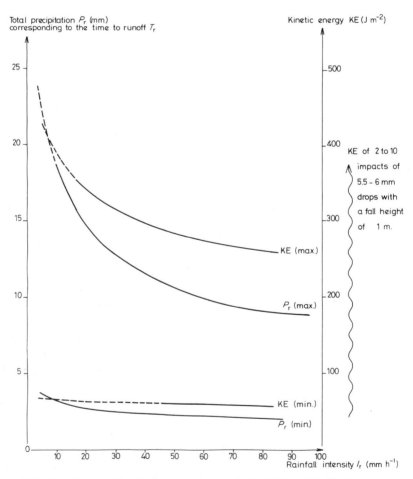

Figure 5.2   Time to runoff in relation to total amount of rainfall and its kinetic energy.

correlation between P_r and I_r. Initial infiltrability is high on dry topsoils and may also be affected by progressive disaggregation and crusting. The fact that the critical KE values vary between 250 and 400 J m^{-2} is remarkable. The Low drop-test for aggregate stability measurements (Low 1954) uses 5.5–6 mm drops, the impact of which provokes sealing by the breakdown of fairly stable loamy aggregates. Imeson and Jungerius (1976) and Bergsma and Valenzuela (1981) applied this test to measure the stability of 4–5 mm prewetted loamy aggregates, discovering that more than 50% of these aggregates are reduced to less than 3 mm in size after 2 − 10 impacts, with a fall height of 1 m. Calculations show that the corresponding KE necessary to initiate sealing then varies between 65 and 380 J m^{-2}, the same order of magnitude as the KE(max) values[*] in Figure 5.2. This suggests that crusting may be primarily a control on the proposed T_r(max) and P_r(max) values. In addition, however, the principle of convergent causalities must be considered, in that threshold conditions for saturation overland flow, starting on a poorly crusted topsoil, may be quite similar to Equation 5.9.

Interrill erosion

Interrill erosion is defined as erosion occurring on soil surfaces where interrill flow, i.e. thin film flow, occurs. This concept was introduced by Meyer et al. (1975) in order to separate the sources of eroded soil into the portion coming from the very top of the soil surface on the interrill areas and that originating from concentrated flow in the relatively deeper rills. Such a division is very useful in mathematical simulation of the soil erosion process, as shown by Foster and Meyer (1975).

On the interrill areas, the several effects of raindrop impact are usually the dominant factors influencing erosion rates. This was demonstrated by Young and Wiersma (1973) who observed that decreasing rainfall impact energy by 89%, without reducing rainfall intensity, decreased soil losses from a 6.9 m^2 laboratory plot by at least 90%. The most important effects of raindrop impact on interrill areas are: (a) destruction of soil aggregates, (b) compaction of the soil top-layer and inwashing of the finest soil fractions which together include surface sealing and crusting, (c) detachment of soil particles, (d) transport of detached particles, and (e) increasing the sediment load. Since (a) and (b) have been discussed previously, more attention is now directed to the remainder.

If we examine the action of a falling drop striking a solid surface, two modes of action are noted. One is the compressive stress from the impact, and the other is the shearing from the lateral jetting water (Huang et al. 1982). The mean stress (P) produced by raindrop impact on a horizontal rigid surface equals ρv^2 with ρ being the density of the water and v equalling the initial drop velocity (Moskovkin & Gakhov 1979). For a drop with a diameter of 3.5 mm and an impact velocity of 7.8 m s^{-1}, $P = 61$ kPa. Results of several investigators indicate that this impact pressure is neither constant nor uniform, with extremely high values at the very instant of impact and on the perimeter of a circle corresponding with the shape of the initial rebound corona (Huang *et al.* 1982). For example, given a drop diameter of 3.5 mm and $v = 7.8$ m s^{-1}, Ghadiri and Payne (1981) measured maximum stresses of 4.4 MPa which they attributed to a transient and localised water hammer effect caused by the initial shock wave. The foregoing calculations and measurements hold for impacts on a solid plane, but on a soil surface one must allow for the effects produced by the porous, loose structure of the soil, which, in general, reduces the impact stress (Moskovkin & Gakhov 1979).

In any event, the high stresses produced by impacting raindrops could partly explain the aggregate destruction and the observed compaction of interrill areas situated on very fine sandy, silty or clayey soils with an unstable structure. Using a strain gauge apparatus, Palmer (1963) measured maximum drop impact stress when a water layer with a depth equalling drop diameter was present, suggesting that under these conditions the soil compaction would also be maximal. However, similar stress measurements undertaken by Ghadiri and Payne (1981) contradict these observations.

Immediately after impact, the vertical force is transformed to lateral shear caused by the high radial flow velocities. This shear stress is an erosive agent that works against the resisting forces of gravity and cohesion to detach and move soil particles. At this stage, soil deformation characteristics and surface microrelief play an important role (Huang *et al.* 1982). During drop impact, the air and water trapped in the soil pores are compressed to high pressures, reducing or even annihilating the shear strength of the soil top layer (De Ploey 1971). This mechanism facilitates detachment of soil particles by the lateral jetting water.

The amount of soil particles detached on a unit interrill area depends on two main factors: (a) the kinetic energy produced by the impacting raindrops and (b) the resistance of the soil surface to raindrop detachment. Poesen (1983) proposed the following relation, enabling an approximate calculation of the weight of detached particles:

$$S = R^{-1} \, (\mathrm{KE})^x \qquad\qquad (5.10)$$

where S is the weight of soil detached per unit surface area per unit time (kg m^{-2} time^{-1}), R is the resistance of the soil surface to detachment, expressed as the kinetic energy required to detach 1 kg of material (J kg^{-1}), KE is the rainfall kinetic energy (J m^{-2} time^{-1}) and x is a coefficient depending on material properties. For loose soil material $x = 1$.

Values for R have been calculated using data obtained during laboratory experiments (Poesen & Savat 1981, Savat & Poesen 1981) and field measurements. The latter have been carried out using small funnels embedded in the soil, field splash cups (Morgan 1978) or small bottles (Poesen in preparation b). Figure 5.3 illustrates the influence of median grain size on R for loose sediments. These R-values agree well with R-values for soils with a similar texture (see Table 5.1), which suggests at least for soils with a low organic matter content, that particle size plays a dominant role in determining soil resistance to splash detachment. As a result of variations in resistance to detachment caused by the initial state and sorting of the sediments, it is necessary to represent the relation between R and the median grain size of the loose sediments not as a single, well-defined curve, but rather as a broad band, showing the possible variations. R-values for different initial states (wet, dry, loose, crusted, etc.) will be published elsewhere (Poesen in preparation a). For soils, it has been generally observed that an increasing content of clay, silt or coarse fragments decreases detachability, while sand content has the opposite effect (Bubenzer & Jones 1971, Bryan 1974). Partially decomposed plant residues act as a mulch and tend to reduce splash detachment (Singer *et al.* 1981). Although several investigators (Adams *et al.* 1958, Bryan 1974) have shown that an increasing aggregate stability decreases the susceptibility of soils to detachment by raindrop impact, others record a positive relation between aggregate stability and

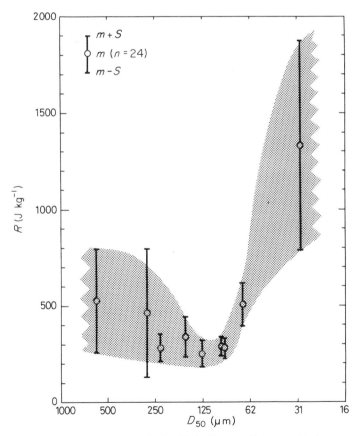

Figure 5.3   Relation between median grain size (D_{50}) of loose sediments and their mean resistance to raindrop detachment (R). The standard deviation (s) of experimental values is shown with each mean (m) (after Poesen in preparation a).

Table 5.1 Resistance to raindrop detachment for bare interrill soil surfaces, based on field data.

Investigator(s)	Soil texture	KE[1] (J m^{-2})	S[2] (kg m^{-2})	R[3] (J kg^{-1})
Sreenivas *et al.* (1947) (Texas, USA)	clayey soil	6 417	6.42	1000
Fournier (1967) (Upper Volta)	sandy ferruginous tropical soil	1 886	2.32	813
Bollinne (1975) (Belgium)	loamy soil	7 375 3 700	41.5 2.4	178 1542
Morgan (1977) (UK)	sandy soil (D_{50} = 0.120 mm)[4]	17 130 17 130	103.67 50.0	165 343
Froehlich & Slupik (1980) (Poland)	loamy soil	6 073 6 073	10.74 3.99	565 1522
Morgan (1981) (UK)	sandy soil (D_{50} = 0.120 mm)[4]	6 938 6 938	37.0 29.0	188 239
Soyer *et al.* (1982) (Zaire)	clayey soil	31 654	28.4	1115
Seiler (1983) (Switzerland)	sandy loam soil	12 080 14 720	8.9 14.3	1357 1029
Poesen (in prep. b) (Belgium)	loamy soil sandy soil (D_{50} = 0.192 mm)[4]	2 940 3 194	1.85 5.00	1587 638
J. M. Roels (pers. comm.) (France)	stony clayey soil (stone cover = 60%)	6 796	0.89	7636

[1] KE = kinetic rainfall energy.
[2] S = weight of splashed soil material.
[3] R = resistance of the soil surface to detachment by raindrop impact (R = KE S^{-1}).
[4] D_{50} = median grain size of the soil material.

detachability (e.g. McIntyre 1958, Mazurak *et al.* 1975, Bollinne 1978, De Ploey 1979, Gasperi-Mago & Troeh 1979).

During rainfall, water content of the topsoil changes continuously and this implies that the apparent cohesion and hence detachability of the soil also change (Fig. 5.4). The appearance of a thin water film on the interrill soil surface further influences detachability as, according to several investigators (Palmer 1963, Mutchler & Young 1975), a very thin water film will accelerate the detachment of particles from a soil surface. These observations, however, are not confirmed by

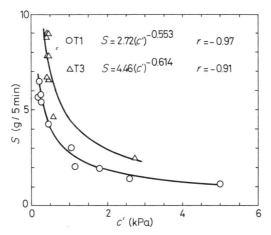

Figure 5.4 Relation between apparent cohesion (c') and detachability (S) of a sandy sediment ($D_{50} = 0.283$ mm). T_1 refers to an initial air-dry loose state, T_3 to an initial desiccated state obtained after a simulated rainstorm (after Poesen 1981).

Ghadiri and Payne 1979, Moss *et al.* 1979 and Poesen 1981). All authors, however, emphasise the time-dependent character of detachability.

After drop impact, detached soil particles are entrained in splash droplets or ejected from the soil surface close to the point of impact. This mode of transport is generally termed splash transport. For loose sediments varying between coarse sand and silty loam, Poesen and Savat (1981) measured mean projected splash distances ranging between 10 and 30 cm. Moeyersons and De Ploey (1976) observed that under simulated rainfall coarse particles with a diameter between 10 and 20 mm were not projected by saltation, but moved owing to the shocks caused by lateral drop strikes. They termed this mode of transport 'splash creep'.

Under vertical rainfall and on a horizontal soil surface it may be assumed that particles are splashed equally in all directions. However, under oblique rainfall, or on a sloping soil surface, there is a net splash transport downwind or down slope De Ploey & Savat 1968, Moeyersons 1983). Several investigators have attempted to model this process (Moeyersons & De Ploey 1976, Kirkby 1980, Poesen & Savat 1981), and recently Poesen (1983) has proposed an improved parametric model enabling net splash transport on a bare smooth interrill surface to be calculated:

$$q_s = \frac{(KE) \cos (a + b)}{R} [(0.301 \sin (a \pm b) + 0.019 (D_{50})^{-0.220}) \tag{5.11}$$
$$(1 - \exp \{ - 2.42 \sin | a \pm b | \})]$$

where q_s is the net splash transport discharge across a unit contour line (kg m^{-1} time^{-1}), KE is the rainfall kinetic energy (J m^{-2} time^{-1}), R is the resistance of the soil to detachment by raindrop impact (J kg^{-1}), e.g. see Figure 5.3 and Table 5.1, D_{50} is the median grain size of the soil material (m), a is the slope angle (degrees), and b is the rainfall obliquity (degrees). If one looks at the net downslope or upslope splash transport, b equals the angle between a vertical line and the mean trajectory of the raindrops projected onto a vertical plane parallel to a slope line. $b = 0°$ for vertical rainfall, $(a + b)$ for downslope-directed rainfall and $| a - b |$ for upslope-directed rainfall. b can be determined using a vectopluviometer (e.g. Reeve 1983) or it can be deduced from $q_s - a$ relationships (Poesen in preparation b). Boundary conditions for the use of this splash transport equation are: (a) inter-*rill soil surfaces must be smooth and bare; (b) 0°* $< (a \pm b)$ *<20°*; (c) 0.000020 m $< D_{50} < 0.000700$ m. Following validation by comparing predicted splash transport and that measured using Ellison-type splash boards (Poesen in preparation b), it may be concluded that the proposed splash transport equation provides a reliable estimate of the order of magnitude with respect to net splash transport discharges on bare slopes (see Fig. 5.5).

As an example, q_s is calculated for a true field situation. During a certain period 255.5 mm of rainfall was recorded. This rainfall amount represents a KE value of 3194 J m^{-2}. Rainfall was directed upslope with a mean obliquity (b) of 20.9°. Slope angle equalled 5.5° and so $| a - b | = 15.4°$. The resistance (R) of the sandy soil ($D_{50} = 0.192$ mm) to raindrop detachment, measured in the field, equalled 664 J kg^{-1} (Table 5.1). Using these data as an input, it can be calculated that approximately 4.8 kg m^{-2} ($=S$) was detached on this interrill surface, while net upslope splash transport equalled 0.64 kg m^{-1} ($=q_s$). From these data it can be computed that mean projected upslope splash distance equalled roughly 0.1 m ($= q_s \, S^{-1}$).

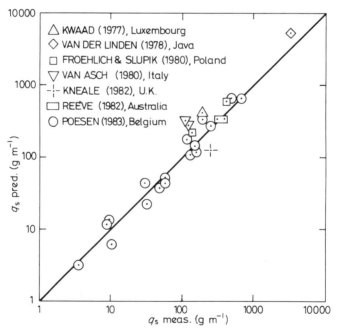

Figure 5.5. Measured compared to predicted net splash (after Poesen, in preparation b). References: see Poesen (1983).

On interrill soil surfaces detached particles are also transported by thin film flow, although in the absence of raindrop impact, its capacity to detach and transport particles is very low (e.g. Savat 1976, Walker et al. 1977). However, in the presence of rainfall, the transport capacity of the interrill flow greatly increases on the condition that its thickness remains smaller than a critical value. In drop-disturbed, shallow interrill flow, soil particles are transported by suspension and saltation or by contact bed load (Moss et al. 1979). As compared with rill flow, interrill flow transports a smaller proportion of larger particles, such as soil aggregates and primary particles, because of basic differences in the detachment and transport mechanisms (Alberts et al. 1980).

Susceptibility of soils to erosion by drop-disturbed interrill flow has been measured in the field on micro-runoff plots (Meeuwig 1970, Blackburn 1975, De Meester et al. 1979) and in the laboratory with the aid of a micro-plot box and simulated rainfall (e.g. Moldenhauer 1965, Epstein & Grant 1967, Farmer & Van Haveren 1971, Poesen 1981, Bryan & De Ploey 1983). Young and Wiersma (1973) and Young and Onstad (1978) developed a laboratory technique in order to measure concurrently interrill and rill erodibility of soils.

A number of investigators have tried to relate the measured interrill erodibility of soils to a number of soil properties. With respect to physical soil properties, the texture and structure play a dominant role. An increasing percentage of gravel, for example, tends to reduce interrill erosion (Meeuwig 1970). Soils with a high content of very fine sand or silt are highly susceptible to interrill and rill erosion (Wischmeier et al. 1971), while an increasing clay content generally lowers the susceptibility of soils to interrill erosion (Meyer 1981). With reference to soil structure, size and water stability of soil aggregates play an important role. Sood and Chaudhary (1980) demonstrated that reducing initial clod size increased interrill soil loss. Many investigators have demonstrated conclusively that aggregate stability is negatively correlated with interrill soil loss (Adams et al. 1958, Bryan 1968, Farmer & Van Haveren 1971, De Vleeschauwer et al. 1978). Interrill erosion usually increases if soil bulk density increases, except on sites with vesicular horizons where bulk density is negatively related to interrill sediment production (Blackburn 1975).

A strong emphasis in the literature would suggest that organic matter content has often been considered to be the dominant chemical soil property affecting interrill soil loss. Many investigators report negative correlations between soil loss due to drop-disturbed interrill flow on the one hand and decomposed organic matter content or organic carbon content of the soil on the other (Bryan 1974, Luk 1978). Partly decomposed organic matter on the soil surface acts as a mulch and tends to reduce interrill erosion (Lattanzi et al. 1974). The presence of sodium in the adsorption complex of the soil has a significant effect on interrill erodibility. Singer et al. (1980) found a strong positive association between interrill soil loss and the sodium adsorption ratio, $SAR = (Na^+) [0.5(Ca^{2+} + Mg^{2+})]^{-0.5}$. Furthermore, these investigators also showed that the percentage of dithionite-extractable iron was inversely proportional to interrill erodibility.

Bryan (1979) and Singer and Blackard (1982) found that the relation between surface slope angle and interrill wash is best described by a polynomial function, the form of which they consider to be dependent on antecedent moisture condition and soil type. Furthermore, De Ploey (1981) has emphasised the time-dependency of interill wash mechanisms.

Conclusion

Soil surface sealing and crusting are two of the key processes controlling runoff generation and rill and interrill erosion, and they are also closely related to aggregate stability. Despite investigation of this topsoil characteristic by an increasing number of researchers, there are still no parametric models available which accurately describe the time-dependent nature of the crusting process, although several indices have been proposed which predict the crustability of a topsoil. Rainfall simulator experiments, both in the field and the laboratory, are essential to further our understanding and modelling capability of runoff generation, a threshold phenomenon intimately connected with topsoil sealing and crusting. Hitherto more attention has been directed towards the analysis of splash erosion and the elaboration of a splash transport model. This is a first step towards the understanding of the rill—interrill system which often governs slope wash.

References

Adams, J., D. Kirkham and W. Sholtes 1958. Soil erodibility and other physical properties of some Iowa soils. *Iowa State Coll. J. Sci.* **32**, 485–540.

Agassi, M., I. Shainberg and J. Morin 1981. Effect of electrolyte concentration and soil sodicity on infiltration rate and crust formation. *Soil Sci. Soc. Am. J.* **45**, 848–51.

Alberts, E., W. Moldenhauer and G. Foster 1980. Soil aggregates and primary particles transported in rill and interrill flow. *Soil Sci. Soc. Am. J.* **44**, 590–5.

Baver, L., W. H. Gardner and W. R. Gardner 1972. *Soil physics.* New York: Wiley.

Bergsma, E. and C. Valenzuela 1981. Drop testing aggregate stability of some soils near Merida, Spain. *Earth Surf. Proc. Landforms* **6**, 309–18.

Blackburn, W. 1975. Factors affecting infiltration and sediment production of semiarid rangelands in Nevada. *Water Res. Res.* **6**, 929–37.

Bollinne, A. 1975. La mésure de l'intensité du splash sur sol limoneux. Mise en point d'une technique de terrain et premiers résultats. *Pédologie* **25**, 199–210.

Bollinne, A. 1978. Study of the importance of splash and wash on cultivated loamy soils of Hesbaye (Belgium). *Earth Surf. Proc.* **3**, 71–84.

Bonell, M., D. A. Gilmour and D. S. Cassells 1983. A preliminary survey of the hydraulic properties of rainforest soils in tropical north-east Queensland and their implications for the runoff process. In *Rainfall simulation, runoff and soil erosion*, J. De Ploey (ed.), 57–78. *Catena* Suppl. 4, Braunschweig.

Bork, H. R. and H. Bork 1981. *Oberflächenabfluss und Infiltration-Ergebnisse von 100 Starkregensimulationen im Einzugsgebiet der Rambla del Santo (SE-Spanien).* Tech. Univ. Braunschweig, Landschaftsgenese und Landschaftsökologie No. 8.

Brewer, R. 1964. *Fabric and mineral analysis of soils.* New York: Wiley.

Bryan, R. 1968. Development, use and efficiency of indices of soil erodibility. *Geoderma* **2**, 5–26.

Bryan, R. 1973. *Surface crusts formed under simulated rainfall on Canadian soils.* Report to Conference held at Pisa, Consiglio Nazionale delle Ricerche 2, 1–30.

Bryan, R. 1974. Water erosion by splash and wash and the erodibility of Albertan soils. *Geogr. Ann.* **56A**, 159–81.

Bryan, R. 1976. Considerations on soil erodibility indices and sheetwash. *Catena* **3**, 99–111.

Bryan, R. 1979. The influence of slope angle on soil entrainment by sheetwash and rainsplash. *Earth Surf. Proc.* **4**, 43–58.

Bryan, R. and J. De Ploey 1983. Comparability of soil erosion measurements with different laboratory rainfall simulators. In *Rainfall simulation, runoff and soil erosion*, J. De Ploey (ed.), 33–56. *Catena* Suppl. 4, Braunschweig.

Bryan, R., A. Yair and W. Hodges 1978. Factors controlling the initiation of runoff and piping in Dinosaur Provincial Park badlands, Alberta, Canada. *Zeitschr. Geomorph. Suppl.* **29**, 151–68.

Bubenzer, G. and B. Jones 1971. Drop size and impact velocity effects on the detachment of soils under simulated rainfall. *Trans Am. Soc. Agric. Engrs* **14**, 625–8

Chorley, R. J. 1978. The hillslope hydrological cycle. In *Hillslope hydrology*, M. J. Kirkby (ed.), 1–42. Chichester: Wiley.

De Boodt, M., L. De Leenheer and D. Kirkham 1961. Soil aggregate stability indexes and crop growth. *Soil Sci.* **91**, 138–46.

De Meester, T., A. Imeson and P. Jungerius 1979. Some problems in assessing soil loss from small-scale field measurements. In *Soil physical properties and crop production in the Tropics*, R. Lal and D. Greenland (eds), 466–73. Chichester: Wiley.

De Ploey, J. 1971. Liquefaction and rainwash erosion. *Zeitschr. Geomorph.* **15**, 491–6.

De Ploey, J. 1977. Some experimental data on slopewash and wind action with reference to Quaternary morphogenesis in Belgium. *Earth Surf. Proc.* **2**, 101–15.

De Ploey, J. 1979. A consistency index and the prediction of surface crusting on Belgian loamy soils. In *Seminar on agricultural soil erosion in temperate non-mediterranean climate*, H. Vogt and T. Vogt (eds), 133–7. Strasbourg.

De Ploey, J. 1980. Some field measurements and experimental data on wind-blown sand. In *Assessment of erosion*, M. De Boodt and D. Gabriels (eds), 541–52. Chichester: Wiley.

De Ploey, J. 1981. Crusting and time-dependent rainwash mechanisms on loamy soils. In *Soil conservation. Problems and prospects*, R. P. C. Morgan (ed.), 139–52. Chichester: Wiley.

De Ploey, J. 1983. Runoff and rill generation on sandy and loamy topsoils. *Zeitschr. Geomorph. Suppl.* **46**, 15–23.

De Ploey, J. and H. Mücher 1981. A consistency index and rainwash mechanisms on Belgian loamy soils. *Earth Surf. Proc. Landforms* **6**, 319–30.

De Ploey, J. and J. Savat 1968. Contribution à l'étude de l'érosion par le splash. *Zeitschr. Geomorph.* **12**, 174–93.

De Vleeschauwer, D., R. Lal and M. De Boodt 1978. Comparison of detachability indices in relation to soil erodibility for some important Nigerian soils. *Pedologie* **28**, 5–20.

Dunin, F. 1976. Infiltration: its simulation for field conditions. In *Facets of hydrology*, J. C. Rodda (ed.), 199–227. Chichester: Wiley.

Emerson, W. W. 1959. The structure of soil crumbs. *J. Soil Sci.* **10**, 235–44.

Emerson, W. W. 1977. Physical properties and structure. In *Soil factors in crop production in a semi-arid environment*, I. G. Russell and E. L. Greacen (eds), 120–9. Brisbane: University of Queensland Press.

Epstein, E. and W. Grant 1967. Soil losses and crust formation as related to some soil physical properties. *Soil Sci. Soc. Am. Proc.* **31**, 547–50.

Epstein, E., W. Grant and R. Struchtemeyer 1966. Effects of stones on runoff, erosion and soil moisture. *Soil Sci. Soc. Am. Proc.* **30**, 638–40.

Farmer, E. and P. Van Haveren 1971. *Soil erosion by overland flow and raindrop splash on three mountain soils.* USDA Forest Serv. Res. Paper INT–100.

Farres, P. 1978. The role of time and aggregate size in the crusting process. *Earth Surf. Proc.* **3**, 243–54.

Foster, G. and L. Meyer 1975. Mathematical simulation of upland erosion by fundamental erosion mechanics. In *Present and prospective technology for predicting sediment yields and sources*, 190−207. Agric. Res. Serv. Rep. ARS-S-40.

Fournier, F. 1967. La recherche en érosion et conservation des sols dans le continent africain. *Sols Africains* **12**, 5−51.

Froelich, W. and J. Slupik 1980. Importance of splash in erosion process within a small flysch catchment basin. *Studia Geomorph. Carpatho-Balcanica* **14**, 77−112.

Gardner, W. R. 1956. Representation of soil aggregate-size distribution by a logarithmic-normal distribution. *Soil Sci. Soc. Am. Proc.* **20**, 151−3.

Gasperi-Mago, R. and F. Troeh 1979. Microbial effects on soil erodibility. *Soil Sci. Soc. Am. J.* **43**, 765−8.

Ghadiri, H. and D. Payne 1979. Raindrop impact stress and soil splash. In *Soil physical properties and crop production in the Tropics*, R. Lal and D. Greenland (eds), 95−104. Chichester: Wiley.

Ghadiri, H. and D. Payne 1981. Raindrop impact stress. *J. Soil Sci.* **32**, 41−9.

Harris, R., G. Chesters and O. Allen 1966. Dynamics of soil aggregation. *Advances in agronomy* **18**, 107−69.

Hartmann, R. and M. De Boodt 1974. The influence of the moisture content, texture and organic matter on the aggregation of sandy and loamy soils. *Geoderma* **11**, 53−62.

Hazelhoff, L., P. Van Hoof, A. Imeson and F. Kwaad 1981. The exposure of forest soil to erosion by earthworms. *Earth Surf. Proc. Landforms* **6**, 235−50.

Hillel, D. 1971. *Soil and water.* New York: Academic Press.

Huang, C., J. Bradford and J. Cushman 1982. A numerical study of raindrop impact phenomena: the rigid case. *Soil Sci. Soc. Am. J.* **46**, 14−19.

Imeson, A. 1983. Studies of erosion thresholds in semi-arid areas: field measurements of soil loss and infiltration in Northern Morocco. In *Rainfall simulation, runoff and soil erosion*, J. De Ploey (ed.), 79−89. *Catena* Suppl. 4, Braunschweig.

Imeson, A. and P. Jungerius 1976. Aggregate stability and colluviation in the Luxembourg Ardennes; an experimental and micromorphological study. *Earth Surf. Proc.* **1**, 259−71.

Imeson, A. and F. Kwaad 1980. Field measurements of infiltration in the Rif Mountains of Morocco. *Studia Geomorph. Carpatho-Balcanica* **15**, 19−30.

Imeson, A. and J. Verstraten 1981. Suspended solids concentrations and river water chemistry. *Earth Surf. Proc. Landforms* **6**, 251−63.

Imeson, A., F. Kwaad and J. Verstraten 1982. An examination of the relationship between soil physical and chemical properties and the development of badlands in Morocco. In *Badland geomorphology and pipe erosion*, R. Bryan and A. Yair (eds), 47−70. Norwich: GeoBooks.

Imeson, A. and M. Vis in press. Assessing soil aggregate stability by ultrasonic dispersion and water-drop impact. *Catena*.

Johnson, C., J. Mannering and W. Moldenhauer 1979. Influence of surface roughness and clod size and stability on soil and water losses. *Soil Sci. Soc. Am. J.* **43**, 772−7.

Kemper, W. and E. Koch 1966. Aggregate stability of soils from Western United States and Canada. *USDA Tech. Bull.* No. 1355. Washington: US Dept. of Agriculture.

Kirkby, M. J. (ed.) 1978. *Hillslope hydrology.* Chichester: Wiley.

Kirkby, M. J. 1980. Modelling water erosion processes. In *Soil erosion*, M. J. Kirkby and R. P. C. Morgan (eds), 183−216. Chichester: Wiley.

Lattanzi, A., L. Meyer and M. Baumgardner 1974. Influences of mulch rate and slope steepness on interrill erosion. *Soil Sci. Soc. Am. Proc.* **38**, 946−50.

Lemos, P. and J. Lutz 1957. Soil crusting and some factors affecting it. *Soil Sci. Soc. Am. Proc.* **21**, 485−91.

Low, A. 1954. The study of soil structure in the field and the laboratory. *J. Soil Sci.* **5**, 57−74.

Luk, S. 1978. Soil erodibility in southern Alberta. *Geogr. Ann.* **60A**, 143−9.

Luk, S. 1979. Effect of soil properties on erosion by wash and splash. *Earth Surf. Proc.* **4**, 241−55.

McIntyre, D. 1958. Soil splash and the formation of surface crusts by raindrop impact. *Soil Sci.* **85**, 261−6.

Marshall, T. and J. Holmes 1979. *Soil physics.* Cambridge: Cambridge University Press.

Mazurak, A., L. Chesnin and A. Triarks 1975. Detachment of soil aggregates by simulated rainfall from heavily manured soils in eastern Nebraska. *Soil Sci. Soc. Am. Proc.* **39**, 732−6.

Meeuwig, R. 1970. *Sheet erosion on intermountain summer ranges.* USDA Forest Serv. Res. Paper INT−85.

Meyer, L. 1981. How rain intensity affects interrill erosion. *Trans Am. Soc. Agric. Engrs* **24**, 1472−5.

Meyer, L., G. Foster and M. Römkens 1975. Source of soil eroded by water from upland slopes. In *Present and prospective technology for predicting sediment yields and sources*, 177−89. Agric. Res. Serv. Report ARS−S−40.

Moeyersons, J. 1975. An experimental study of pluvial processes on granite grus. *Catena* **2**, 289−308.

Moeyersons, J. 1983. Measurements of splash-saltation fluxes under oblique rain. In *Rainfall simulation, runoff and soil erosion*, J. De Ploey (ed.), 19−31. *Catena* Suppl. 4, Braunschweig.

Moeyersons, J. and J. De Ploey 1976. Quantitative data on splash erosion, simulated on unvegetated slopes. *Zeitschr. Geomorph. Suppl.* **25**, 120−31.

Moldenhauer, W. 1965. Procedure for studying soil characteristics by using disturbed samples and simulated rainfall. *Trans Am. Soc. Agric. Engrs* **8**, 74−5.

Morgan, R. P. C. 1977. *Soil erosion in the United Kingdom: field studies in the Silsoe area, 1973−75.* Nat. Coll. Agric. Engng Silsoe, Occ. Paper No. 4.

Morgan, R. P. C. 1978. Field studies of rainsplash erosion. *Earth Surf. Proc.* **3**, 295−9.

Morgan, R. P. C. 1981. Field measurement of splash erosion. *Int. Assoc. Hydrol. Sci. Publn.* **133**, 373−82.

Morin, J., Y. Benyamini and A. Michaeli 1981. The effect of raindrop impact on the dynamics of soil surface crusting and water movement in the profile. *J. Hydrol.* **52**, 321−35.

Moskovkin, V. and V. Gakhov 1979. Physical aspects of raindrop erosion. *Soviet Soil Sci.* **11**, 716−20.

Moss, A., P. Walker and J. Hutka 1979. Raindrop-stimulated transportation in shallow water flows: an experimental study. *Sed. Geol.* **22**, 165−84.

Mücher, H. 1981. Soil properties influencing resistance to erosion. In *Compendium of Geomorphological Methods: Part A*, P. van Olm and P. Jungerius (eds), 97−101. In progress.

Mücher, H. and T. Morozova 1983. The application of soil micromorphology in Quaternary geology and geomorphology. In *Soil micromorphology*, P. Bullock and C. Murphy (eds), Vol. 1, 151−94. Berkhamsted: Academic Publications.

Mutchler, C. and R. Young 1975. Soil detachment by raindrops. In *Present and prospective technology for predicting sediment yields and sources*, 113−17. Agric. Res. Serv. Report ARS−S−40.

Nortcliff, S., J. B. Thornes and M. Waylen 1979. Tropical forest systems: a hydrological approach. *Amazoniana* **6**, 557−68.

North, P. 1976. Towards an absolute measurement of soil structural stability using ultrasound. *J. Soil Sci.* **27**, 451−9.

Palmer, R. 1963. The influence of a thin water layer on water drop impact forces. *Int. Assoc. Hydrol. Sci. Publn.* **65**, 141−8.

Philip, J. 1957. The theory of infiltration, 1: the infiltration equation and its solution. *Soil Sci.* **83**, 345−57.

Poesen, J. 1981. Rainwash experiments on the erodibility of loose sediments. *Earth Surf. Proc. Landforms* **6**, 285−307.

Poesen, J. 1983. *Regenerosiemechanismen en bodemerosiegevoeligheid.* Unpublished Ph.D thesis, Catholic University of Leuven.

Poesen, J. in preparation (a). An improved splash transport model.

Poesen, J. in preparation (b). Field measurements of splash erosion to validate a splash transport model. Paper presented at IGU symposium on The role of geomorphological field experiments in land and water management, Aug.−Sept. 1983, Bucharest.

Poesen, J. and J. Savat 1980. Particle-size separation during erosion by splash and runoff. In *Assessment of erosion*, M. de Boodt and D. Gabriels (eds), 427−439. Chichester: Wiley.

Poesen, J. and J. Savat 1981. Detachment and transportation of loose sediments by raindrop splash. Part II: Detachability and transportability measurements. *Catena* **8**, 19−41.

Reeve, I. 1983. *An analysis of the transport of soil by rainsplash and an evaluation of methods of measurement.* Unpublished Master of Nat. Res. thesis, University of New England.

Savat, J. 1976. Discharge velocities and total erosion of a calcareous loess: a comparison between pluvial and terminal runoff. *Rev. Géomorph. Dyn.* **24**, 113–22.

Savat, J. and J. Poesen 1981. Detachment and transportation of loose sediments by raindrop splash. Part I: The calculation of absolute data on detachability and transportability. *Catena* **8**, 1–17.

Scoging, H. and J. B. Thornes 1979. Infiltration characteristics in a semiarid environment. *Int. Assoc. Hydrol. Sci. Publn.* **128**, 159–68.

Seiler, W. 1983. Bodenwasser und Nährstoffhaushalt unter Einfluss der rezenten Bodenerosion am Biespiel zweier Einzugsgebiete im Basler Tafeljura bei Rothenfluh und Anwil. *Physiogeographica* **5.**

Singer, M., and J. Blackard 1982. Slope angle–interrill soil loss relationships for slopes up to 50%. *Soil Sci. Soc. Am. J.* **46**, 1270–3.

Singer, M., J. Blackard and P. Janitsky 1980. Dithionite iron and soil cation content as factors in soil erodibility. In *Assessment of erosion*, M. De Boodt and D. Gabriel (eds), 259–67. Chichester: Wiley.

Singer, M., Y. Matsuda and J. Blackard 1981. Effect of mulch rate on soil loss by raindrop splash. *Soil Sci. Soc. Am. J.* **45**, 107–10.

Smith, R. E. and J. Parlange 1978. A parameter efficient hydrologic infiltration model. *Water Res. Res.* **14**, 533–8.

Sood, M. and T. Chaudhary 1980. Soil erosion and runoff from sandy loam soil in relation to initial clod size, tillage time, moisture and residue mulching under simulated rainfall. *J. Indian Soc. Soil Sci.* **28**, 24–7.

Soyer, J., T. Miti and K. Aloni 1982. Effets comparés de l'érosion pluviale en milieu peri-urbain de région tropicale (Lubumbashi, Shaba, Zaire). *Rev. Géomorph. Dyn.* **31**, 71–80.

Sreenivas, L., J. Johnston and H. Hill 1947. Some relationships of vegetation and soil detachment in the erosion process. *Soil Sci. Soc. Am. Proc.* **12**, 471–4.

USDA 1951. *Soil survey manual.* USDA Handbook no. 18. Washington: US Dept. of Agriculture.

Walker, P., J. Hutka, A. Moss and P. Kinell 1977. Use of a versatile experimental system for soil erosion studies. *Soil Sci. Soc. Am. J.* **41**, 610–12.

Whipkey, R. and M. J. Kirkby 1978. Flow within the soil. In *Hillslope hydrology*, M. J. Kirkby (ed.), 121–44. Chichester: Wiley.

Wischmeier, W., C. Johnson and B. Cross 1971. A soil erodibility nomograph for farmland and construction sites. *J. Soil Water Cons.* **26**, 189–93.

Yair, A. and H. Lavee 1976. Trends of sediment removal from arid scree slopes under simulated rainstorm experiments. *Hydrol. Sci. Bull.* **12**, 379–91.

Yair, A. and J. Rutin 1981. Some aspects of the regional variation in the amount of available sediment produced by isopods and porcupines, northern Negev, Israel. *Earth Surf. Proc. Landforms* **6**, 221–34.

Yair, A., D. Sharon and H. Lavee 1978. An instrumented watershed for the study of partial area contribution of runoff in the arid zone. *Zeitschr. Geomorph. Suppl.* **29**, 71–82.

Young, R. and C. Onstad 1978. Characteristics of rill and interrill eroded soils. *Trans Am. Soc. Agric. Engrs* **21**, 1126–30.

Young, R. and J. Wiersma 1973. The role of rainfall impact in soil detachment and transport. *Water Res. Res.* **9**, 1629–36.

Zaslavsky, D. and G. Sinai 1981. Surface hydrology: IV – Flow in sloping layered soil. *J. Hydraul. Div. Am. Soc. Civ. Engrs* **107**, 53–64.

Zeman, L. and H. O. Slaymaker 1975. Hydrochemical analysis to discriminate variable runoff source areas in an alpine basin. *Arctic Alpine Res.* **7**, 341–51.

6

Soil properties and subsurface hydrology

M. McCaig

Introduction

The soil toposequence or 'catena' is a familiar concept in pedology which recognises the interrelationship between soils developed in different positions on a slope. Different soils develop partially in response to downslope (lateral) movement of drainage water within the slope soil cover which carries with it soil solutes and suspensions. Excessively well-drained and saturated soils experience completely different physicochemical conditions and these in turn encourage either the mobilisation or precipitation/crystallisation of ions within the developing soil profile. Pedologists have made wide use of the toposequence concept to explain the spatial pattern of soils over slopes both in temperate (e.g. Walker 1966) and semi-arid (e.g. Dan *et al.* 1968) environments. Since the first development of the concept, over 40 years ago (Milne 1936), the nature of the subsurface hydrological processes that are of primary importance in the development of slope soil catenas (Ruhe & Walker 1968) has been the subject of a great deal of study.

However, the recent explosion of interest and research into hillslope hydrology has concentrated upon the way these processes affect the hydrological response of slopes and catchments to storm rainfall. Whilst the importance of soil factors in determining both the occurrence and nature of the dominant processes of hillslope drainage and runoff generation has been widely commented upon (e.g. Dunne *et al.* 1975, Dunne 1978) relatively little progress has been made toward linking the results of these process observations with the toposequence concept. Some geomorphological investigations have been made into both the pattern of solutional removal from slope soils (Burt *et al.* 1981, Crabtree & Burt 1983) and the changing chemical state of the soil solution during recharge and drainage (e.g. Trudgill *et al.* 1980a), but the observations and reports available to date only begin to demonstrate how hillslope hydrology may make a significant contribution to the formation of different soil characteristics in toposequences.

In the following review of relationships between soil properties and subsurface flow, two distinct areas are considered. First, there is the question of the significance of pedological characteristics inherited from parent lithology in determining the occurrence of specific flow processes. Second, there is the complex role which subsurface flow processes play in the development of profile characteristics of slope soils, which owing to the small body of available data, can only be treated in an interpretative and general fashion.

Subsurface flow processes: the context of pedological relationships

Following the general recognition that in humid lands rainfall intensities are generally too low and soil surface infiltration rates too high for the widespread generation of surface runoff by the classic Horton (Horton 1933) 'infiltration excess' mechanism, the possibility of storm runoff on slopes being generated by other mechanisms has been widely investigated. These enquiries have recorded the significance of subsurface flow processes, in terms of their contributions to storm and total runoff, in a number of different climatic and geomorphological milieux. The largest set of observations has been made in humid temperate climates and reports are available which detail the occurrence of various processes on hillslopes (Whipkey 1965, 1967), channel sideslope (Beven *et al.* 1977) and in headwater catchments (Weyman 1970, 1974, Dunne & Black 1970a, 1970b, McCaig 1980).

Leaving aside for the present at-a-site variability in specific processes, it would, intuitively, seem probable that the pedological characteristics inherited from local lithology might account for the occurrence of different subsurface drainage processes. The physical property which leads to the occurrence of lateral subsurface drainage in the soil is decreasing permeability with depth. Zaslavsky and Rogowski (1969) held such vertical anisotropy to be among the most important of soil properties. Sharp decreases in vertical permeability commonly occur above the textural B horizon of a clay-rich soil, or in sandy soils at the B−C horizon interface. Significant changes in vertical permeability may also occur elsewhere in soil profiles, e.g. the A horizon, or within organic matter, between amorphous and fibrous peat layers, for example (Dasberg & Neuman 1977).

Soil anisotropy alone, however, is not sufficient to ensure that significant volumes of lateral subsurface slope drainage occur. A supply of water and a gradient or hydraulic head are required for flow to occur. In addition, rapid matrix flow in capillary pores can only occur when the soil becomes saturated, saturation freeing interstitial water from capillary suction. Given these conditions, some proportion of infiltrated rainfall will drain from a slope as lateral subsurface flow.

A number of different subsurface flow processes have been described. From the range of observations available, it would appear that in anisotropic slope soils capable of generating subsurface flow there is a potential continuum of change in the type and relative importance (dominance) of the subsurface flow processes occurring. This continuum ranges between soil matrix 'throughflow' in capillary pores, 'percoline' flow in both capillary and non-capillary macropores and 'pipeflow' in natural pipes that may be several tens of centimetres in diameter. The question of exactly how interactions occur between various site factors, principally pedology, topography and vegetation, has not been rigorously explored, but the hydrological importance of soil macropore networks has recently been stressed by a number of workers (see below).

Throughflow

Throughflow has been defined by Kirkby and Chorley (1967) as the 'downslope flow of water occurring physically within the soil profile, usually under unsaturated conditions except close to flowing streams, occurring where permeability decreases with depth.'

Field studies of hillslope hydrology have shown, however, that saturated throughflow, although spatially non-uniform in its occurrence on slopes, is quite common and an important contributor to stream runoff. In the United Kingdom, the majority of early work on hillslope hydrology (e.g. Weyman 1970, 1974) was carried out in study areas where the local soil conditions were such that matrix throughflow was by far the most important process of subsurface drainage. As a result, there has been a continuing tendency amongst new students of the subject of hillslope hydrology to think of matrix throughflow as the 'normal' process of slope drainage. However, published reports describing throughflow, or subsurface flow more loosely defined as 'interflow' (Amerman 1965), frequently mention that saturated flow in non-capillary pores, structural voids, decayed root channels and tension cracks is a significant component of the 'stormflow' component of total lateral subsurface drainage (e.g. Hursh & Hoover 1941, Whipkey 1965, 1967, Finlayson 1977) and one of the main factors accounting for the spatial variability in measured flow volumes on slopes (e.g. Arnett 1974).

Recently, a good deal of research demonstrating the function of macropores as pathways both for infiltrating and laterally draining soil water has been carried out (e.g. Bouma et al. 1977, Bouma et al. 1981, Bouma et al. 1982). In addition, several studies of the movement and leaching of soluble salts from clayey soils have concluded that a considerable proportion of the total slope drainage within saturated soils bypasses capillary soil pores leaving areas of near-stagnant soil water (e.g. Wild 1972, Quinsberry & Phillips 1976, Kissel et al. 1973). The existence of interconnected preferential flow paths or 'pipes' within soil sections that can lead to this sort of bypass flow has been vividly demonstrated by the use of X-ray microscopy (Anderson 1979).

In the context of examining hydrological relationships with soil characteristics, these observations appear to suggest that it is both inappropriate and practically impossible to consider matrix throughflow as a single, isolated process of soil-water and sediment transportation. It is more appropriate to view the matrix and macropore flow mechanisms within a slope soil as an interrelated group, acting (in broad terms) as a 'sheet' process over the slope in which the vertical and downslope vectors of flow concentration are generally larger than the lateral ones.

Percolines and pipeflow

In contrast to the case of throughflow type processes, percolines (Bunting 1961, 1964) and natural pipe systems are expressions of slope drainage that occur when there is lateral concentration of drainage down slope. Such concentration may occur, for example, in an apparently random fashion as a consequence of the development of networks of macropores, tension cracks, structural voids or decayed root systems within the soil. The most intensively studied systems of natural pipes in the UK have been developed in upland soils with thick organic (peaty) horizons or in blanket peat deposits. In this sort of environment it has been suggested that pipe systems develop owing to the formation of desiccation cracks (Gilman & Newson 1980). The nature of superficial deposits and irregularities in the basal weathering profile (due to jointing patterns, etc.) are alternative factors which might cause lateral concentration of subsurface slope drainage.

Seepage lines or subsurface flow in discrete lines not occupied by pipes (percolines) have not been studied in any great detail since Bunting's description of this form of drainage. It would seem probable that percolines occur without the development of pipes when the texture of the local soil is such that the maximum stable size of macropores is relatively small.

For some time there has been a consensus regarding the factors which may give rise to extensive natural pipe networks in hillslope soils. Many workers, in a diverse range of climates (e.g. Fletcher et al. 1954, Parker & Jenne 1967, Crouch 1976), have produced lists of factors thought to be significant controls of the occurrence of piping. In reviewing previous research Jones (1971) produced a list of such soil and site factors arranged in assumed order of importance:-

(a) susceptibility of soil to cracking in dry periods, high silt-clay content and high percentage of swelling clay (especially montmorillonite);
(b) periodic high intensity rainfall and devegetation;
(c) biotic break-up of the soil and a relatively impermeable basal horizon;
(d) an erodible layer above this base, high exchangeable sodium and high base exchange capacity or high soluble salts (i.e. alkaline soils);
(e) steep hydraulic gradient.

Later studies of piping (e.g. Barendregt & Ongley 1977) have re-emphasised the importance of the occurrence in the soil of a high percentage of swelling clay (montmorillonite) and seasonally dry periods which desiccate the upper soil horizons.

In some respects, however, Jones's list and those produced by the original workers on which it is based are misleading since they do not differentiate between the conditions *necessary* for specific processes of pipe erosion and the general pedogeomorphic setting in which natural pipes are liable to form and play a significant role in hillslope hydrology. Re-examination of the lists of 'causal factors' identified in studies of piping (Table 6.1) shows that the most significant features that define the pedogeomorphic setting are:

(a) vertical soil anisotropy;
(b) a sufficient hydrostatic head/hydraulic gradient;
(c) the occurrence of periods of desiccation;
(d) a downslope outlet (gully, stream channel, etc.).

It can be readily seen that (a) and (b) are the same conditions that are described as sufficient to ensure that some form of lateral subsurface drainage occurs on a slope. The occurrence of periods of desiccation appears, therefore, to be the principal factor that differentiates sites that experience piping from those that do not. Hence it can be suggested that *sufficient* conditions for piping (periodic desiccation of the soil) are determined by climate (or, more generally, variability in the supply of slope drainage water) which acts indirectly through its influence on soil structure. However, certain soil chemical and textural characteristics can be considered as *necessary* conditions for the erosive development of natural pipe networks. Published studies give examples of specific

Table 6.1 Sufficient conditions for piping.

	Soil anisotropy	Hydrostatic head/ hydraulic gradient	Periods of desiccation	Downslope outlet	Other
Fletcher *et al.* 1954	X	X		X	erodible soil layer
Parker and Jenne 1967	X	X	X	X	high % exchangeable sodium
Crouch 1976	X	X	X	–	change in vegetation
Barendregt and Ongley 1977	X	X	X		saturation in basal soil horizons
McCaig 1980	X	X	X	–	

mechanisms of pipe erosion, particularly in semi-arid lands and in soils with high exchangeable sodium content (e.g. Sherard *et al.* 1972).

Relationships with climatic and geomorphological variables

Before considering details of relationships between slope soils and the processes of subsurface drainage described in the preceding sections, it is worthwhile examining the main factors (other than pedology) that appear to affect the occurrence of these processes at any site. This should help to draw clearer distinctions between those process—soil correlations that relate to such primary factors and those that are due to process—response 'feedback' mechanisms in the soil landscape system.

Microclimate   The preceding discussion suggests that, in general terms at least, the dominant subsurface flow process on a slope is perhaps best considered to be a reflection of soil structure. Soil structure, whilst being influenced by the textural characteristics of the local substrate, is also highly dependent upon such factors as the general climatic regime, local microclimate and other factors that influence the frequency and degree of wetting and drying within the soil (Finney *et al.* 1962).

Field studies of subsurface flow processes in the UK have usually given detailed consideration only to the studied sites' dominant drainage process. Most commonly, these dominant processes have been the two 'end members' of the range of flow processes, namely saturated soil matrix throughflow (e.g. Anderson & Burt 1978) and natural pipeflow (e.g. Jones 1978, Gilman 1972). The questions of dominance and relationships with site conditions have therefore received very little attention. As a result the role of subordinate processes in hillslope hydrological systems has been generally underemphasised, and factors that determine the occurrence of specific processes at a site have not been clearly differentiated from those accounting for their local variability. In particular, whilst the effect of microclimate on soil conditions has been studied (Reid 1975, 1977), the effects of local variations in factors such as rainfall receipt and evapotranspiration on the nature of subsurface flow processes has been largely ignored. This omission is somewhat surprising since differences between drainage processes and soil types between slopes of different aspects can be quite striking. For example, in a study of erosion processes around an upland stream head in the Yorkshire Pennines (McCaig 1980) it was noted that whilst natural pipes were extensively developed on the south-west-facing slope studied, on nearby north-facing slopes of similar general physiography and lithology (Millstone Grit), pipe systems were much more limited in extent. In the sandy soil of the mid-section of the south-facing slope relatively drier conditions during late summer led to more frequent and intense desiccation of the upper peaty horizon. This produced a closely spaced network of tension cracks, which were presumably exploited by drainage water, and eventually led to the erosive development of the natural pipes observed. The difference in character between these slopes would seem to be due, at least in part, to their aspect.

Geomorphology   The topographic position of the soil on a slope and its relation to zones of soil moisture storage and supply can, potentially, have a greater total effect on the moisture regime of a soil than any microclimatic differences due to aspect. The importance of slope position, in terms of the frequency and volume of low draining through a soil section, has been widely noted (Troeh 1964) and can be seen from the equations that describe the throughflow process.

In homogeneous porous media, saturated matrix flow can be described by Darcy's Law (Darcy 1856):

$$Q = K \ d\phi/dx \qquad (6.1)$$

where Q is the discharge rate, K is the saturated hydraulic conductivity and, $d\phi/dx$ is the combined moisture and topographic gradient.

At any point on a slope the depth to which saturation occurs is determined by the mass balance between influent drainage from up slope, vertical infiltration and downslope discharge. Thus, for any point on a slope the topographic parameters of slope gradient and area drained per unit contour length are of key importance in determining the volume of discharge passing that point and the frequency with which saturated conditions, at any depth within the soil profile, become established. The relationship between throughflow, surface topography, subsurface flow volumes and spatial variations in areas of soil surface saturation around a headwater hillslope hollow have been described in detail by Burt (1978) and Anderson and Burt (1978).

One possible approach towards determining how geomorphology may affect the nature of subsurface flow processes is to compare sites where different processes occur in terms of some conceptual model of hillslope form. The nine-unit landform model of Dalrymple *et al.* (1968; see also Conacher & Dalrymple 1977) is perhaps the best known of this type of model, which seeks to set 'catenary' sequences of relationships between geomorphological processes and soil conditions on a slope into a general framework (Fig. 6.1). The elements of the slope profile which Anderson and Burt studied can be broadly described as belonging to units 2, 3, 5 and 6 of the model (Fig. 6.2). A noteworthy feature of the profile of the site studied by these authors is the relatively narrow, flat, interfluve above the summital convexity which offers little potential for soil-water storage between storms and is unlikely to generate surface runoff under normal rainfall conditions, a point of some difference between this site and sites where extensive pipe networks have been observed.

Whilst links between the occurrence of laterally concentrated subsurface slope drainage (piping) and soil conditions have been explored in detail in both humid temperate and semi-arid lands (e.g. Parker & Jenne 1967, Jones 1971), relationships between these processes and geomorphology have not been a subject of such intense interest. Jones (1981), however, has reviewed the pedo-geomorphic environment in which piping occurs, using the nine-unit landform model described above, and identifies (Jones 1981, p.39) the landform units which can be regarded as 'preferred locations' for piping as (a) the seepage slope

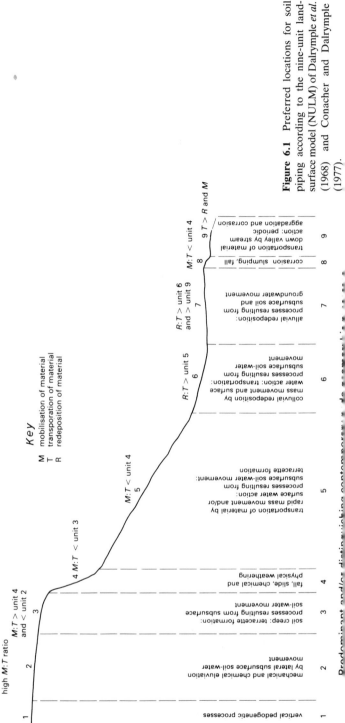

Figure 6.1 Preferred locations for soil piping according to the nine-unit land-surface model (NULM) of Dalrymple et al. (1968) and Conacher and Dalrymple (1977).

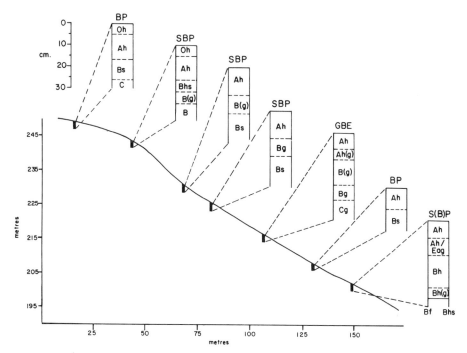

Figure 6.2 Generalised pattern of downslope change in soil characteristics in a hillslope hollow (composite redrawn from Burt *et al.* 1981, Figures 1 and 6). BP brown podzolic soil, SBP stagnogleyic brown podzolic soil, GBE gleyed brown earth, S (B) P stagnogleyic brown podzolic/stagnopodzol, BE brown earth.

(unit 2), (b) the transportational mid-slope (unit 6) and (c) the alluvial toe slope (unit 7). It is apparent from the previous discussion that these units are the same as those on which throughflow-type processes have been observed. It is suggested, therefore, that it is the degree of development of different slope facets and their arrangement (rather than simply their occurrence), in the slope profile which influences the relative importance of different subsurface drainage processes. For example, the geomorphology of the gritstone moorlands of Yorkshire, with broad and flat, semi-permanently saturated, peat-covered summits and relatively steep mid-slope sections subject to seasonal desiccation, provides conditions that favour piping rather than laterally unconcentrated slope drainage as throughflow. In contrast, the more distinctly defined divides and convexo-concave slopes of the Mendip and Quantock hills of south-west England offer less opportunity for lateral concentration of drainage, and here throughflow is the dominant process.

Relationships between processes and soil formation

In Figure 6.1 the soil conditions described for each of the slope unit associations are produced by a balance between the processes of: (a) solution of material *in situ*, (b) transport and deposition of chemical ions, and (c) transport and deposition of solids. In humid temperate climates, at least, it seems likely that other mechanical and gravitational processes of soil erosion exceed, in their effects on soil development, hydrologically driven processes only on convex slope segments which experience very little infiltration and drainage or on bare 'free face' segments (units 3 and 4 of the nine-unit model).

Solution of material in situ

The suggestion has frequently been made that the deeper soils found in topographic hollows around stream heads as well as the hollows themselves result from locally high rates of solutional removal by lateral subsurface drainage (e.g. Cleaves 1970). It is not until recently, however, that attempts have been made to determine whether such suggestions are accurate. Burt *et al.* (1981) and Crabtree and Burt (1983) used the microweight loss tablet technique to assess the solutional denudation around a hillslope hollow in which soil moisture conditions were monitored. Their results indicated that the thickest soils and highest rates of solutional removal from the unweathered rock tablets did indeed occur where the highest volume of subsurface flow is recorded and saturated conditions are most prolonged. The development of deeper soils in areas of flow convergence would appear to be the consequence of a greater intensity, in these sites, of the combined processes of 'weathering' (the physical disintegration of the rock structure and the mobilisation of ions under high negative Eh conditions) and the removal of weathering products, in the soil solution, from the profile. It should be noted, however, that the solute concentration in slope drainage water is controlled by a number of closely interrelated factors in addition to flow volume and the soil's physicochemical state.

One such important factor is vegetation which, in the case of different species of moorland grass communities, has been found to be closely correlated with soil hydrological conditions (Rutter 1955, McCaig 1980). The effects of vegetation on soil- and surface-water solute concentrations range between those due to plant physiology and those due to the physical presence of a high proportion of organic material in those soil horizons which carry the bulk of the subsurface and surface flow. McCaig (1980) noted a persistent positive difference between the solute concentration in water draining through natural pipe networks cutting through mineral soil and that of flow from a boggy 'flush' area. Whilst direct calculations from continuously measured discharge data were not possible, periodic recording of flow rates from the different runoff source areas throughout the year suggested that ephemeral saturation of the mineral soil layers in which the pipes were formed and the different vegetation communities combined to give a higher rate of solutional denudation (per unit area drained) from the piped areas than the more frequently flowing flush area. However, Trudgill *et al.* (1980b) have presented results which show that the displacement and dilution of long residence time soil water by macropore flow may produce an initial flushing of solute-rich water followed by a progressive decrease in solute concentrations.

Transport and deposition of chemical ions

The process of chemical removal, or eluviation, from a soil profile requires two basic conditions. First, chemical ions within the soil must be in a mobile state and, second, there must be an effective vertical or lateral mechanism for the removal of mobilised ions from the soil. Both the frequency and extent of eluviation within the soil have long been recognised as being closely correlated with slope drainage conditions. Certain ions, notably iron and manganese, are mobilised within the soil as a result of the establishment of reducing conditions when a soil horizon becomes saturated. When saturated, oxygen within the soil atmosphere is consumed by soil organisms and anaerobic micro-organisms convert available oxidised elements into their reduced, more soluble, form. Iron and manganese in their oxidised state are important soil colouring agents, giving the soil a brightish hue. Removal of these elements from parts of the soil profile leads to the progressive development of the characteristic pallid grey/brown mottled morphology of gleyed horizons.

Gleying in soils has been used as a criterion of drainage conditions in soil studies and classification (e.g. Canadian Department of Agriculture 1970). The degree of mottling developed in a soil horizon is dependent upon the length of time for which saturated conditions exist in that horizon. For example, Veneman *et al.* (1976) found that the chroma values of mottles in a soil toposequence were related to the duration of saturation. Mottles with chromas of less than 2 occurred only in soils experiencing periods of saturation exceeding several days each year, while mottles with chromas of less than 1 were present only in soils which were saturated for several months. Several other studies have established empirical correlations between soil-water regime and quantitative indices of soil morphology (e.g. Latshaw & Thompson 1968, Crown & Hoffman 1970, Simonsen & Boersma 1972). The rate of development of gley characteristics varies with factors such as soil depth and parent lithology, development being slower in the deeper parts of a profile owing to the decrease in soil organic matter with depth. However, it has been suggested that, as a broad rule of thumb, recognisable gleying occurs at and below soil depths that experience saturation by drainage water for approximately two months of the year (Dunne *et al.* 1975, Daniels *et al.* 1971). The depth at which gleying is found in the soil can be expected, therefore, to correlate over a hillslope with the spatial pattern of areas of surface saturation that develop during and after storm rainfall. On a rectilinear slope drained by throughflow, gley morphology will become more distinctly developed and possibly occur at a shallower depth with increasing distance down slope. On irregular slopes, gleying will be most strongly developed within hollows and around stream heads, producing the so-called bog-gley soils (see, for example, Fig. 6.2). In these areas soil saturation occurs owing to the convergence of subsurface flow, and surface runoff during drainage is generated by the 'return flow' (Dunne & Black 1970a, 1970b) from up slope. The morphology of the 'stagno' or bog-gley soils developed in these sites can be expected to be somewhat different from that of surface water or pseudo-gley soils formed where saturation occurs owing to rainfall exceeding the soil infiltration or water storage capacity. These differences arise since, in return flow sites, there is little or no downward movement of drainage water within the soil and so the down-profile illuviation of mobilised ions which produces the thick iron pans found in some pseudo-gley soils is limited.

The alternation of oxidising and reducing conditions in slope soils up slope of areas of perennial saturation leads to the development of quite distinct gley morphologies in different slope positions. In particular, ions mobilised from surface horizons may be oxidised and partially fixed within the horizons in which ephemeral subsurface drainage occurs. For example, Gerasimova (1968) detailed different types of gley morphology developed in the brown forest soils of the Carpathians. Here, gleying occurred in the lowest (B−C horizon) part of the soil profile. At the slope base 'pseudo-gley' soils were found whilst on the mid-slope 'ecigley' soils characterised by weakly crystallised illuvial iron oxides were developed. The genesis of these 'ecigley' soils was described as due to the lateral movement of soil solutes and suspensions of clay and iron oxide as well as 'seasonal over-wetting'. This form of gley morphology has been termed a 'typical' gley (Bunting 1968, p. 126).

The extent to which illuvial features are developed in the profiles of soils at the base of slopes, it is suggested, is determined in a rather complex fashion by the magnitude, frequency and sequence of flow on a slope. In the case where influent drainage from upslope becomes 'ponded' at the surface or impeded within the subsoil and fails to drain as surface runoff, evapotranspiration during dry spells may lead to the concentration of salts within the profile of the soil. Subsurface flow during the next rainfall will deliver more salts to the soil and a cycle of illuviation due to lateral flow is established. This simplistic model of the salt enrichment process has frequently been used to explain salt-rich slope-base soils. However, studies of hillslope hydrology allow consideration of rather more complex situations. For example, when describing the percolines developed on gritstone moorland on Blackbrook Moor (Derbyshire) and Tinghede (West Jutland), Bunting (1961) noted that the thickest development of semi-indurated, illuvial soil layers on the slopes occurred in their central parts and was not associated with topographic features such as breaks in slope. It would seem, therefore, that in the simple surface depression of the percolines, lateral flow was in some way impeded. The growth of areas of surface soil saturation around stream heads and the mechanism of 'return flow' runoff generation may together account for this impedence of flow. A hypothetical model of the relationships between flow, saturated conditions and illuviation during a sequence of wetting and drying is shown in Figure 6.3.

Following the onset of rainfall, the initial growth of the area of surface soil saturation within the percoline is due to infiltration, subsurface flow being too slow to participate in this preliminary expansion. During the infiltration phase, down-profile illuviation of mobilised ions and organo-chelated compounds is possible (Fig. 6.3a). As rainfall and drainage proceed, lateral flow becomes more important, the rate of subsurface inflow to the percoline increases beyond the capacity of the soil matrix, and a further expansion of the area of surface

(a)

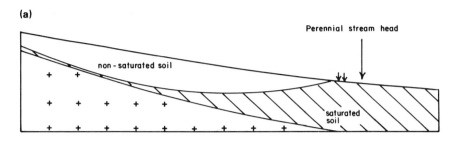

(b)

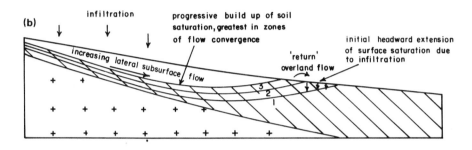

(c)

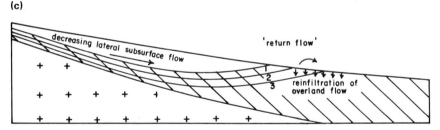

Figure 6.3 Generalised soil-water movements during a wetting and drying cycle in a percoline/stream head hollow. (a) dry conditions; (b) wetting; (c) drying.

saturation occurs (Fig. 6.3b). During this phase of drainage, down-profile movement of the soil solution is limited and the infiltrated water from the earlier phase of drainage may be bypassed by the laterally moving 'return flow'. During the drying phase, subsurface runoff rates are initially maintained by continued lateral inflow from up slope. As the volume of this flow decreases, further drainage is fed by a reduction of the saturated area within the percoline. If the rate of drying is most rapid in those areas proximal to the surface drainage channel, it is possible that re-infiltration of surface water within the percoline may occur during this phase (Fig.6.3c). This may lead to some further illuviation within the percoline soil. Bunting's description of the percolines on Blackbrook Moor gives some evidence that more rapid, or more complete, drying does occur in the lower parts of percolines where open stream channels provide a ready route

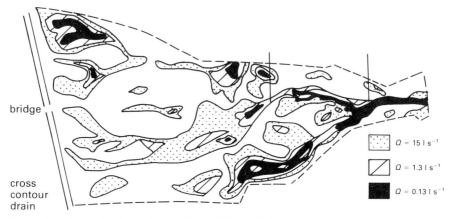

Figure 6.4 Areas of surface soil saturation at different discharges in the Slithero Clough catchment (from McCaig 1980).

for drainage water. Bunting (1961) noted that the brightest coloured (reddish-yellow) sands in the C horizons of the soils developed in the percolines occurred in the lower parts nearest to the perennial stream channel, suggesting that oxidising conditions occurred in this part of the slope.

The examples given above refer specifically to the case of subsurface drainage in percolines. However, the sorts of relationships suggested between the frequency and sequence of hydrological conditions, and the development of certain soil characteristics, should be quite similar in the case of slopes drained by throughflow type processes. That is to say, the main areas in which illuvial features will develop can be expected to be ones of ephemeral saturation, for example, those peripheral to a stream head hollow.

A slightly different and more variable case can be expected where slope drainage is through natural pipe networks. In slopes drained by pipes, the contribution to storm runoff from upslope areas is, generally, more rapid than drainage by throughflow or percoline drainage. However, pipe networks are commonly discontinuous and soil moisture conditions over a piped slope tend to be 'patchy', areas of surface saturation occurring both in pipe source areas and below pipe effluxes (e.g. see Fig. 6.4). In consequence, a more complex pattern of soil features, due to the precipitation, oxidation and reduction of ions moved in the drainage water, can be expected.

Mechanical transport and erosion

The transport of fine particles within the soil profile by both vertically infiltrating and laterally draining soil water has been widely reported (e.g. Bunting 1961, Jones 1971). The size grades of material moved are commonly clays to medium silts. The name, often rather loosely applied, to the process of particle translocation within the soil profile is 'lessivage'.

The processes of particle translocation lead to the development of relatively clay- or silt-rich soil horizons, and when some of this movement is lateral, in association with subsurface flow, the soils in areas of drainage convergence (slope-base hollows) may show particularly well-developed lessivage features. A characteristic soil micromorphological feature where clay translocation occurs is the development of clay skins or 'cutans' on ped surfaces.

The mechanics of fluid flow in porous media are well known and it is possible to calculate the pore sizes and flow velocities which have the capacity to transport material within soil pores. Ingles (1964) showed, for example, that a clay particle of 2 μm diameter would require a flow velocity of 3.6×10^{-6} m s^{-1} (assuming laminar flow) to prevent it settling from that flow. This result suggests that the transport of clay particles in throughflow draining in most loamy soils is possible. However, the above calculations assume laminar flow in straight tubes of equal size, which is a rather unrealistic assumption for natural soils. Resistance to flow and the trapping of particles in bends and on other surface irregularities all serve to inhibit lateral translocation of particles. There is evidence to suggest also that clay translocation is controlled by a rather complex set of both physical and chemical conditions. For example, Mel'nikov and Kovenya (1974) showed, using radioactive tracing techniques, that the movement of fines (particles <0.5 μm) by drainage water within the soil is affected by surface roughness and soil pH (which affects the degree to which clay particles are dispersed). Clay particles were found to move most freely in soils of neutral reaction. In media with a high surface area, particle movement was relatively slow, suggesting repeated adhesion of particles as they moved through soil pores.

It has been suggested that macropores and pipes themselves develop in soils as a result of the mechanical eluviation of material from within the soil matrix (Aitchison & Wood 1965), the process of soil 'boiling' (the development of soil pipes due to instability resulting from seepage pressure) being described by Terzaghi and Peck (1948) over 30 years ago. This latter process has been of particular concern to civil engineers in the context of earth fill dam failure but would seem to be of minor significance on natural slopes.

The association of piping with alkaline soils has been closely studied and the process of pipe formation through clay dispersion appears to be important in such soils (Cole & Lewis 1960, Wood *et al.* 1964, Ingles & Aitchison 1970). Clay dispersion occurs in soil where clay and/or highly concentrated salt solutions bond the soil together. Lateral influent drainage into the soil weakens these bonds, either because of the disequilibrium between the high salt concentrations of the interstitial soil water and much lower concentrations in the drainage water or by cation exchange (replacement of divalent by monovalent ions) on the surfaces of clay molecules. A comprehensive review of the literature on this subject is provided by Jones (1981, p.96), who makes the general observation that the critical or necessary conditions under which this type of erosion may occur is a complex function of the soluble sodium percentage and clay minerals present in the soil. The chemical nature, particularly the total cation content, of rainfall and influent subsurface flow are also important factors in this process. The effect of decreasing electrolyte concentrations in soil water is generally to decrease soil permeability (Quirk & Schofield 1955, Christansen & Ferguson 1966).

Turbulent flow in macropores and pipes, once formed, is capable of a much greater amount of both transportation and physical erosion than matrix through-flow in capillary pores. Pilgrim and Huff (1983) observed that rapid subsurface stormflow in macropores carried material originally suspended within the flow by the action of splash erosion. This form of erosion offers a mechanism for the translocation of soil surface materials to lower parts of the soil profile.

At the extreme, pipe roof collapse following growth (in radius) of a pipe leads to the removal of large amounts of sediment. McCaig (1980) noted that whilst soil depths in a drainage line depression occupied by a discontinuous and partially collapsed pipe system were generally deeper than on the adjacent slopes (Fig. 6.5), soil depth in this depression was irregular and the soils devel-

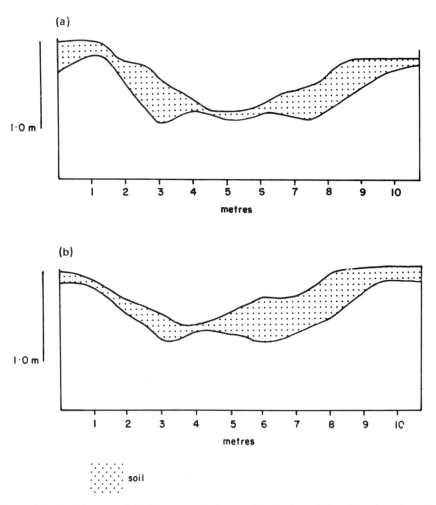

Figure 6.5 Variations in soil depth across a piped stream head hollow, Slithero Clough catchment, Yorkshire (from McCaig 1980). See Figure 6.4 for locations.

oped showed only poor horizonation. The depth of soil within the depression was generally observed to be inversely related to the steepness of the drainage line's 'side slopes', the deepest soils being bounded by the shallowest sides. This relationship suggests that soil depth within the drainage line was controlled by a sequence of pipe growth, collapse and infill by erosion of the pipe's sides, aided by more rapid solutional weathering of the bedrock in the wetter floor of the drainage line. A sequence of several such cycles would cause the depression to grow in amplitude, until eventually it formed a topographically recognisable 'stream-head hollow'. This sequence of development would be self-terminating since, once the depression grows beyond a certain size, the volume of sediment delivered by the infilling processes will be insufficient to keep pace with the downslope removal by the 'pipeflow' and so an open stream channel would be formed.

Summary

In the preceding sections an account has been given of both the context in which soil characteristics affect the occurrence of specific subsurface flow processes and the role these processes play in the formation of certain soil characteristics. This review is, however, by no means comprehensive, this being partially due to the limited amount of published data directed toward investigating these relationships. Nevertheless, two general summary observations can be made. First, in terms of controlling the occurrence of different subsurface flow processes it is recognised that soil anisotropy is a basic prerequisite for any form of lateral subsurface flow to occur. One of the most important factors determining the nature of these processes is the structure of the soil, which is influenced at least as much by the climate, moisture regime and physiography of a slope as by its parent lithology. In this sense, therefore, pedological controls on subsurface flow processes are indirect ones. The pedological characteristics which determine the nature of subsurface drainage processes are, themselves, a reflection of the influence of climate, topography and lithology. Second, it is evident that sub-surface drainage plays both a passive and an active role in the formation of certain soil characteristics. Passive effects can be considered to be those (mobilisation/precipitation of ions) that result from the changing soil moisture and aeration conditions within a soil profile that experiences lateral subsurface drainage, while soil solutes and suspensions are actively transported by these processes.

Acknowledgements

I am indebted to Susan Dale and Clive Smith of Geomorphological Services Ltd for their help in producing the manuscript and diagrams of this text. I would also like to thank the editors for their efforts in improving this paper.

References

Aitchison, G. D. and C. C. Wood 1965. Some interactions of compaction, permeability and post-construction deflocculation affecting the probability of piping failure in small earth dams. In *Proceedings of the 6th International Conference on Soil Mechanics and Foundation Engineering*, Canada, Vol. 2, 442–6.

Amerman, C. R. 1965. The use of unit source watershed data for runoff prediction. *Water Res. Res.* **1** (4), 499–508.

Anderson, M. G. 1979. On the potential of radiography to aid studies of hillslope hydrology. *Earth Surf. Proc.* **4**, 77–83.

Anderson, M. G. and T. P. Burt 1978. The role of topography in controlling throughflow generation. *Earth Surf. Proc.* **3**, 331–44.

Arnett, R. R. 1974. Environmental factors affecting the speed and volume of topsoil interflow. In *Fluvial processes in instrumented watersheds*, K. G. Gregory and D. E. Walling (eds) 7–23. IBG Spec. Publn No. 6.

Barendregt, R. W. and E. D. Ongley 1977. Piping on the Milk river canyon, southwestern Alberta – a contemporary dryland geomorphic process. In *Erosion and solid material transport in inland waters*, Proceedings of the Paris symposium. 233–243. IAHS–AISH Publn.

Beven, K. J., R. Iredale and M. G. Kew 1977. *Hydrologic response of headwater and sideslope areas in the Crimple Beck catchment*. Working Paper 184, School of Geography, Leeds University.

Bouma, J., C. F. M. Belmas and L. W. Dekker 1982. Water infiltration and redistribution in a silt loam subsoil with vertical worm channels. *J. Soil Sci. Soc. Am.* **46**, 917–21.

Bouma, J., L. W. Dekker and C. J. Muilwijk 1981. A field method of measuring short circuiting in clay soils. *J. Hydrol.* **52**, 347–54.

Bouma, J., A. Jongerius, O. Boersma, A. Jager and D. Schoonderbeek 1977. The function of different types of macro-pores during saturated flow through four swelling soil horizons. *J. Soil Sci. Soc. Am.* **41**, 945–50.

Bunting, B. T. 1961. The role of seepage moisture in soil formation, slope development and stream initiation. *Am. J. Sci.* **259**, 503–18.

Bunting, B. T. 1964. Slope development and soil formation on some British sandstones. *Geogr. J.* **130**, 73–9.

Bunting, B. T. 1968. *The geography of soil*. London: Hutchinson University Library.

Burt, T. P. 1978. *Runoff processes in a small upland catchment with special reference to the role of hillslope hollows*. Unpublished Ph D. thesis, Bristol University.

Burt, T. P., R. W. Crabtree and N. A. Fielder 1981. *Patterns of hillslope solutional denudation in relation to the spatial distribution of soil moisture and soil chemistry over a hillslope hollow and spur*. Paper presented at the Annual Conference of the IGU Commission for field experiments in geomorphology, University of Exeter, 1981.

Canadian Department of Agriculture 1970. *A system of soil classification for Canada*. Ottawa: Department of Agriculture.

Christansen, D. R. and H. Ferguson 1966. The effect of interactions of salts and clays on unsaturated water flow. *Soil. Sci. Soc. Am. Proc.* **30**, 549.

Cleaves, E. T. 1970. Geochemical balance of a small watershed and its geomorphic implications. *Bull. Geol Soc. Am.* **81**, 3015–31.

Cole, D. C. H. and J. G. Lewis 1960. Piping failure of earthen dams built of plastic materials in arid climates. In *Proceedings of the 3rd Australia – New Zealand Conference on Soil Mechanics and Foundation Engineering*, 93–9.

Conacher, N. J. and J. B. Dalrymple 1977. The nine-unit land surface model: an approach to pedogeomorphic research. *Geoderma* **18**, 1–54.

Crabtree, R. W. and T. P. Burt 1983. Spatial variation in solutional denudation and soil moisture over a hillslope hollow. *Earth Surf. Proc. Landforms* **8**, 151–61.

Crouch, R. O. 1976. Field tunnel erosion – a review. *J. Soil Conserv. Service, New South Wales* **32**, 98–111.

Crown, P. H. and D. W. Hoffman 1970. Relationships between water table levels and type of mottles in four Ontario grey soils. *Canadian J. Soil Sci.* **50**, 453−5.

Dalrymple, J. B., R. J. Blong and A. J. Conacher 1968. A hypothetical nine-unit land surface model. *Zeitschr. Geomorph.* **12**, 60−76.

Dan, J., D. M. Yaalon and M. Koyundjisky 1968. Catenary soil relationships in Israel 1. The Netanya Catena on coastal dunes of Sharon. *Geoderma* **2**, 95−120.

Daniels, R. B., E. E. Gamble and L. A. Nelson 1971. Relations between soil morphology and water table levels on a dissected North Carolina coastal plain surface. *Soil Sci. Soc. Am. Proc.* **35**, 781−4.

Darcy, M. 1856. *Les fontaines publiques de la Ville de Dijon.* Paris: Dalmont.

Dasberg, S. and S. P. Neuman 1977. Peat hydrology in the Mula basin, Israel 1. Properties of peat. *J. Hydrol.* **32**, 219−39.

Dunne, T. 1978. Field studies of hillslope flow processes. In *Hillslope hydrology*, M. J. Kirkby (ed.) 177−222. London: Wiley.

Dunne, T. and R. D. Black 1970a. An experimental investigation of runoff production in permeable soils. *Water Res. Res.* **6**, 478−90.

Dunne, T. and R. D. Black 1970b. Partial area contributions to storm runoff in a small New England watershed. *Water Res. Res.* **6**, 1296−311.

Dunne, T., T. R. Moore and G. M. Taylor 1975. Recognition and prediction of runoff producing zones in humid regions. *Hydrol. Sci. Bull.* **20**, 305−27.

Finlayson, B. 1977. *Runoff contributing areas and erosion.* Res. Pap. 18, School of Geography, University of Oxford.

Finney, H. R., N. Molowaychuk and M. K. Meddeson 1962. The influence of microclimate on the morphology of certain soils of the Allegheny Plateau of Ohio. *Soil Sci. Soc. Am. Proc.* **26**, 287−92.

Fletcher, J. E., E. Harris, M. D. Peterson and V. N. Chandler 1954. Piping. *Trans Am. Geophys. Union* **35**, 258−62.

Gerasimova, M. I. 1968. Types of gley phenomena in Carpathian foothill soils. In *Transactions of the International Congress on Soil Science*, Adelaide, Vol. IV, 433−9.

Gilman, K. 1972. *Pipe flow studies in the Nant Gerig.* NERC Institute of Hydrology, Subsurface Section, Internal Report 50.

Gilman, K. and M. Newson 1980. *Soil pipes and pipeflow − a hydrological study in upland Wales.* BGRG Res. Monog. No. 2, Norwich: GeoBooks.

Horton, R. E. 1933. The role of infiltration in the hydrological cycle. *Trans Am. Geophys. Union* **14**, 446−60.

Hursh, C. R. and M. D. Hoover 1941. Soil profile characteristics pertinent to hydrologic studies in the Southern Appalachians. *Soil Sci. Soc. Am. Proc.* **6**, 414−22.

Ingles, O. G. 1964. *The effect of lime treatment on permeability−density−moisture relationships for three montmorillonitic soils.* Paper No. 30, Water Research Foundation of Australia Ltd and soil mechanics section CSIRO, Colloquium on failure of small earth dams, 16−19 Nov.

Ingles, O. G. and G. D. Aitchison 1970. Soil−water disequilibrium as a cause of subsidence in natural soils and earth embankments. In *IASH-UNESCO, International symposium on land subsidence*, 1969, Tokyo, 342−53. IASH Publn. 89.

Jones, J. A. A. 1971. Soil piping and stream channel initiation. *Water Res. Res.* **7**, 602−10.

Jones, J. A. A. 1978. Soil pipe networks: distribution and discharge. *Cambria* **5**, 1−21.

Jones, J. A. A. 1981. *The nature of soil piping − a review of research.* BGRG Res. Monogr. No 3. Norwich: GeoBooks.

Kirkby, M. J. and R. J. Chorley 1967. Throughflow, overland flow and erosion. *Bull. Int. Assoc. Sci. Hydrol.* **12**, 5−21.

Kissel, D. C., J. T. Ritchie and E. Bugnett 1973. Chloride movement in undisturbed swelling clay soil. *Soil Sci. Soc. Am. Proc.* **37**, 21−4.

Latshaw, G. J. and R. F. Thompson 1968. Water table study verifies soil interpretations. *J. Soil Water Conserv.* **23**, 65−7.

McCaig, M. 1980. *Erosion of stream head hollows*. Unpublished Ph.D. thesis, University of Leeds.

Mel'nikov, M. K. and S. V. Kovenya 1974. Simulation studies of lessivage. In *10th International Congress on Soil Science*, Moscow 1974, Vol. VI, Commission V, Pt 2, 600–9.

Milne, G. 1936. A provisional soil map of East Africa. *Amani Memoirs* No. 28.

Parker, G. G. and E. A. Jenne 1967. *Structural failure of western US highways caused by piping*. Washington: US Geol. Survey Water Resources Division.

Pilgrim, D. H. and D. D. Huff 1983. Suspended sediment in rapid subsurface stormflow on a large field plot. *Earth Surf. Proc. Landforms* **8**, 451–63.

Quinsberry, U. L. and P. E. Phillips 1976. Percolation of simulated rainfall under field conditions. *Soil Sci. Soc. Am. Proc.* **40**, 484–90.

Quirk, J. P. and R. K. Schofield 1955. The effect of electrolyte concentration on soil permeability. *J. Soil Sci.* **6**, 163.

Reid, I. 1975. Seasonal variability of rainwater redistribution by field soils. *J. Hydrol.* **25**, 71–80.

Reid, I. 1977. Soil environment and the hydrometeorological mosaic. *Agric. Meteorol.* **18**, 425–33.

Ruhe, R. U and D. H. Walker 1968. Hillslope models and soil formations I. Open systems. In *Transactions of the 9th International Congress on Soil Science* Vol. IV, 551–60.

Rutter, A. T. 1955. The composition of wet heath vegetation in relation to the water table. *J. Ecol.* **43**, 507–43.

Sherard, J. L., R. S. Decker and N. L. Ryber 1972. Piping in earth dams of dispersive clay. In *Proceedings ASCE special conference on performance of earth and earth supported structures*, Vol. 1, 589–626.

Simonsen, G. M. and L. Boersma 1972. Soil morphology and water table relations II. Correlation between annual water table fluctuations and profile features. *Soil. Sci. Soc. Am. Proc.* **36**, 649–53.

Terzaghi, K. and R. B. Peck 1948. *Soil mechanics in engineering practice*. New York: Wiley.

Troeh, F. R. 1964. Landform parameters correlated to soil drainage. *Soil. Sci. Soc. Am. Proc.* **28**, 808–12.

Trudgill, S. T., I. M. S. Laidlaw and P. L. Stuart 1980a. Soil water residence times and solute uptake on a dolomite bedrock – preliminary results. *Earth Surf. Proc.* **5**, 91–100.

Trudgill, S. T., I. M. S. Laidlaw and P. J. C. Walker 1980b. Chemical erosion in soils in relation to soil water residence time. In *Assessment of erosion*, M. De Boodt and D. Gabriels (eds) 361–8. Chichester: Wiley.

Veneman, P. L. M., M. J. Vepraskas and J. Bouma 1976. The physical significance of soil mottling in a Wisconsin toposequence. *Geoderma* **15**, 103–118.

Walker, P. M. 1966. Postglacial environments in relation to landscape and soils on the Cary drift Iowa. *Iowa State Univ. Agric. Home Econom. Exp. Station Res. Bull.* **549**, 835–75.

Weyman, D. R. 1970. Throughflow on hillslopes and its relation to the stream hydrograph. *Bull. Int. Assoc. Sci. Hydrol.* **15**, 25–33.

Weyman, D. R. 1974. Runoff process, contributing area and streamflow in a small upland catchment In *Fluvial processes in instrumental watersheds*, K. J. Gregory and D. E. Walling (eds), 33–45. IBG Spec. Publn. 6.

Whipkey, R. Z. 1965. Subsurface stormflow on forested slopes. *Bull. Int. Assoc. Sci. Hydrol.* **10**, 74–85.

Whipkey, R. Z. 1967. Theory and mechanics of subsurface stormflow. In *Proceedings of the International Symposium on Forest Hydrology (1965)*. 255–9.

Wild, A. 1972. Nitrate leaching under bare fallow at a site in northern Nigeria. *J. Soil Sci.* **23**, 315–25.

Wood, C. C., G. D. Aitchison and O. G. Ingles 1964. *Physicochemical and engineering aspects of piping failures in small earth dams*. Paper presented at CSIRO Colloquium on Failure of Small Earth Dams, 16–19 Nov, water research foundation of Australia Ltd and soil mechanics section.

Zaslavsky, D. and A. S. Rogowski 1969. Hydrologic and morphologic implications of anisotropy and infiltration in soil profile development. *Soil Sci. Soc. Am. Proc.* **33**, 594.

7

Soil creep: a formidable fossil of misconception

B. L. Finlayson

Introduction

Soil creep became a necessary process in geomorphology to explain denudation on slopes that are too gentle for mass failure, and that show no evidence of erosion by running water. Although several specific processes contribute to total soil creep, process modelling has concentrated on relatively predictable processes such as freeze–thaw and moisture change, especially since this allowed the adoption from engineering science of ideas relating to an essentially different kind of creep. Furthermore, field measurements of creep have been largely interpreted in terms of expansion and contraction in the soil and results not fitting this model have generally been ignored.

Associated observations of soil properties have been mainly limited to grain size distributions and Atterberg limits. However, a detailed record of soil behaviour in response to the whole range of processes associated with creep can be found in the soil fabric at both the macro- and microscale. This pedo-geomorphic approach is already well known (Conacher & Dalrymple 1977, Paton 1978), though little attempt has yet been made to reconcile traditional ideas of soil creep within this framework.

Soil creep in geomorphology

Soil creep was recognised as an important hillslope process when, for example, W. M. Davis (1892) invoked a 'creeping process' to explain the summit convexities of badland divides. He later described the process (Davis & Snyder 1898) as the result of temperature fluctuations (including freezing and thawing), growth and decay of plant roots, and the action of earthworms, ants and various animals. He also suggested that these causes are strongest near the soil surface where the creeping waste mantle would accordingly move faster, and, finally, that the rate of creep increased with slope angle. Davis refined his ideas on the role of creep in the landscape in a paper on rock floors in arid and humid climates (Davis 1930). In humid regions, soil creep was described as the process responsible for development of low-angle summit convexities where, although still a very slow process, it would be more active than other forms of erosion.

Gilbert (1909) ascribed the convexity of hilltops to creep, and demonstrated how creep could produce a convex profile, by initially assuming that creep rate is a function of slope angle. The responsibility of creep for the form of a particular slope element (the summit convexity) was therefore well established in geomorphological literature early in the 20th century. On slopes where other processes operate, soil creep may still be active but its morphological effect will still be slight compared with processes such as slope failure and surface wash.

The most widely quoted geomorphological definition of soil creep is that of Sharpe (1938), who defined creep as 'the slow downslope movement of superficial soil or rock debris, usually imperceptible except to observations of long duration' (p. 21). Sharpe also emphasised its distinctly surface character, with movement dying away with depth and only affecting a 'few feet' below the soil surface. The moving material is not bounded by a plane of sliding, since creep is a 'flow process', and in his classification, Sharpe identified five types of slow flow (Table 7.1). He drew a distinction between creep in temperate and tropical climates and creep in cold climates controlled by frost. True soil creep is viewed as a process of temperate and tropical climates, and it is the field evidence and processes of this category of soil creep that are considered in this Chapter.

Geomorphological techniques for measuring soil creep (Anderson & Finlayson 1975) monitor the behaviour of soil creep *en masse*. However, many causes of soil creep, according to Davis, Gilbert and Sharpe, do not involve movement of the soil *en masse* but rather the movement of individual grains or soil aggregates by, for example, growth and decay of roots, expansion of vegetable matter by wetting, swaying of trees in the wind, burrowing or excavating by animals, and removal of soluble fractions of rocks or minerals. Sharpe also anticipated that not all movements would be downhill, a matter discussed further below.

Two types of soil creep have been recognised more recently, following work by Terzaghi (1950, 1953). They are 'seasonal' or 'skin' creep, and 'continuous'

Table 7.1 The five classes of creep movement recognised by Sharpe (1938).

Movement		Earth or rock	Earth or rock,	Earth or rock
Kind	Rate	plus ice	dry or with minor amounts of ice or water	plus water
flow usually imperceptible		rock glacier creep solifluction	rock creep talus creep soil creep	solifluction

or 'mass' creep. Most modern discussions of creep are classified thus, although with occasional modifications (Kirkby 1967, Carson & Kirkby 1972, Statham 1977). Hutchinson (1968) adds a third category of 'progressive' creep, and Statham (1977) also adds a third category, 'random' creep. Seasonal creep is essentially soil creep as understood by earlier writers, and Statham's 'random' creep belongs in this category. Kirkby (1967) has modelled seasonal creep due to expansion and contraction, following ideas developed by Davison (1889).

Continuous, or mass, creep has entered geomorphological discussion (Carson 1971) from engineering geology. This is creep in the rheological sense, where movement occurs when the shear stress is less than the strength of the soil and strain is both time- and stress-dependent. In the engineering context it includes soil compaction under building foundations and slow movement of soil under applied loads. Hutchinson (1968) apparently distinguished continuous and progressive creep because he considered there was little evidence for continuous creep on natural slopes, although Skempton (1964) had suggested that progressive creep occurred prior to failure. However, Kojan (1967) has been widely quoted as providing evidence for continuous creep on natural slopes, an enthusiasm which is considered below to be premature. Carson and Kirkby (1972) have argued that seasonal creep is continuous creep with a seasonal component related to soil conditions. Alternatively, Young (1972) dismisses continuous creep as of minor geomorphological importance except as a precursor of rapid mass movements, and this is supported by Lohnes and Handy (1968). They studied creep movement in glacial till in Iowa and found that a soil with 38% clay and a unit weight of 80 lb ft^{-3} (1280 kg m^{-3}) on a slope of 35° required 4 ft (1.2 m) of overburden to initiate continuous creep. Culling (1983a) also distinguishes engineering and geomorphological approaches, noting that soil creep as understood by the engineer is directed, persistent and strong and is essentially a molecular process, while soil creep as understood by the geomorphologist is random, intermittent and weak and is a particulate phenomenon.

The evidence for soil creep

The evidence for soil creep includes that based on both field measurements and environmental indicators.

Field measurements of soil creep

Technical details of field measurement methods are provided in Selby (1966) and Anderson and Finlayson (1975). Davison (1889) appears to have initiated the direct measurement of creep in experiments on soil movement specifically produced by freezing and thawing. His methods provide the conceptual basis for development of both measuring techniques and theory, which have been applied even in areas where freezing is unimportant, and which emphasise movement that is down slope only and that dies away with depth.

The Young Pit (Young 1960) is specifically suited to measuring downslope

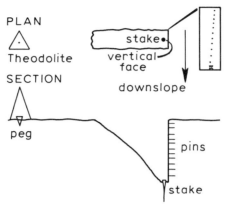

Figure 7.1 Layout of a Young Pit. Alternative ways of measuring the pin positions are given in Young (1960).

movements below the soil surface (Fig. 7.1). Young (1960) reported results from nine pits which had been remeasured at intervals of 6−12 months over two years. He stated that 'disturbances of rods during burial and excavation were distinguishable as random movements, contrasted with systematic movement of adjacent rods due to soil movements' (p. 121). Later (Young 1963), he reported Young Pit measurements made over a four-year period, 'after rejecting rods which showed inconsistent movements, assumed to be caused by disturbance during re-excavation' (p. 129). The criteria upon which individual pins were accepted or rejected are unclear but the decisions may in part have been prompted by preconceived ideas about the way the soil should behave.

Results of creep observations in 52 Young Pits in south-west Scotland were reported by Kirkby (1967). Downslope transport was found in 34 pits, while in 18, net transport was up slope. This division was described as 'probably significant' at the 95% level. The large scatter in the data concealed any possible correlation with slope angle, although like Young (1963), Kirkby (1967) believed that there was a possible correlation with slope angle which was obscured by 'noise' in the data. Slope angle explained very little of the variance in Kirkby's results, but he nevertheless developed a theory of soil creep based on the assumption that creep rate is proportional to the sine of slope angle. This theory is internally consistent but is based on an assumption which lacks the support of Kirkby's own, or any other, field data set.

Soil movements measured in 15 Young Pits in a small catchment in the Mendip Hills, Somerset, are shown as velocity profiles in Figure 7.2 (Finlayson 1976). Volumetric rates for the same pits are shown in Table 7.2. Sites are defined by soil type. The Maesbury series soils are free-draining brown earths and the Ashen series soils are peaty podzols. Ashen series soils are on gentle slopes (< 6°) while Maesbury soils occur on steeper slopes (> 9°). Slope angles for individual sites are shown in Table 7.2, and it is obvious that there is no relationship between creep and slope angle. Upslope movements in these Young Pits are not just occasional aberrant pins which may have been disturbed during

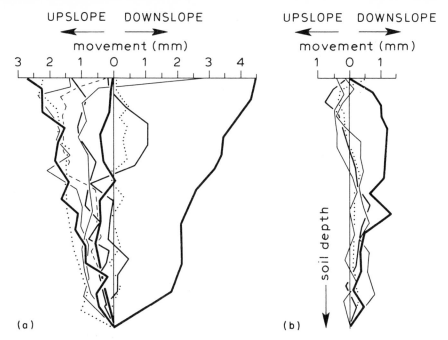

Figure 7.2 Profiles of movement in Young Pits; vertical scales are standardised proportional soil depths. All sites located in the catchment of East Twin Brook, Mendip Hills, Somerset, UK; (a) Maesbury series soils, (b) Ashen series soils.

Table 7.2 Soil creep measurements in a small catchment in the Mendip Hills, using Young Pits and a travelling telescope that measures the profiles of flexible tubes. Negative values indicate upslope movement (after Finlayson 1976).

Site	Soil series	Slope angle (degrees)	Volumetric movement (cm³ cm⁻¹ yr⁻¹)		
			Young Pits	Flexible tubes Total	Net downslope
2	Maesbury	14.0	−5.2	3.7	3.7
4	Maesbury	26.5	18.3	12.7	−4.2
5	Maesbury	28.0	1.5	7.5	7.1
6	Ashen	5.5	6.1	9.6	−3.9
7	Ashen	6.0	1.0	12.5	−10.9
8	Ashen	1.0	2.8	3.9	−0.9
9	Ashen	5.5	−0.8	3.1	−1.3
10	Maesbury	19.5	−6.1	2.5	2.1
11	Maesbury	25.5	−1.5	1.0	0.9
12	Maesbury	27.5	1.4	3.5	3.1
13	Maesbury	29.0	−13.7	14.9	12.2
18	Maesbury	9.5	−10.7	2.4	2.2
19	Maesbury	9.0	−13.1	2.8	0.4
21	Maesbury	12.0	−8.1	6.9	6.2
23	Ashen	6.0	−0.2	13.9	10.9

excavation. Neither are the upslope movements found only on gentle slopes, as the steepest site in the group, pit 13, showed a very large upslope movement. These results and those of Kirkby (1967) are the only published Young Pit results showing upslope movement, yet in each case this is frequent and substantial.

Troeh (1975), using a different measurement technique but one which, like the Young Pit, only detects movement on a predetermined plane parallel to the direction of maximum slope, published only ·the downslope movements. However, in correspondence he has written that some data do show small amounts of upslope rebound, usually during the summer period. At first these were thought to be inaccurate readings, but there are now too many for that to be the explanation (F. R. Troeh personal communication).

Alternative measurement techniques are available which can detect movement in any direction. One such instrument, the strain gauge inclinometer (Wilson & Mikkelson 1978), has been used in field measurements of soil creep on natural slopes (Kojan 1967, Fleming & Johnson 1975). In this method, the profile of a flexible tube inserted into a borehole is measured by lowering the inclinometer down the tube. Successive readings with the inclinometer rotated through 90° allow the three-dimensional form of the tube to be calculated, and movement to be measured in any direction. Since readings are considered not to disturb the site seriously, they can be repeated frequently.

The widely quoted work of Kojan (1967) reports first-year results from 12 strain gauge inclinometer tubes at 4 sites in the northern coastal ranges of California. Very large displacements were noted, including one case of 14 cm of surface movement in 6 months, and one of movement extending to 10.5 m below the soil surface. The tubes also moved in various directions, not just down slope. Ziemer (1977) presented a new analysis of the data reported by Kojan (1967) and included an additional 7 years of data collected by Kojan at 35 access casings at 7 sites. Based on the 8 years of borehole measurement, Ziemer concluded that the displacement rates reported by Kojan were excessively high. In fact, no consistent direction of soil movement was detected for the duration of the study and no significant rate of soil creep was found during the 8 years in which the boreholes were observed. Three reasons were given for rejecting the data (Ziemer 1977, Ziemer & Strauss 1978). First, the holes in which the casings were installed were not backfilled, and so recorded movements were partly differential movements between the casings and the boreholes. Second, careful analysis of the instrument showed that instrument error was larger than the reported creep, and third, the instrument calibration was not consistent. One is left to wonder if other soil creep data sets would withstand a similar rigorous analysis by someone not indoctrinated with geomorphological ideas on the subject.

Fleming and Johnson (1975) used a tiltmeter (which also employs a strain gauge to measure deformation of a flexible tube) in expansive silty clay soils in central California. Their results show movements, related to soil moisture changes, occurring in various directions, and not only down slope. These results could not be reconciled with the conventional theories of soil creep.

The major advances in the measurement systems used by Kojan (1967) and Fleming and Johnson (1975) are (a) that site disturbance during installation and reading can be minimised, and (b) that they are not constrained by prior assumptions about the direction of movement. However, if geomorphological interest is centred on near-surface movements in the top 20 cm (Kirkby 1967) where bulk densities are low, even the insertion of a strain gauge inclinometer into a flexible tube could cause tube displacement of the same order as that expected in the soil. A measuring system which avoids this problem of movement during reading is the travelling telescope (Finlayson 1976, Finlayson & Osmaston 1977). Flexible tubes are used that contain a sealed light source at their base and cross-wires set at intervals up the tube. A travelling telescope mounted on a tripod is used to measure the profiles of the tubes by sighting on the cross-wires when lit from below. Profiles of movement are relative to the lowest cross-wire.

Results from 15 tubes located close to the Young Pits for which data are quoted above have been reported (Finlayson 1976, 1981). The tube readings are for slightly different time periods. Examples of profiles of movement for three tubes are shown in Figure 7.3, and rates of total movement and net downslope movement are given in Table 7.2. These results confirm that movement can be in any direction, supporting the above critique of techniques that presuppose directly downslope movement. In geomorphology, soil creep is generally defined as a downslope transport process, but it would appear that much of the movement in soils is recoverable and not simply directed. Thus a serious question arises as to when in a monitoring programme it may be assumed that net non-recoverable soil creep has occurred. Much longer series of measurements will be necessary before this issue can be resolved.

The elusive relationship between the downslope component of soil creep and slope angle has been investigated using the tube/telescope data. For individual tubes, a regression of downslope movement on the sine of slope angle is not statistically significant ($F = 2.87$ with 1 and 11 degrees of freedom, $p = 0.11$). When median rates for slope classes are used, the regression explains 72% of the variance in median creep rates ($F = 7.80$ with 1 and 3 d.f., $p = 0.06$). The tube results in Table 7.2 therefore actually provide the closest example yet published of a relationship between creep rate and slope angle.

Although the geomorphological theory of soil creep has been developed for soil movements with a strong vertical component (Kirkby 1967), little attempt has been made to measure movements in a vertical direction. Only Young (1963, 1978) has attempted this, in Young Pits where pins were observed to move towards the bedrock and slightly down hill, with the inward movement exceeding that down slope. Spotts and Brown (1975) have described an instrumental technique for monitoring vertical movements in the soil using induction coils, but few results have been reported.

This review of measurement evidence for soil creep reveals cause for concern. Field measurements are often carried out during a postgraduate degree programme, and have therefore been for rather short periods (e.g. Kirkby 1967,

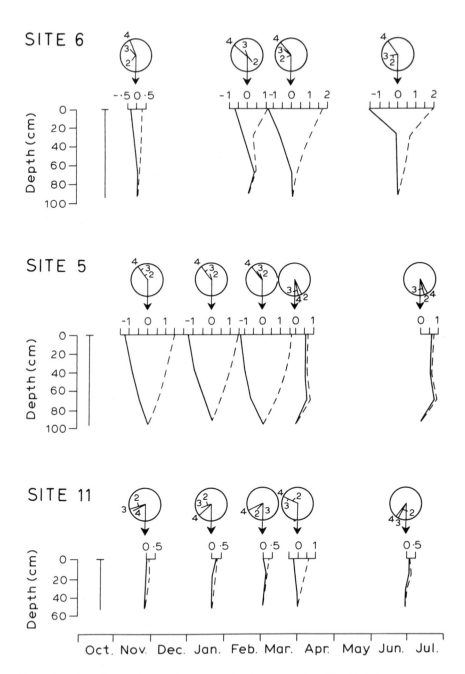

Figure 7.3 Directions and rates of movement for three examples of flexible tubes measured by travelling telescope. Directions of movement are shown on the polar graphs. The arrow indicates the downslope direction and the bars are measured at the cross-wires, no. 4 being at the soil surface. The lengths of the bars indicate relative amounts of movement. On the velocity profiles the broken lines show total movement and the solid lines net downslope movement.

Slaymaker 1972, Williams 1973, Finlayson 1976, Anderson 1977), although Kojan (1967) and Young (1963, 1978) have reported longer-term records. Thus excessive generalisation based on limited data often occurs. The design of measuring systems has largely reflected a single theory which actually originated from consideration of freeze—thaw effects rather than any of the other processes thought to contribute to soil creep. There has accordingly been a somewhat unproductive cross-fertilisation from periglacial work, as noted by Fleming and Johnson (1975). The reliability of certain field data is also in doubt. Most researchers seem to believe that the process behaves as Davison (1889) and Kirkby (1967) have suggested, but that their measurements contain a lot of 'noise'. The noise : signal ratio is at least unity for such short periods of observation. This ratio could be expected to improve in studies of longer duration, and it is certainly time to consider more seriously the sources of noise and the processes responsible for it.

Despite lack of convergence between theory and field measurement, the theory persists. Sharpe (1938) drew attention to the additional evidence for creep provided by a range of environmental indicators, which therefore also require evaluation.

Environmental indicators of creep
Sharpe (1938) believed that 'evidences of creep of the soil mantle are to be found on almost every soil-covered slope' (p. 22). Figure 7.4 and its legend are taken from Sharpe (1938, Figure 2, p. 22) and summarise these evidences. Many subsequent discussions of creep have reproduced this evidence in the absence of

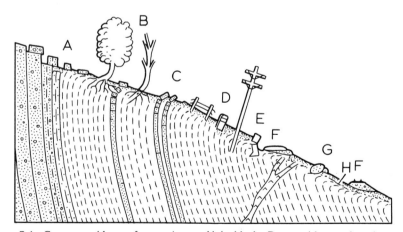

Figure 7.4 Common evidence of creep: A moved joint blocks; B trees with curved trunks concave upslope; C downslope bending and drag of bedded rock, weathered veins, etc. also present beneath soil elsewhere on the slope; D displaced posts, poles, and monuments; E broken or displaced retaining walls and foundations; F roads and railroads moved out of alignment; G turf rolls downslope from creeping boulders; H stone line at approximate base of creeping soil. A and C represent rock creep; all other features are due to soil creep. Rather similar effects may be produced by some types of landslides (from Sharpe 1938).

support (e.g. Selby 1982), despite the substantial case made against much of it by Parizek and Woodruff (1956, 1957b) and Phipps (1974). Only Young (1972) has cited evidence in support of the tree curvature hypothesis by quoting Schmid (1955), who found that curvature develops during the first ten years of tree growth, and that older trees are strong enough to withstand creep pressure.

The case against tree curvature as an indicator of creep is convincing (Parizek & Woodruff 1957b, Phipps 1974). While geomorphologists cite curved tree trunks as evidence of creep, botanists never cite creep to explain deformed or curved trees, there being sufficient and more convincing plant physiological and environmental causes of curvature. The straight trunks on steep mountain slopes of the mountain ash (*Eucalyptus regnans*) and alpine ash (*E. delegatensis*) that support the hardwood industry in Victoria, Australia, suggest either that soil creep does not operate on mountain slopes in Victoria, or that it does not affect the trees!

Other evidence cited by Sharpe has also been questioned. The displacement of posts, poles and monuments, and outcrop curvature beneath the soil provide at best qualified evidence of creep. Statham (1977) has suggested that man-made objects may be tilted because they impose an extra load on a small area of otherwise static soil. Outcrop curvature may reflect soil creep (Young 1972), or cambering or collapse (Statham 1977). The origin of stone lines in the soil is also a subject of dispute. Paton (1978) and Young (1972) present examples which they attribute to soil creep. However, Parizek and Woodruff (1956) re-investigated the site of Sharpe's (1938) stone line evidence and concluded that it did not support the creep hypothesis. Until this issue has been clarified, the danger remains that the use of stone lines as evidence of soil creep is a circular argument.

Alternative approaches to soil creep

The preceding sections show that there is substantial confusion and misconception surrounding the phenomenon of soil creep. The use of the word 'creep' to describe such a wide range of processes creates confusion, but renaming one group of those processes 'soil melt' (Parizek & Woodruff 1957a) does not seem a viable solution. The evidence for creep provided by field measurements is confused and ambiguous, some is unreliable, and often it does not support theory. The relationship between creep rate and slope angle is indicative of divergence between theory and measurement. Environmental indicators of creep are invalid or lead potentially to circular argument. Nevertheless the Davison—Kirkby theory persists, suggesting either the need for new methods of testing it or the need for new theory. Overall, geomorphologists seem to have sought predetermined evidence rather than critically examining that evidence.

An alternative theory of soil creep has been developed by Culling (1963, 1965, 1981, 1983a, 1983b). It suffers because field testing is difficult, although the approach described by Coutts *et al.* (1968) using radioactive ^{59}Fe may have some potential. Initially, Culling (1963) claimed to be developing the ideas of Strahler (1952) and therefore of Sharpe (1938) and Gilbert (1909). As noted above, these early workers placed more emphasis on forces tending to move individual grains or small aggregates than is true of more recent approaches. While most writers mention factors such as growth and decay of plant roots, trees swaying in the wind, and the effect of burrowing animals, these are usually relegated to minor roles, and the measurement techniques commonly used cannot respond to movements on such a small scale.

Culling, however, has attempted to consider creep as the sum total of the movements of individual soil particles. The soil is envisaged as an aggregate of particles which usually act as individual bodies (Culling 1981). His conceptual model is one where particles are constrained into a 'cell' by surrounding particles, and each is subject to frequent small random forces to which it can react, with varying amounts of freedom because of constraints imposed by the surrounding soil structure (the 'cell'). If both activity and opportunity coincide, a particle can escape from one cell into another ('translation'). Particle activity is generated by 'turbation' (cf. 'pedoturbation', Hole 1961, Buol *et al.* 1973) which can be (a) physical, as in thermal expansion and contraction, freeze and thaw of soil water, variations in pore-water pressure, local displacement by throughflow, electrical and magnetic phenomena, Van der Waals' forces and gravity; (b) chemical, as in solution and precipitation, hydration and desiccation, crystal-lisation and reactions leading to rearrangement, changes in temperature, volume and the nature of chemical bonds; and (c) biological, which is harder to assess but no less significant. In short, particles execute a three-dimensional random walk upon which is superimposed any drift due to the presence of a general directed force, presumably gravity (Culling 1983b). These movements have been analysed in various ways, using diffusion analysis (Culling 1981), blurred crystal lattices (Culling 1983b) and rate process theory (Culling 1983a). Although presently untestable in the field, Culling's theories lead in focusing on individual soil particles.

Paton (1978) has also suggested this approach to the study of downslope soil movement, recommending the detailed investigation of soil fabric. The existence of pedological evidence for soil creep is implicit in the pedogeomorphic approach adopted by Conacher and Dalrymple (1977). Unit 3 of their nine-unit land-surface model is, by definition, the product of soil creep, but unfortunately lacks pedological criteria for its identification. Pedological evidence of creep does occur, however, in the work of Beinroth *et al.* (1974), Chandler and Pook (1971) and Crampton (1974), who interpret small shear planes in soils as the result of soil creep. Given the range of processes thought to contribute to creep, it is un-necessarily restrictive to require that microshearing should be present to provide evidence of creep.

Pedological evidence of creep on slopes in mid-Wales is presented by Adams and Raza (1978). They argue that if a given block of soil does not contain an equivalent amount of any element derived from the bedrock, then it has experienced relative depletion or enrichment. Loss or gain of Fe was found to correlate with slope angle (Fig. 7.5). Soil profiles showing net loss of Fe were either at the slope crest where there is no upslope source of Fe, or in the valley bottom where the soils consist of Fe-depleted material gained from up slope. Arguing that vertical processes in the soil should be constant over the slope since they are climatically controlled, but that the creep rate should increase with slope angle, they suggested that the rapid downslope movement of eluvial horizons at the soil surface should leave behind a subsoil that is relatively Fe-rich (see also Finlayson 1977). The creep mechanism is not considered, but the pedological evidence may represent the net result of creep over a time span much longer than anything achieved by direct measurement.

Hole (1961) has defined nine categories of pedoturbation (soil mixing) which he considers to be a 'special case of erosion which succeeds in prolonging the long-term process of seaward erosion' (p. 375; cf. the parable on erosion given by Leopold 1949 p.104): 'prolonging' because presumably pedoturbation may locally include upslope movements, and only in the long term will there be net downslope transport. His nine categories are:

(a) Faunalpedoturbation: soil mixing by animals such as ants, earthworms, moles, rodents, wombats, lyrebirds and man.
(b) Floralpedoturbation: mixing by plants as in tree-throw.
(c) Congelipedoturbation: mixing by freeze—thaw cycles (cf. Peltier 1950).
(d) Argillipedoturbation: mixing of materials in the solum by movements of expansive clays.
(e) Gravipedoturbation: mixing by non-catastrophic mass-wasting, such as creep.
(f) Aeropedoturbation: mixing by movement of gases in the soil during and after rains.
(g) Aquapedoturbation: mixing by upwelling water currents in the solum.
(h) Crystallpedoturbation: mixing by the growth and wasting of crystals in the soil.
(i) Seismipedoturbation: mixing by vibrations, notably earthquakes.

Many of these are recognisable as processes commonly held to be responsible for soil creep: argilli- and congelipedoturbation have attracted most attention. It is, however, clearly unreasonable to expect pins in a Young Pit to move down slope in a regular velocity profile when subject to floral- or faunalpedoturbation, for example.

Floralpedoturbation has been investigated by Lutz and Griswold (1939) and Pawluk and Dudas (1982). The latter concentrated on tree-throw and list criteria for its effect on soils (see Fig. 7.6). Pedorelicts derived from B and C horizon materials in all Ah horizon fabrics (both turbated and adjacent unturbated soils) were considered to reflect considerable mixing and shifting across the landscape surface. Lutz and Griswold (1939) considered additional effects. When tree roots decay the channels thereby left in the soil are filled by material from the

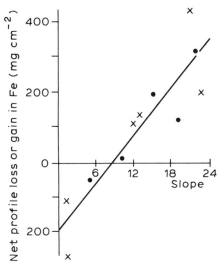

Figure 7.5 The relationship between net profile loss or gain in Fe and slope of profile site for the soils in two slope sequences in west mid-Wales (from Adams & Raza 1978).
Foel Wyddon $- y = 25 x - 207, r = 0.92$ $(p = 0.05)$; Tarren Cadian $- y = 19 x - 157, r = 0.91$ $(p = 0.05)$;
All data $- y = 23 x - 195$, $r = 0.91$ $(p = 0.001)$.

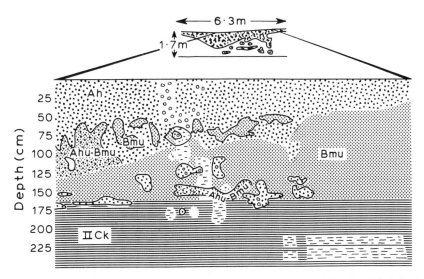

Figure 7.6 Pedological evidence of floralpedoturbation by tree-throw (after Pawluk & Dudas 1982).

surface or side walls. During growth, tree roots push material aside and roots agitate the soil as trees sway in strong winds.

An excellent review of the effects of animals on soils is provided by Hole (1981), who lists 12 activities of animals in soils including mounding, mixing, forming and filling voids, forming and destroying peds, and regulating soil erosion, plant and animal litter, nutrient cycling and the movements of air and water in the soil. Many of these processes, by moving soil particles or creating voids into which adjacent soil material can move, actively generate the interparticle motions which Culling (see above) has attempted to model.

Several studies of the contributions of animal activity to slope erosion in Luxembourg have been published (e.g. Imeson 1976, 1977, Imeson & Kwaad 1976, Hazelhoff *et al.* 1981). Since overland flow does not occur on these slopes, the main sediment transport process is considered to be splash erosion, which requires bare soil at the surface. These studies document the effects of earth-worms and larger burrowing and mounding animals in exposing bare soil to rainsplash. Subsurface processes such as the excavation and filling of animal burrows therefore also occur (cf. Moeyersons 1978), and the slopes are affected by a 'creeping process' in the Davisian (1892) sense, which includes faunal-pedoturbation and surface rainsplash. Perhaps significantly, although rainsplash is usually considered distinct from soil creep, it can also develop slope convexity (Mosley 1973).

Micromorphological study of soils (Brewer 1964) offers a potential tool for in-depth investigation of the subsurface processes (Mücher 1974). Many charac-teristics of soils as seen in thin section are only explicable in terms of differential movements within the soil (e.g. cutans, tubules and plasma separations, Brewer 1964). Three examples of soil microfabrics illustrating these features are shown in Figure 7.7. Cady (1974) has also described microfabrics indicating movement and pressure in the soil. The efforts of Mücher (1974) and others in applying micromorphological analysis in studies of geomorphological processes is a highly desirable development which could improve understanding of soil creep.

There is likely to be considerable spatial variation of both rates and types of pedoturbation processes. Tree-throw, for example, will be absent from grass-lands while argillipedoturbation will be at a maximum in areas with a markedly seasonal rainfall regime and expansible clay soils, such as gilgai soils (Hallsworth *et al.* 1965). Soil creep may be seen to have a multiplicity of causes with different combinations present in different areas and with individual factors varying in intensity from place to place or from time to time, or being absent altogether. There is a danger of circular argument regarding the role of creep in soil studies. Hole (1961) cites creep as a process of gravipedoturbation, and Buol *et al.* (1973) claim that creep, by removing surficial material as fast or faster than most pedo-genic horizons can form, may be an important factor in the formation of entisols. Thus soil scientists are adopting a geomorphological view of soil creep, the 'catch-all term' criticised by Parizek and Woodruff (1957a), rather than recognising that all pedoturbation processes are actually responsible for soil creep.

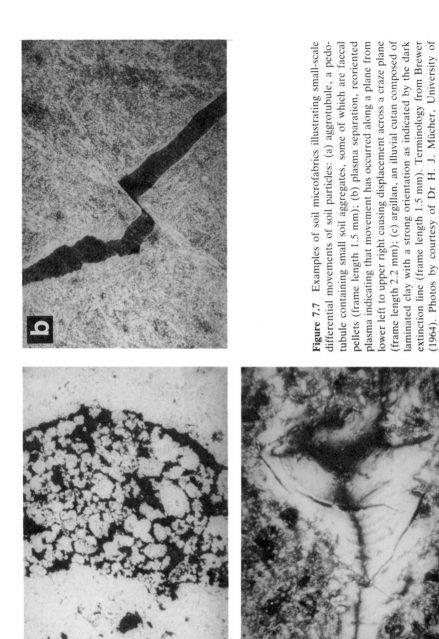

Figure 7.7 Examples of soil microfabrics illustrating small-scale differential movements of soil particles: (a) aggrotubule, a pedotubule containing small soil aggregates, some of which are faecal pellets (frame length 1.5 mm); (b) plasma separation, reoriented plasma indicating that movement has occurred along a plane from lower left to upper right causing displacement across a craze plane (frame length 2.2 mm); (c) argillan, an illuvial cutan composed of laminated clay with a strong orientation as indicated by the dark extinction line (frame length 1.5 mm). Terminology from Brewer (1964). Photos by courtesy of Dr H. J. Mücher, University of Amsterdam.

Conclusion

In this review it has been shown that soil creep was initially viewed as the consequence of numerous contributory processes. Subsequently, however, field measurement techniques have responded only to bulk movements, with theoretical attention focusing on expansion and contraction processes. The measurements have shown some movements of this type, but also seemingly 'random' components, or 'noise', not explicable by this theory. Some field data, and environmental indicators of soil creep, have been shown to be unreliable. More complete understanding of soil creep processes requires greater attention to the sources of 'noise' in the field data. The theoretical framework developed by Culling (1963) could help explain the 'noise', and the evidence preserved in the soil fabric offers a solution to the present difficulties of testing this theory. This is essentially the pedogeomorphic approach stressed by Mücher (1974) and Conacher and Dalrymple (1977), among others. Pedological evidence shows that movement in the soil (pedoturbation) has many causes, with complex interactions of biological, physical and chemical processes which are not readily accommodated within a rigidly deterministic framework such as that used by Carson (1971) in discussing soil creep. More relevant is the 'science of complexity' (Nicolis & Prigogine 1977), in which the laws of strict causality are limiting situations applicable only to idealised cases. A revised approach to soil creep, incorporating micro- and macropedological techniques and a plurality of causation and interaction would seem the best way forward.

References

Adams, W. A. and M. A. Raza 1978. The significance of truncation in the evolution of slope soils in mid-Wales. *J. Soil Sci.* **29**, 243–57.

Anderson, E. W. 1977. *Soil creep. An assessment of certain controlling factors with special reference to upper Weardale, England.* Unpublished Ph.D. thesis, University of Durham.

Anderson, E. W. and B. L. Finlayson 1975. *Instruments for measuring soil creep.* BGRG Tech. Bull. 16. Norwich: Geo Abstracts.

Beinroth, F. G., G. Uehara and H. Ikawa 1974. Geomorphic relationships of oxisols and ultisols on Kauai, Hawaii. *Soil Sci. Soc. Am. Proc.* **38**, 128–31.

Brewer, R. 1964. *Fabric and mineral analysis of soils.* New York: Wiley.

Buol, S. W., F. D. Hole and R. J. McCracken 1973. *Soil genesis and classification.* Ames: Iowa State University Press.

Cady, J. G. 1974. Applications of micromorphology in soil genesis research. In *Soil microscopy,* Proc. 4th Int. Working Meeting on Soil Micromorphology, G. K. Rutherford (ed.), 20–7. Kingston: Limestone Press.

Carson, M. A. 1971. *The mechanics of erosion.* London: Pion.

Carson, M. A. and M. J. Kirkby 1972. *Hillslope form and process.* Cambridge: Cambridge University Press.

Chandler, R. J. and M. J. Pook 1971. Creep movement in low gradient clay slopes since the late glacial. *Nature* **229**, 399–400.

Conacher, A. J. and J. B. Dalrymple 1977. The nine-unit landsurface model. An approach to pedogeomorphic research. *Geoderma* **18**, 1–154.

Coutts, J. R. H., M. F. Kandil, J. L. Nowland and J. Tinsley 1968. Use of radioactive ^{59}Fe for tracing soil particle movement. 1. Field studies of splash erosion. *J. Soil Sci.* **19**, 311−41.

Crampton, C. B. 1974. Micro-shear fabrics in soils of the Canadian north. In *Soil microscopy*, Proc. 4th Int. Working Meeting on Soil Micromorphology, G. K. Rutherford (ed.), 655−64. Kingston: Limestone Press.

Culling, W. E. H. 1963. Soil creep and the development of hillside slopes. *J. Geol.* **71**, 127−61.

Culling, W. E. H. 1965. Theory of erosion on soil covered slopes. *J. Geol.* **73**, 230−54.

Culling, W. E. H. 1981. New methods of measurement of slow particulate transport processes on hillside slopes. In *Erosion and sediment transport measurement*, 267−74. Proc. Florence Symp., IAHS Publn. 133.

Culling, W. E. H. 1983a. Rate process theory of geomorphic soil creep. In *Rainfall simulation, runoff and erosion*, J. de Ploey (ed.), 191−214. *Catena* Suppl. 4. Cremlingen: Catena Verlag.

Culling, W. E. H. 1983b. Slow particulate flow in condensed media as an escape mechanism: mean translation distance. In *Rainfall simulation, runoff and erosion*, J. de Ploey (ed.), 161−90. *Catena* Suppl. Cremlingen: Catena Verlag.

Davis, W. M. 1892. The convex profile of bad-land divides. *Science* **20**, 245.

Davis, W. M. 1930. Rock floors in arid and in humid climates. *J. Geol.* **38**, 1−27, 136−58.

Davis, W. M. and W. H. Snyder, 1898. *Physical geography*. Boston: Ginn.

Davison, C. 1889. On the creeping of the soilcap through the action of frost. *Geol. Mag.* **6**, 255−61.

Finlayson, B. L. 1976. *Measurements of geomorphic processes in a small drainage basin*. Unpublished Ph.D. thesis, University of Bristol.

Finlayson, B. L. 1977. *Runoff contributing areas and erosion*. School of Geography, University of Oxford, Research Paper 18.

Finlayson, B. L. 1981. Field measurements of soil creep. *Earth Surf. Proc. Landforms* **6**, 35−48.

Finlayson, B. L. and H. A. Osmaston 1977. *An instrument system for measuring soil creep*. BGRG Tech. Bull. 19, Norwich: Geo Abstracts.

Fleming, R. W. and A. M. Johnson, 1975. Rates of seasonal creep of silty clay soil. *Q. J. Engng Geol.* **8**, 1−29.

Gilbert, G. K. 1909. The convexity of hilltops. *J. Geol.* **17**, 344−50.

Hallsworth, L., G. R. Robertson and F. R. Gibbson 1955. Studies in pedogenesis in New South Wales. VII The 'gilgai' soils. *J. Soil Sci.* **6**, 1−31.

Hazelhoff, L., P. van Hoof, A. C. Imeson and F. J. P. M. Kwaad 1981. The exposure of forest soil to erosion by earthworms. *Earth Surf. Proc. Landforms* **6**, 235−50.

Hole, F. D. 1961. A classification of pedoturbation and some other processes and factors of soil formation in relation to isotropism and anisotropism. *Soil Sci.* **91**, 375−7.

Hole, F. D. 1981. Effects of animals on soil. *Geoderma* **25**, 75−112.

Hutchinson, J. N. 1968. Mass movement. In *The encyclopedia of geomorphology*, R. W. Fairbridge (ed.), 688−95. New York: Reinhold.

Imeson, A. C. 1976. Some effects of burrowing animals on slope processes in the Luxembourg Ardennes. I The excavation of animal mounds in experimental plots. *Geogr. Ann.* **58A**, 115−25.

Imeson, A. C. 1977. Splash erosion, animal activity and sediment supply in a small forested Luxembourg catchment. *Earth Surf. Proc.* **2**, 153−60.

Imeson, A. C. and F. J. P. M. Kwaad 1976. Some effects of burrowing animals on slope processes in the Luxembourg Ardennes. II The erosion of animal mounds by splash under forest. *Geogr. Ann.* **58A**, 317−28.

Kirkby, M. J. 1967. Measurement and theory of soil creep. *J. Geol.* **75**, 359−78.

Kojan, E. 1967. Mechanics and rates of natural soil creep. In *Proceedings of the 5th Annual Engineering Geology and Soils Engineering Symposium*, Idaho Dept. Highways, Idaho State Univ., Pocatello, 233−53.

Leopold, A. 1949. *The Sand County Almanac*. Oxford: Oxford University Press.

Lohnes, R. A. and R. L. Handy 1968. Rheological approach to soil creep. *Geol Soc. Am. Spec. Paper* **115**, 132−3.

Lutz, H. J. and F. S. Griswold 1939. The influence of tree roots on soil morphology. *Am. J. Sci.* **237**, 389–400.

Moeyersons, J. 1978. The behaviour of stones and stone implements, buried in consolidating and creeping Kalahari sands. *Earth Surf. Proc.* **3**, 115–28.

Mosley, M. P. 1973. Rainsplash and the convexity of badland divides. *Zeitschr. Geomorph. Suppl. Bd* **18**, 10–25.

Mücher, H. 1974. Micromorphology of slope deposits: the necessity of a classification. In *Soil microscopy*, Proc. 4th Int. Working Meeting on Soil Micromorphology, G. K. Rutherford (ed.), 553–66. Kingston: Limestone Press.

Nicolis, G. and I. Prigogine 1977. *Self organization in non-equilibrium systems: from dissipative structures to order through fluctuations.* New York: Wiley.

Parizek, E. J. and J. F. Woodruff 1956. Apparent absence of soil creep in the east Georgia Piedmont. *Geol. Soc. Am. Bull.* **67**, 1111–16.

Parizek, E. J. and J. F. Woodruff 1957a. A clarification of the definition and classification of soil creep. *J. Geol.* **65**, 653–6.

Parizek, E. J. and J. F. Woodruff 1957b. Mass wasting and deformation of trees. *Am. J. Sci.* **255**, 63–70.

Paton, T. R. 1978. *The formation of soil material.* London: George Allen & Unwin.

Pawluk, S. and M. J. Dudas, 1982. Floral pedoturbation in black chernozemic soils of the Lake Edmonton Plain. *Can. J. Soil Sci.* **62**, 617–29.

Peltier, L. C. 1950. The geographic cycle in periglacial regions as it is related to climatic morphology. *Ann. Assoc. Am. Geogs.* **40**, 214–36.

Phipps, R. L. 1974. The soil creep – curved tree fallacy. *J. Res. US Geol Surv.* **2**, 371–7.

Schmid, J. 1955. *Der bodenfrost als morphologischer factor.* Heidelberg: Huthig.

Selby, M. J. 1966. Methods of measuring soil creep. *J. Hydrol.* (NZ) **5**, 54–63.

Selby, M. J. 1982. *Hillslope materials and processes.* Oxford: Oxford University Press.

Sharpe, C. F. S. 1938. *Landslides and related phenomena.* New York: Columbia University Press.

Skempton, A. W. 1964. Long term stability of clay slopes. *Geotechnique* **14**, 77–101.

Slaymaker, H. O. 1972. Patterns of present sub-aerial erosion and landforms in mid-Wales. *Trans Inst. Brit. Geogs.* **55**, 47–68.

Spotts, J. W. and K. W. Brown 1975. A technique for installing induction coils in a profile with minimum soil disturbance. *Soil Sci. Soc. Am. Proc.* **39**, 1006–7.

Statham, I. 1977. *Earth surface sediment transport.* Oxford: Clarendon Press.

Strahler, A. N. 1952. Dynamic basis of geomorphology. *Geol Soc. Am. Bull.* **63**, 923–38.

Terzaghi, K. 1950. Mechanism of landslides. *Geol. Soc. Am. Bull.*, Berkey Vol., 83–122.

Terzaghi, K. 1953. Some miscellaneous notes on creep. In *Proceedings of the 3rd International Conference on Soil Mechanics and Foundation Engineering*, Vol. 3, 205–6.

Troeh, F. R. 1975. Measuring soil creep. *Soil Sci. Soc. Am. Proc.* **39**, 707–9.

Williams, M. A. J. 1973. The efficacy of creep and slopewash in tropical and temperate Australia. *Austral. Geogr. Stud.* **11**, 62–8.

Wilson, S. D. and P. E. Mikkelson, 1978. Field instrumentation. In *Landslides analysis and control*, Transport Research Board, Special Report No. 176, R. L. Schuster and R. J. Krizek (eds), 112–38. Washington: National Academy of Science.

Young, A. 1960. Soil movement by denudational processes on slopes. *Nature* **188**, 120–2.

Young, A. 1963. Soil movement on slopes. *Nature* **200**, 129–30.

Young, A. 1972. *Slopes.* Edinburgh: Oliver & Boyd.

Young, A. 1978. A twelve year record of soil movement on a slope. *Zeitschr. Geomorph. Suppl. Bd.* **29**, 104–10.

Ziemer, R. R. 1977. *Measurement of soil creep by inclinometer.* Engineering Tech. Report ETR–7100–4. Washington: USDA Forest Service.

Ziemer, R. R. and D. Strauss 1978. *A statistical approach to instrument calibration.* Special Report EM–7100–11. Washington: USDA Forest Service.

8

Soil mechanics and natural slope stability

C. Rouse and A. Reading

Introduction

This chapter reviews the study of slopes in the context of soil mechanics and mass movement processes. The term 'soil' is throughout used in the engineering sense, that is, to refer to the regolith or loose detrital material on any hillslope. A convenient and frequently used subdivision of 'soil' is between cohesive (clay) soils exhibiting interparticle attraction, and non-cohesive or granular soils which have cohesion intercepts of zero (see below). A full account of the range of mass movement processes on slopes is provided by Brunsden and Prior (1984).

Shear strength

Mohr–Coulomb criteria
Any soil mass will become unstable and liable to failure if shear stress exceeds the available shear strength. The factor of safety for a slope is defined as the ratio of shear strength to shear stress, so that failure occurs when this factor is just below unity. The failure criterion is expressed as

$$\tau \leq s \text{ for stability}$$

where τ is shear stress and s is shear strength. Coulomb (1776) first suggested that the shear strength of a soil might be expressed in the form:

$$s = c + \sigma_n \tan \varphi \tag{8.1}$$

where σ_n is the normal stress on the failure surface, φ is the angle of shearing resistance, and c is the cohesion. Mohr (1914) showed that the failure condition

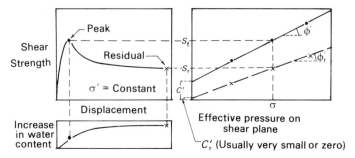

Figure 8.1   Relationship between 'peak' and 'residual' shear strength from direct shear tests (after Skempton 1964).

may be defined in terms of a Mohr envelope which is tangential to all such failure circles. In practice this envelope is found to be a straight line over a considerable range of normal stresses, and may therefore be expressed in the same way as the Coulomb strength criterion; hence, the Mohr–Coulomb failure theory. The envelope represents the upper limit of stable stress states (Fig. 8.1). Any attempt to impose a stress state represented by a point above the envelope causes the soil to yield and become unstable. The Mohr–Coulomb criterion is a hypothesis of value only in so far as it correctly describes the behaviour of real soils. In this respect it has certain defects, but routine tests for shear strength are almost always interpreted in terms of this criterion.

Drained saturated samples

When an excavation is made in a saturated soil, the resulting stress relief produces an immediate drop in pore-water pressure in the sides and below the excavation. In this 'short-term' condition the strength parameters are those defining the undrained shear strength. With time, the low pore-water pressures will rise towards an equilibrium state, accompanied by a corresponding increase in the water content of the soil around the excavation. The time-dependency of this swelling process may be the cause of many delayed or 'long-term' failures, and is considered in detail for the London Clay by Vaughan and Walbancke (1973). Their data suggest that at least 60 years may be required before equilibrium long-term pore-water pressures are reached in London Clay cutting slopes.

Slopes composed of granular soils have high enough permeabilities for long-term equilibrium pore pressures to be reached so rapidly that short-term conditions never apply, even under rapid excavation. Slopes of cohesive soils, if excavated or eroded slowly as in natural slopes, may also not experience short-term conditions if pore pressures equilibrate faster than the erosion rate. Thus in natural slopes the long term is critical for slope stability.

Coulomb's expression must generally be restated in the form (Terzaghi 1925):

$$s = c' + \sigma'_n \tan \varphi' \tag{8.2}$$

where c' and φ' are approximately constant parameters (Fig. 8.1) expressing the shear strength in terms of effective stress, and σ'_n is the effective normal stress. For all practical purposes, it may be assumed that:

$$\sigma'_n = \sigma_n - u \tag{8.3}$$

where u is the pore-water pressure. This relationship between shear strength and effective stress is found to be approximately valid for a wide variety of materials and conditions of loading.

Residual shear strength

It is commonly found that long-term failures in cohesive and granular soils exhibit apparent strengths at failure which are lower than the effective strengths defined by Equation 8.2, and it becomes important to differentiate between 'first time' failures and reactivation of old failure surfaces (Vaughan & Walbancke 1973, Rouse 1981). The apparent strengths will be overestimated by laboratory tests unless the residual strength is determined (Skempton 1964).

Skempton performed direct shear tests to large displacements (see Rouse 1981 for explanation of the apparatus) on samples of overconsolidated clays (Fig. 8.1), to determine the peak strength at failure (s_f) and the residual shear strength (s_r). Both peak and residual strengths defined by tests over a range of normal pressures accorded with the Coulomb−Terzaghi model (Fig. 8.1):

$$s_f = c' + \sigma_n \tan \varphi' \tag{8.4}$$

$$s_r = c'_r + \sigma'_n \tan \varphi'_r \tag{8.5}$$

where c'_r is approximately zero, and does not equal c', and $\varphi' \geq \varphi'_r$.

As a result of work by Petley on Walton Wood Clay, Skempton (1964, p. 83) suggested that the residual strength of a clay is the same whether the clay has been overconsolidated or normally consolidated. The value of φ'_r should then only depend on the soil properties, and when plotted against clay content a general decrease in φ'_r with increasing clay content is apparent (Skempton 1964, Rouse 1967, 1975). A full explanation of residual strength in clay soils is provided in Rouse (1981).

In granular soils the density at the residual state reaches a constant value for a given normal stress, irrespective of the initial density of the soil. The soil continues shearing without further volume change once it has reached the residual strength, which is therefore sometimes called the 'constant volume' strength, and φ'_r is equivalent to φ'_{cv}. The residual strength of clay soils is probably close to the magnitude of interparticle friction, while for granular soils it is significantly higher, for although shearing occurs at constant overall volume, the individual grains must still interact, and extra work is done in excess of that required to overcome interparticle friction.

The values of φ'_r for clay samples obtained using ring shear apparatus (Bishop et al. 1971) clearly show that multiple reversal direct shear tests give results generally higher than the true residual strength. Also the ratio τ/σ_n' (i.e. tan φ_r' for $c_r' = 0$) generally decreases with increasing normal stress (see below). At very low stresses (i.e. σ_n' less than 50 kPa) the ring shear apparatus gives values of φ_r' similar to the reversal direct shear box, which cannot be used satisfactorily below 50 kPa. As normal stress in ring shear tests increases, φ_r' drops rapidly to a constant value at high normal stress.

The value of φ_r' estimated from field observation is very sensitive to the accuracy of pore-water pressure measurement and the positioning of piezometers. Some classic studies underestimated effective normal stresses by incorporating excessively high pore-water pressures (Hutchinson *et al.* 1973, Chandler 1976). Thus, high values of φ_r' determined from reversal direct shear box tests have been used, and since field normal stresses are generally lower than 50 kPa the results have been fortuitously close. With thin regoliths developed by weathering, the possible range of normal stress will be small and envelope curvature will not be important. For clay slopes where deep-seated landslides may occur, the range of normal stress will be large and a constant value of φ_r' will not apply.

Slope stability models

The choice of an appropriate failure model is a major area of soil mechanics which can only be briefly reviewed here. Rotational slips are analysed by various circular arc methods, one of the commonest being Bishop's method of slices (Bishop & Morgenstern 1960) or one of its modified forms (Spencer 1967). More sophisticated methods are also available, unconstrained by the assumption of a circular arc shear plane (Morgenstern & Price 1965, Sarma 1973). Detailed study of landslides demands considerable data on slide geometry, pore-water pressure and material strength parameters, which may be difficult to obtain. However, geomorphologists may be more concerned with answering general questions about the relationships between hillslope angles and landslide processes in an area, and simpler methods of analysis such as the infinite planar slide model have been used, as outlined below.

On straight hillsides which are essentially rock slopes mantled by weathered debris, the failure surface is confined to the regolith mantle and may be considered almost planar and parallel to the ground surface (Skempton & De Lory 1957, Carson & Petley 1970, Rouse 1975, Rouse & Farhan 1976, Chandler 1977). Fluctuations in water-table level (i.e. the phreatic surface) or the creation of temporary perched water systems will markedly affect the factor of safety of a hillslope. Thus the maximum likely pore-water pressures must be assessed, and Skempton and De Lory (1957) suggest that this occurs when the phreatic surface coincides with the ground surface and subsurface flow is parallel to the ground slope. Chandler (1970a) identified this condition in the Uppingham slide. This does not imply that the regional ground flow net extends to the slope surface; 'throughflow' may provide a suitable mechanism (Rouse 1975).

The slope stability model of Figure 8.2 represents the case of plane failure

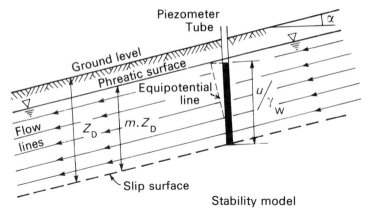

Stability model

Figure 8.2 The 'infinite slope' stability analysis (Z_D is depth to failure plane, u is pore-water pressure).

and parallel flow, and when the residual strength can be represented by φ_r' alone (since c_r' approximates to zero):

$$\alpha_L = \arctan\left[(1 - (m\,\gamma_w/\gamma_s))\tan\varphi_r'\right] \tag{8.6}$$

where (Fig. 8.2) α_L is the limiting angle of valley side slope (the minimum angle on which landsliding can occur), γ_w is the unit weight of water, γ_s is the saturated unit weight of soil, and m is the position of phreatic surface with respect to ground level (i.e. $m = 1$ when the phreatic and ground surfaces coincide, and $m = 0$ when the phreatic surface is at or below the failure surface).

Many assumptions are made in the above model, but the analysis seems to explain the limiting stability of a wide variety of natural hillslopes when used prudently. The approach is directly pertinent to interpretation of process–form relationships on slopes and landscape explanation.

Stability of slopes in granular soils

Most early studies of long-term slope stability on granular lithologies were dominated by the 'threshold slope' concept. Soil mechanical principles were applied to slopes over relatively large areas and, when used sensibly, satisfactorily explained the limiting stability of a range of slopes. Carson and Petley (1970) found close agreement between predicted limiting angles and peaks of the frequency distributions of hillside slope angles in the Pennines and Exmoor, although their postulated trimodal frequency distribution is statistically weak. 'Peaks' of $20-21°$, $25-27°$ and $32-34°$ were taken as valid despite rejection in a non-parametric test. The $37°$ slopes were observed to be rock scree slopes and the remaining 'peaks' were found to agree closely with predicted limiting angles, suggesting a marked change in φ'_r as particle size decreases during the course of weathering.

This argument was developed further by Carson (1971) in a study of slopes in the Laramie Mountains, Wyoming. The frequency distribution of angles of surveyed rectilinear slopes showed a pronounced modal group at 25–28° and upper and lower limits of 33° and 18°. The detrital mantle of 19–23° slopes was found to have $\varphi'_r = 33°$, and substituting this into the limiting equilibrium equation (Eqn 8.6) with the appropriate saturated bulk density produces $\alpha_L = 17.5°$. The value of $\varphi'_r = 33°$ is therefore taken to explain the upper and lower limits to the frequency distribution. Carson's explanation of the modal group (1971, p. 44) is, however, confused and his quoted calculations and limits are incorrect; in fact, the limiting angle lies 2° lower than the limits of the modal group, which is not therefore adequately explained in terms of φ'_r.

Rouse (1975) extended the application of soil mechanical approaches to studies of slope stability in the valleys of West Glamorgan, South Wales. The residual shear strength of the granular regoliths on these slopes varied very little about a mean of $\varphi'_r = 36°$, based on numerous tests. The limiting equilibrium model for infinite planar slides (Eqn 8.6) gave threshold slopes of 36° (dry state with $m = 0$) and 22° (saturated state with $m = 1$). Slopes in the study area were randomly sampled to define a frequency distribution (Fig. 8.3) with population mean $19.4° \leq \mu \leq 21.3°$ (95% confidence), and the 'unstable' slopes had a population mean of $21.5° \leq \mu \leq 24.7°$. Since the limiting angle is 22° the unstable slopes have a central tendency equivalent to the calculated limiting angle. The frequency distribution, irrespective of apparent stability (Fig. 8.3), was shown to have a population mean close to the limiting angle, demonstrating that rapid mass movement was important in terms of evolution of these slopes. A similar analysis was performed on the fossil cliff line of the Gower Peninsula, which has scree slopes grading into solifluction benches. The various slope facets were explained in terms of the engineering properties and palaeoenvironment (Rouse 1975, p. 232). For example, the detrital slopes had a population mean (Fig. 8.4) of $20° \leq \mu \leq 22°$ (95% confidence) and with $\varphi'_r = 34°$ for crushed limestone and drift, and $m = 1$ in Equation 8.6, α_L is 21°. The detrital slopes were assumed to have been controlled by Late Devensian rapid mass movement when water tables were at the surface of what are now fossil slopes.

Rouse and Farhan (1976) also differentiated the slope angle frequency distribution for the Lliw Valley, West Glamorgan, into subjective groups. The mean residual strength ($\varphi'_r = 35°$) and mean saturated unit weight ($\gamma_s = 17.4$ kN m^{-3}), based on numerous sampled debris mantles, were incorporated in the shallow planar slide model (Eqn 8.6) to predict threshold slopes of 35° (dry state) and 17.2° (saturated state). The slopes were randomly sampled but measured in the field to define a frequency distribution (Fig. 8.5) with a population mean for valley side slopes of $15.5° \leq \mu \leq 17.8°$ equivalent to the limiting angle, so rapid mass movement was considered important in interpreting valley side slopes. The multimodal distribution was also analysed in terms of the soil mechanical properties. For example, between the two limiting angles of 17.2° and 35° (Fig.

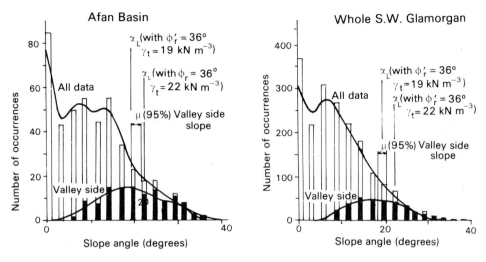

Figure 8.3 Frequency distribution of angles of slope segments in West Glamorgan (after Rouse 1975).

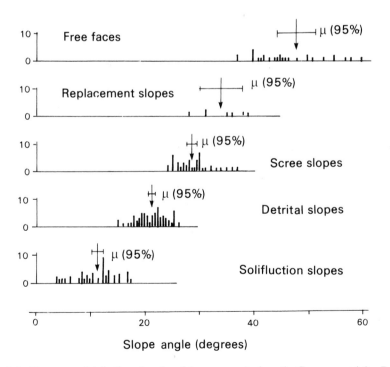

Figure 8.4 Frequency distribution of angles of slope segments along the Gower coast (after Rouse 1975).

8.5) there are two important peaks at $18-22°$ and $28-30°$. These are the rectilinear fossil detrital slopes developed on shale and sandstone respectively; the φ_r' values do not differ between the two types of site. This was interpreted as a result of variation in pore-water pressure creating different threshold values (Rouse & Farhan 1976, p. 334).

Similar conclusions were drawn regarding the role of periglacial conditions for colluvial (detrital) masses in the Appalachian Plateau province of the United States (Philbrick 1961, D'Appolonia *et al.* 1967, Gray *et al.* 1977, Gray & Gardner 1977). However, large-scale movement of the type previously assigned to periglacial conditions occurred in a colluvial mass in that region at McMechan, West Virginia, in March 1975, indicating that such movements may not be solely related to past climatic conditions. Detailed study indicated that movements may have been triggered by a period of abnormally high precipitation preceded by at least 30 years of low precipitation (Gray & Gardner 1977). Large-scale movements are thought to occur infrequently between long intervals dominated by smaller-scale superficial mass movements similar to those occurring in South Wales. In the periods between large-scale events, colluvium builds up and equilibrium is disturbed in portions of the larger mass, until a trigger mechanism initiates another large-scale movement.

Richards and Anderson (1978) attempted to interpret the origin of fossil dry valleys in fluvioglacial sands and gravels in north Norfolk. Mean maximum valley side slope angles were $16.7°$ and $19.5°$ at two locations studied, although modal values of $20.5°$ and $23.5°$ respectively are more appropriate definitions of the threshold or limiting slopes. Secondary peaks in the slope angle histograms occurred at $13-15.9°$ For the two locations, the average φ_r' values of $34.2°$ and $36.1°$ gave predicted maximum stable angles of $19.0°$ and $19.2°$ respectively, assuming saturated conditions. The φ_r' values therefore give predicted limiting angles less than the modal groups for the two valleys. This may reflect the accuracy of unit weights used in Equation 8.6; the authors do not explain their method of estimation (Dunkerley 1982). However, they suggest that the assumption that $m = 1$ under periglacial conditions might not be valid. If m is lower, this would allow a steeper limiting angle, a possibility which runs counter to arguments used elsewhere (Dunkerley 1976, Rouse & Farhan 1976, Gray & Gardner 1977).

It is evident from comparison of these studies that, while early investigations employed very little data and less rigorous forms of analysis, more recent work has considerably developed the application of the infinite planar slide model to the interpretation of natural threshold slopes.

The mechanics of scree slopes

Engineering research into the strength of rockfill suggests that the minimum angle of shearing resistance at constant volume (φ_{cv}', which is similar to φ_r'; see above) of most scree materials is likely to be about $39-40°$ for angular rockfill, $32-33°$ for rounded sands and gravels, and $35-37°$ for poorly sorted gravels. However, the inclination of well-drained scree slopes will depend on φ_r' only if

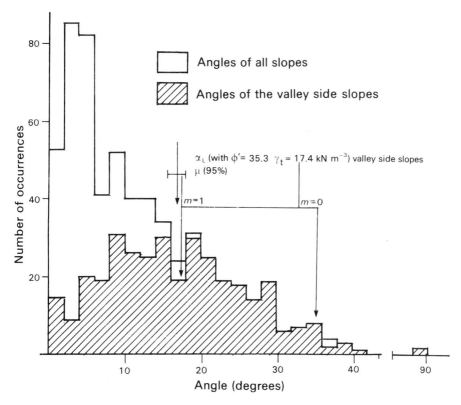

Figure 8.5 Frequency distribution of slope angles for the Lliw Valley (after Rouse & Farhan 1976).

they are being rapidly eroded or if deposition rates are high compared with denudation rates on the scree, and their typical 35° angle is therefore not a limiting angle (Chandler 1973), which would be approximately 39° for angular scree materials. Scree slopes in Spitzbergen were characteristically 35° but were sometimes up to 39° where basal erosion occurred (Chandler 1973). Shallow landslides in screes on slopes less than 39–40° were attributed to impeded drainage due to permafrost at shallow depth. The frequent occurrence of talus slopes at 35° remained unexplained. The upper limit of 39–40° for angular scree is also evident in Figure 8.5, and Rouse and Farhan (1976) note fresh scree accumulations at angles greater than 36° while Statham (1973) measured maximum angles of approximately 39° on scree slopes at Cader Idris in mid-Wales.

Statham (1973, 1976) maintained that scree slope development could not be explained by an angle of repose model when rockfall operates at the scree head. Then, scree slopes have a straight upper section at an angle less than the angle of repose, and a basal concavity. Particles are well sorted, with the largest at the scree base. Evidence from the Isle of Skye and Cader Idris supported a model developed as an alternative to limiting equilibrium, and based on an energy input from rockfall (Statham 1976). This model predicts that, as the height of the headwall decreases, the basal concavity becomes shallower, until the headwall disappears and the scree adopts a rectilinear form at the maximum angle of stability. Sorting is assumed to reflect trapping of smaller particles in surface depressions, and strong downslope orientation of particles occurs because particles align themselves to offer the least resistance to motion (Statham 1973). This idea of dynamic input from rockfall led to a dispute between Carson (1977) and Statham (1977) over the definitions of angle of repose (φ_{rep}), angles of shearing resistance (φ_r' and φ_{cv}') and the angles of talus slopes. Carson advocates that angles of talus slopes equal φ_{rep} although φ_{rep} is less than φ_{cv}'. He maintains that angles of rest of angular rock fragments measured from tilting box experiments are affected by 'wall effects' and small relative sample size (overestimating φ_{cv}'), and are unrepresentative of field conditions where they are actually found to be $35-36°$. He also highlights the problem of confusing the concepts of the angle of repose and the residual angle of shearing resistance in soil mechanics. Carson's findings in field tests of angles of repose of $35-36°$ led him to dispel the conclusions of Chandler (1973) outlined above. Other research suggests that rounded scree or gravelly mixtures behave differently from angular scree (Rouse & Farhan 1976).

Stability of slopes in clay soils

Temperate environments

Early studies of clay slopes, like those of granular soils, applied laboratory analysis of mechanical properties to areal interpretation of threshold slopes. However, Chandler later (1970a, 1971, 1973, 1974) concentrated on detailed investigation of site failures, and more rigorous explanations of failure processes emerged.

Skempton and De Lory (1957) stated that in several areas of the London Clay, natural hillsides were not in final equilibrium (Fig. 8.6). Where the groundwater level reaches the surface in winter, slips can occur on 10° slopes but all slopes lower than this are stable. This critical slope was found to be in agreement with a laboratory-deduced φ' of 20° assuming c' was zero. Hutchinson (1967) also worked on London Clay slopes, demonstrating the free degradation of fossil cliffs to a stable state. He identified and classified different types of failure (Fig. 8.7a), and compiled a frequency distribution of unstable slopes (classed by landslide type) and stable slopes (Fig. 8.7b). The ultimate angle for unstable coastal slopes was found to be about 8° while the steepest slopes inland were at angles of 10° and were attributed to different groundwater levels. Hutchinson (1969) also re-evaluated the major landslides at Folkestone Warren in Kent, which had not been previously examined in terms of effective stresses. All major slides at this site were shown to have occurred within the period of seasonally high groundwater. Furthermore, three slides investigated using a non-circular slip surface stability analysis indicated that the average strength mobilised on

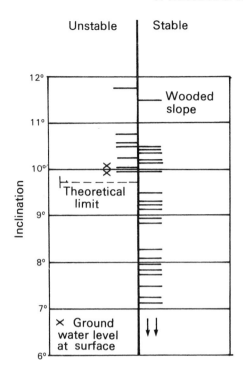

Figure 8.6 Natural slopes in London Clay (after Skempton & De Lory 1957).

the shear surface approximated to the residual strength of the Gault Clay. An understanding of this process allowed Hutchinson (1969) to devise a mechanism for retrogression of the rear scarp.

Chandler identified several important principles relating to long-term stability of natural clay slopes in studies of failures in the Lias Clay in Northamptonshire. At Uppingham (Chandler 1970a) the limiting angle for stability was 8.5−9° with most unstable slopes in the 9−10° range. Natural slopes appear to be flattening in this area as a result of landsliding. The infinite slope stability analysis (Eqn 8.6 and Fig. 8.2) was demonstrated to be an appropriate model for the 1968−9 movement noted on a slope of 8.9° and borings and trial pits confirmed that movements at Rockingham nearby are generally shallow, being less than 4 m in depth (Chandler 1971). At Rockingham the slide was stable when analysed, but movement began in the Lateglacial and was renewed following Iron Age deforestation and since the 17th century. In each of 12 case records of failures in the Lias Clay, Chandler (1974) noted that groundwater conditions had reached the long term steady state, in less than 10 years for brecciated Lias and in no more than 60 years for weathered but unbrecciated Lias, although seasonal fluctuations occur at shallow depths. A final trigger for slope movement occurs when heavy rain causes a rise in pore-water pressure close to the ground surface. The lower bound to the strength data is given by $c' = 0$ and $\varphi'_r = 23°$ (Fig. 8.8). This is slightly higher than values for the London Clay and is possibly associated with the lower plasticity of the Upper Lias Clay. Brecciated

(a)

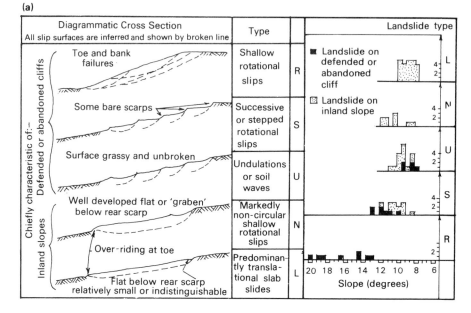

(b)

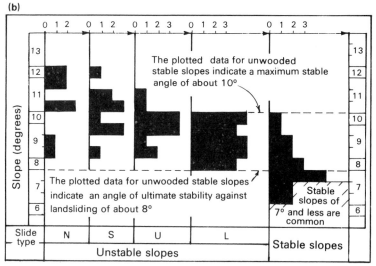

Figure 8.7 Stability of London Clay cliffs: (a) main types and frequencies of landslide slopes; (b) data for stable and unstable inland slopes (after Hutchinson 1967).

samples have a lower strength and reach long-term steady pore pressures more quickly; failures therefore occur at lower strengths, and brecciated slopes have angles of 4°, whereas unbrecciated slopes frequently exceed 8° On these steeper slopes the brecciated clay (a product of periglacial processes) had probably been removed by landsliding. A complication revealed by these detailed studies is that, at the appropriate normal effective stresses, 'back analysis' shows the

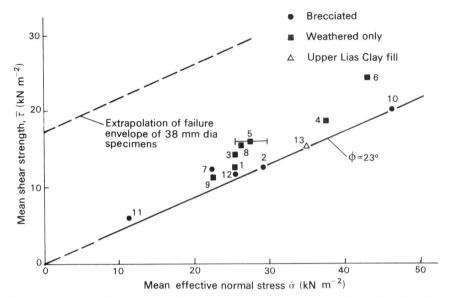

Figure 8.8 Relation between mean shear strength and normal pressure in Upper Lias Clay (after Chandler 1974).

residual shear strength to be lower than indicated by direct shear tests, but higher than the value given by ring shear tests (Chandler 1973, Hutchinson *et al.* 1973).

Periglacial conditions and slope stability

Hutchinson (1967) recognised the possibility that solifluction mantles often determine present slope form and behaviour in temperate environments. The maximum slope angle for stiff fissured clays in areas affected by past periglacial activity is dictated by the presence of pre-existing slip surfaces formed during these periods (Weeks 1969), and these slip surfaces may be found in slopes flatter than the limiting angle of stability against landsliding. For landslides to occur on 3–4° slopes requires considerable artesian pore pressures. This could have resulted from permafrost sealing off an underlying aquifer. In Northamptonshire, Chandler (1970b) identified two sites where failure surfaces existed on slopes of 6.75° and 4° where limiting angles for landsliding should be 12° and 8.75° respectively if $m = 1$ in Equation 8.6. Artesian pore pressures caused by permafrost in the Northampton Sand were invoked. Chandler (1972) confirmed this interpretation by examining an active mudflow site in Vestspitzbergen, where movement of silty material occurred on slopes as low as 8.5–12° when stability analysis predicted limiting angles of 16–18°. High pore-water pressures were observed, where permafrost at shallow depth prevented downward seepage of melt water. Soil water drainage was confined to a shallow layer, and was concentrated by soil banding parallel to the slope surface, probably caused by ice lense segregation. Surface overnight freezing, accumulation of 'silt seeps', and frost shattering of the siltstone bedrock just above the permafrost also encour-

Table 8.1 Selected soil properties for Dominican samples.

Soil type	Sample name		In situ moisture (% dry weight)	In situ unit wt. (kN m⁻³)	Dry unit weight (kN m⁻³)	Sat.unit weight (kN m⁻³)	Void ratio	Degree Saturation	Porosity	Sat.water content (% dry weight)	Liquid limit % dry	Liquid limit % wet	Plastic limit % dry	Plastic limit % wet	Plasticity Index dry	Plasticity Index wet	Activity dry	Activity wet
allophane	Grandfond	Unf	72.0	13.26	7.71	14.8	2.62	78.2	0.72	90	56	101	43	69	13.5	32		2.04
latosolic	Grandfond	Toe	92.0	14.89	7.84	14.89	2.56	100.0	0.72	92	54	80	36	56	18	24	1.91	
	Grandfond	Scar										59		44		15		
allophane	Pichelin	Unf	69.5	13.39	7.90	14.93	1.96	78.1	0.66	69	51		42		8.8		3.17	
latosolic	Pichelin	Toe	54.0	12.17	9.40	15.90	2.52	42.6	0.72	89	77		52		25		9.14	
	Pichelin	Scar									86		57		29			
allophane	Carhome	Unf	90.0	10.42	5.47	13.35	4.09	62.8	0.81	144	83	96	52	55	32	41	1.2	1.54
latosolic	Carhome	Toe	69.0	14.15	8.37	15.23	2.33	84.2	0.70	82	71	93	43	56	28	42		
	Carhome	Scar	65.0	12.92	7.81	14.87	2.57	72.4	0.72	90	65		44		24			
allophane	Bells	Unf	48.0	11.35	7.67	14.78	2.63	51.8	0.72	92	88		54		34		2.4	
latosolic	Bells	Toe	49.0	12.03	8.07	15.04	2.45	56.8	0.71	86	55		34		21		1.72	
allophane	Rosalie	Unf	70.5	11.89	6.98	14.33	2.99	66.8	0.75	105	60		38		21		2.1	
latosolic	Rosalie	Toe	44.5	14.47	10.01	16.30	1.78	70.9	0.64	63	72		51		22		1.32	
allophane	Attley	Unf	73.5	11.62	6.71	14.16	3.15	66.0	0.76	111	73		43		30		5.7	
latosolic	Attley	Toe	65.0	10.43	6.32	13.91	3.41	54.2	0.77	120	85		46		40		1.37	
	Attley	Scar	47.5	13.18	8.94	15.60	2.12	63.7	0.68	74	77		46		32			
kandoid	Vielle Case	Unf	38.5	12.00	8.66	15.42	2.22	49.3	0.69	78	75	60	44	35	31	25	2.3	1.8
	Vielle Case	Toe	39.5	12.34	8.85	15.54	2.15	52.2	0.69	76	71	62	36	39	35	23	2.8	
	Vielle Case	Scar	53.5	13.31	8.67	15.43	2.21	68.7	0.69	78	72		36		36		1.8	1.5
kandoid	Calibishie	Unf	48.0	14.09	9.52	15.98	1.93	70.8	0.66	68	59		31		28		1.0	
	Calibishie	Toe	52.0	9.03	5.94	13.66	3.69	40.0	0.79	130	78	69	36	36	42	33	1.3	1.1
	Calibishie	Scar	55.0	10.68	6.89	14.27	3.04	51.3	0.75	107	57		33		24			
allophane	Freshwater Lake	Unf	88.0	11.98	6.37	13.94	3.37	74.1	0.77	119	50	148	37	85	13	95	4.0	29.1
latosolic	Freshwater Lake	Toe	100.0	11.56	5.78	13.55	3.82	74.3	0.79	134	65	153	47	79	18	74		9.4
	Freshwater Lake	Scar									70		48		21			
Smectoid	Méro	Unf	13.0	13.67	12.09	17.65	1.30	28.3	0.57	46	47	68	21	25	25	43		
	Soufrière	Unf	29.5	14.23	10.98	16.93	1.54	54.6	0.61	54	42		27		non-plastic			
allophane	Galion*	Unf	18.5	21.07	17.81	21.35	0.56	92.0	0.36	20	28	—	18	16	10	–		
allophane podzol	Pont Cassé	Unf	180	11.23														4.5

* Galion is a protosol — when developed it will become a smectoid.

aged artesian pressures. Such observation of active periglacial slope processes considerably aids interpretation of low-angle shear surfaces in southern England.

Stability of slopes in clay soils in tropical environments

The residual soils of the tropics differ dramatically from those of temperate regions, mainly because of the unusual behaviour of allophane latosols, which have extremely high field moisture contents of 42−180% (Wallace 1973). Lower moisture contents are limited to kandoid soils or protosols (immature soils) of a smectoid type (Table 8.1). Values of 20−70% are typical of a wide range of temperate soils. Samples from Dominica, West Indies, analysed by the authors illustrate the low field dry unit weight of allophane soils, ranging from 5.47−10.01 kN m^{-3} (Table 8.1); only smectoid and skeletal soils had high unit weights. Field values for the degree of saturation were 40−100% (Table 8.1), compared with a range of 85−95% noted by Wallace (1973). One low value at Méro reflects the peculiar nature of this skeletal soil. The field moisture content was between 80% and 100% of the liquid limit. Whereas Wallace's (1973) observations were in the continually wet highlands of New Guinea, the authors' data from Dominica relate to a seasonally wet environment; thus there are differences in the range of field moisture contents.

Without exception, these tropical residual soils become non-plastic when air-dried or oven-dried, and do not become plastic when wetted again (Wallace 1973). This affects size analysis and index tests. Wallace was unable to obtain representative measures of size distribution because of aggregation on drying.

Table 8.2   Values of τ'/σ'_n for Dominican soils.   handl

Sample name	Soil type	Normal stress (kN m^{-2})				Mean φ'_r	Date of landslide
		41.802	74.502	134.452	229.827		
Direct shear							
Carhome Unf	allophane	0.9343	0.6985	0.8116	0.7234	35°	1980
Freshwater Lake Unf	allophane	0.6981	0.7681	0.7177	0.5396	26.7°	1979
Rosalie Toe	allophane	0.6380	0.6765	0.6002	0.6535	33°	1982
Bells Toe	allophane	0.5450	0.5610	0.5964	0.6185	32.5°	1982
Galion Unf	protosol	0.7141	0.6534	0.7024	0.7553	37.8°	unfailed
Bells Unf	allophane	0.7882	0.7054	0.7042	0.6648	32.6°	1982
Attley Unf	allophane	0.6212	0.8019	0.7535	0.6726	33.8°	1982
Grandfond Unf	allophane	—	0.7768	0.7176	0.7402	35.6°	1980
						ultimate	
		40.038	69.528	128.508	246.468	φ'_r	
Ring shear							
Attley Unf	allophane	0.8745	—	0.4876	0.4636	25.4°	1982
Grandfond Unf	allophane	0.8737	0.8484	0.7770	0.5912	34.4°	1980

The authors have also found it impossible to deflocculate wet samples; the problem is not in dispersing samples, but in preventing reflocculation (Wesley 1973). Soils with high allophane contents may therefore not be amenable to meaningful particle size analysis. Liquid and plastic limit determinations are also affected by drying. In Table 8.1, 'wet' and 'dry' samples are compared; large discrepancies occur in the index properties for allophane soils (e.g. Grandfond and Freshwater Lake samples), but not for kandoid soils. This has considerable implications for standard methods of testing. The increase of plasticity index with increasing liquid limit is very small (Table 8.1), so that on a plasticity chart the allophanes do not run parallel to the A-line but fall progressively below as the liquid limit increases (Wesley 1973).

Wallace (1973) identified a critical normal stress for allophane soils below which their compressibility is low, but above which it is high. The critical pressure ranged from 107 to 375 kN m^{-3} in tests, and these exceed the present overburden pressure. The wet climate and presence of highly hydrated allophane and halloysite suggests that desiccation has not caused overconsolidation. There is no geomorphological evidence of former burial to explain overconsolidation. This critical compressibility limit is important in laboratory testing. Problems of compressibility and drying are related by Wallace (1973) to the structure of the cementation bonds and viscous gel, a complex structure which also explains anomalies of shear strength. The destruction of undisturbed structure on remoulding the soils generally results in a substantial reduction of permeability, but not of shear strength (Wesley 1977).

Wesley (1977) tentatively suggests that lower shear strengths relate to decreasing water content, after discussing residual shear tests on two samples each of allophane and halloysite (kandoid) soils. Comparisons of data in Tables 8.1 and 8.2 do not substantiate this idea, since φ'_r appears unrelated to the physical properties apart from a general relationship with the index values (particularly the plasticity index). The sample with the lowest φ'_r (26.7°; Freshwater Lake Unf) has the highest plasticity index (95), while that with the highest φ'_r (37.8°, for the protosol Galion Unf) has one of the lowest plasticity indices (10). However, insufficient tests for φ'_r have been conducted for more definite conclusions
characteristic decrease in τ/σ_n (which equals φ'_r when $c'_r = 0$) with increasing σ'_n
loysites with temperate clays, and noted unusually high φ'_r values for tropical allophanes, and very little variation between their peak and residual strengths. This reflects the non-crystalline nature of allophane, and the lack of layer lattice minerals (Wallace 1973, Wesley 1977).

The Dominican samples tested in a ring shear apparatus both reveal the characteristic decrease in τ'/σ'_n (which equals φ'_r when $c'_r = 0$) with increasing σ'_n (Table 8.2). This contrasts with the lack of strength envelope curvature resulting from Wesley's (1977) ring shear tests, but insufficient tests have been performed to allow definite comparison. In both Dominican samples tested by ring shear, the ultimate level of τ'/σ'_n (i.e. of φ'_r) is less than for direct shear tests, for which shear stress–normal stress plots are perfectly linear with c'_r approximately zero. A critical problem is therefore which value of φ'_r is appropriate for studies of limiting slope angles. The low stresses of the ring shear apparatus give values of about 41° while direct shear gives about 34° as the ϕ_r' of allophanes!

These soils have permeabilities 10–1000 times those of temperate sedimentary soils despite their generally higher clay content (Wesley 1977). Groundwater levels on the slopes are usually low, but rise sharply during periods of heavy rainfall, making pore pressure measurement difficult. The slopes are also commonly steeper than those in temperate sedimentary clays, and slope failures tend to be translational surface slides rather than deep-seated rotational failures. The methods of stability analysis developed for temperate slopes may therefore be inappropriate for the different conditions experienced on the steeper tropical slopes, but individual slide analyses are few. Wesley (1977) studied two sites and provided geotechnical depth profiles, but no pore-water pressure data or flow nets. This lack of pore pressure data also characterises the Dominican studies, although the age and geometry of these landslides are known. Inaccessibility through hurricane-damaged forest, and the rapid regrowth of secondary vegetation, inhibit the direct observation that will be required to explain how hillsides in clays can exist at angles of 40° even allowing for the high mean ϕ_r' values of allophanes. The explanation may lie in the higher ϕ_r' values implied by ring shear tests at the low stresses applicable to shallow translational slides, and in the normal maintenance of low pore pressures by the high permeability. Certainly, however, catastrophic landslides may be triggered by heavy rain such as occurred in hurricanes David and Alan in 1979.

Stability analysis and threshold slopes: some limitations and conclusions

The study of threshold slopes has gained in understanding with increasing rigour of analysis; nevertheless, some applications overextend the model. For example, the stability of basal slopes argues against the explanation of basal concavity by invoking increased probability of high pore-water pressures (Carson 1969, 1971, Carson & Petley 1970). Some problems require further evaluation. For example, Anderson et al. (1980, 1982) maintain that slope angle histograms may be composed of many different elements, suggesting the need to differentiate such components (Skempton & De Lory 1957, Rouse 1975, Dunkerley 1976, Rouse & Farhan 1976). Several problems bedevil laboratory determination of shear strength. Shear box tests may give unrepresentative results for granular soils if particles greater than one-tenth of the shear box length are not screened out (Chandler 1973), and in fissured clays the strength of small intact samples differs markedly from that of large bulk samples containing fissures. Ring shear tests indicate a non-linear relationship between τ'/σ_n' and σ_n' (Chandler et al. 1973), although this is less critical in the range of normal stresses known to occur in shallow translational slides. Modelling individual slides introduces additional problems. The appropriate worst pore pressure condition must be defined, and although parallel flow with the water table at the surface may satisfactorily represent the condition for periglacial perched water tables,

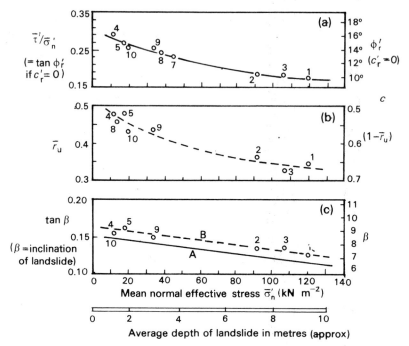

Figure 8.9 (a) Residual strength; (b) average pore pressure; (c) infinite slope analysis, for Upper Lias landslides (after Chandler 1982). A predicted from (a) and (b): $\tan \beta = (1 - \bar{r}_u) \tan \varphi_r'$; B best fit relationship: $\tan \beta = 0.013 + (1 - \bar{r}_u) \tan \varphi_r'$.

alternative flow net models may be preferable for Postglacial failures induced by high rainfall (Anderson *et al.* 1980). The predicted threshold angle is also dependent on the selected stability analysis, and for a specific slide in cohesive materials, the infinite planar model gives a lower threshold angle than the conventional method of slices (Anderson *et al.* 1980, p. 107).

Chandler (1982) gathered case records of several Lias Clay failures, and performed 'back analyses' for the slides. The average shear stress was plotted as a ratio of the average normal stress (Fig. 8.9a), and confirms the conclusion that φ_r' is a function of the effective stress. Also, the average pore-water pressure ratio ($\bar{r}_u$) is plotted against the corresponding normal stress, and varies with the average depth of the landslide (Fig. 8.9b). Since slope angle (β) is less than 10° Equation 8.6 can be simplified to:

$$\tan \beta = (1 - \bar{r}_u) \tan \varphi_r' \qquad (8.7)$$

Chandler (1982, p. 436) takes parameters from Figures 8.9a and b to create the distribution of Figure 8.9c. Observed data from the slides generally agree with Equation 8.7, but underestimate the slope tangent by about 10%. Greater discrepancies would have arisen had laboratory ring shear values and infinite planar slide analyses been employed. Chandler (1982) concludes that three explicit assumptions of the infinite slope analysis are suspect. First, φ'_r is not a constant but is stress-dependent, although for granular soils and shallow surface failures this is not critical. Secondly, he questions the assumption of uniformity in origin and development of slopes throughout an area. Thirdly, the assumption is questioned that $m = 1$ represents the most severe groundwater condition experienced by the slope as it reaches its ultimate angle of stability. Thus, Chandler concludes that there is no unique threshold limit in the Lias Clay, but rather a range from 9° to 7° (possibly as low as 6°) depending on the size of land-slide and depth to the failure plane.

Both in terms of explaining recent, periglacial and Postglacial failure, and in terms of general models of slope development under mass wasting processes, the use of stability models requires a sophistication which cannot readily be matched by field measurement or laboratory determinations. Therefore inter-pretation of agreement between predicted and observed angles requires con-siderable caution (Anderson *et al.* 1982, Dunkerley 1982). It is not necessary to abandon the concept completely; indeed, alternative approaches to slope evo-lution have been fraught with even more conceptual fallacies. What is required is a better appreciation of the variability of field conditions and laboratory-deduced soil parameters, and a move away from broad-brush geomorphologi-cal perspectives towards studies of specific slides which will eventually permit general inference from individual site analyses.

References

Anderson, M. G., K. S. Richards and P. E. Kneale 1980. The role of stability analysis in the inter-pretation of the evolution of threshold slopes. *Trans Inst. Brit. Geogs.* **5**, 100–12.

Anderson, M. G., K. S. Richards and P. E. Kneale 1982. Reply to comment by Dr D. L. Dunkerley. *Trans Inst. Brit. Geogs.* **7**, 236–9.

Bishop, A. W. and N. Morgenstern 1960. Stability coefficients for earth slopes. *Geotechnique* **10**, 129–50.

Bishop, A. W., G. E. Green, V. K. Garga, A. Andresen and J. D. Brown 1971. A new ring shear apparatus and its application to the measurement of residual shear strength. *Geotechnique* **21**, 273–328.

Brunsden, D. B. and D. Prior (eds) 1984. *Mass movements on slopes.* Chichester: Wiley.

Carson, M. A. 1969. Models of hillslope development under mass failure. *Geogr. Anal.* **1**, 76–101.

Carson, M. A. 1971. An application of the concept of threshold slopes to the Laramie Mountains, Wyoming. *Inst. Brit. Geogs Spec. Publn. 3*, 31–47.

Carson, M. A. 1977. Angles of repose, angles of shearing resistance and angles of talus slopes. *Earth Surf. Proc.* **2**, 363–80.

Carson, M. A. and D. J. Petley 1970. The existence of threshold hillslopes in the denudation of the landscape. *Trans Inst. Brit. Geogs.* **49**, 71–95.

Chandler, R. J. 1970a. A shallow slabslide in the Lias Clay near Uppingham, Rutland. *Geotechnique* **20**, 253–60.

Chandler, R. J. 1970b. Solifluction on low-angled slopes in Northamptonshire. *Q. J. Eng. Geol.* **3**, 65–9.

Chandler, R. J. 1971. Landsliding on the Jurassic escarpment near Rockingham, Northamptonshire. *Trans Inst. Brit. Geogs.* **3**, 111–28.

Chandler, R. J. 1972. Periglacial mudslides in Vestspitzbergen and their bearing on the origin of fossil 'solifluction' shears in low angled clay slopes. *Q. J. Eng. Geol.* **5**, 223–41.

Chandler, R. J. 1973. The inclination of talus, Arctic talus terraces and other slopes composed of granular materials. *J. Geol.* **81**, 1–14.

Chandler, R. J. 1974. Lias Clay: long-term stability of cutting slopes. *Geotechnique* **24**, 21–38.

Chandler, R. J. 1976. The history and stability of two Lias Clay slopes in the upper Gwash Valley, Rutland. *Phil Trans R. Soc. Ser. A* **283**, 463–92.

Chandler, R. J. 1977. The application of soil mechanics methods to the study of slopes. In *Applied geomorphology*, J.R. Hails (ed.), 157–82. Oxford: Elsevier.

Chandler, R. J. 1982. Lias Clay slope sections and their implications for the prediction of limiting or threshold slope angles. *Earth Surf. Proc. Landforms* **7**, 427–38.

Chandler, R. J., M. Pachakis, J. Mercer and J. Wrightman 1973. Four long-term failures of embankments founded on areas of landslip. *Q. J. Eng. Geol.* **6**, 405–22.

Coulomb, C. A. 1776. Essai sur une application des regles des maximis et minimis à quelques problémes de statique relatif à l'architecture. *Mem. Acad. R. Pres. Divers Savants* **7**.

D'Appolonia, E., R. Alperstein and D. D'Appolonia 1967. Behaviour of a colluvial slope. *Am. Soc. Civ. Engrs,. J. Soil Mech. Found. Div.* **93**, Paper SM4.

Dunkerley, D. L. 1976. A study of long-term stability in the Sydney Basin, Australia. *Eng. Geol.* **10**, 1–12.

Dunkerley, D. L. 1982. A comment on 'The role of stability analysis in the interpretation of the evolution of threshold slopes'. *Trans Inst. Brit. Geogs.* **7**, 233–5.

Gray, R. E. and G. D. Gardner 1977. *Processes of colluvial slope development, McMechen, West Virginia.* Paper presented at Symp. Int. Assoc. Eng. Geol., Prague.

Gray, R. E., H. F. Ferguson and J. V. Hamel 1977. Slope stability in the Appalachian Plateau of Pennsylvania and West Virginia. In *Rockslides and avalanches*, B. Voight (ed.). New York: Elsevier.

Hutchinson, J. N. 1967. The free degradation of London Clay cliffs. In *Proceedings of the Geotechnical Conference*, Oslo, Vol. 1, 113–18.

Hutchinson, J. N. 1969. A reconsideration of the coastal landslides at Folkestone Warren, Kent. *Geotechnique* **19**, 6–38.

Hutchinson, J. N., S. H. Somerville and D. J. Petley 1973. A landslide in periglacially disturbed Etruria Marl at Bury Hill, Staffordshire. *Q. J. Eng. Geol.* **6**, 377–404.

Mohr, O. 1914. *Abhandlungen aus den Gebiete der Technischer Mechanik.* Berlin: W. Ernst.

Morgenstern, N. R. and V. E. Price 1965. The analysis of stability of general slip surfaces. *Geotechnique* **15**, 79–93.

Philbrick, S. S. 1961. *Old landslides in the Upper Ohio Valley.* Paper presented at Geol Soc. Am. Annual Meeting.

Richards, K. S. and M. G. Anderson 1978. Slope stability and valley formation in glacial outwash deposits, north Norfolk. *Earth Surf. Proc.* **3**, 301–18.

Rouse, W. C. 1967. Residual shear strength of sandstone soils. *Geotechnique* **17**, 298–300.

Rouse, W. C. 1975. Engineering properties and slope form in granular soils. *Engng. Geol.* **9**, 221–35.

Rouse, W. C. 1981. Direct shear. In *Geomorphological techniques*, A. S. Goudie (ed.) 125–8. London: George Allen & Unwin.

Rouse, W. C. and Y. I. Farhan 1976. Threshold slopes in South Wales. *Q. J. Eng. Geol.* **9**, 327–38.

Sarma, S. K. 1973. Stability analysis of embankments and slopes. *Geotechnique* **23**, 423–33.

Skempton, A. W. 1964. Long-term stability of clay slopes. *Geotechnique* **14**, 77–101.

Skempton, A. W. and F. A. De Lory 1957. Stability of natural slopes in London Clay. In *Proceedings of the 4th International Conference on Soil Mechanics*, Vol. 2, 378–81.

Spencer, E. 1967. A method of analysis of embankments assuming parallel interslice forces. *Geotechnique* **17**, 11–26.

Statham, I. 1973. Scree slope development under conditions of surface particle movement. *Trans Inst. Brit. Geogs.* **59**, 41–53.

Statham, I. 1976. A scree slope rockfall model. *Earth Surf. Proc.* **1**, 43–62.

Statham, I. 1977. Angle of repose, angles of shearing resistance and angles of talus slopes – a reply. *Earth Surf. Proc.* **2**, 437–40.

Terzaghi, K. 1925. Principles of soil mechanics. *Eng. News Record* **95**, 742.

Vaughan, P. R. and H. J. Walbancke 1973. Pore-pressure changes and delayed failure of cutting slopes in overconsolidated clay. *Geotechnique* **23**, 531–9.

Wallace, K. B. 1973. Structural behaviour of residual soils of the continually wet highlands of Papua-New Guinea. *Geotechnique* **23**, 203–18.

Weeks, A. E. 1969. The stability of natural slopes in south-east England as affected by periglacial activity. *Q. J. Eng. Geol.* **2**, 49–61.

Wesley, L. D. 1973. Some basic engineering properties of halloysite and allophane clays in Java, Indonesia. *Geotechnique* **23**, 471–94.

Wesley, L. D. 1977. Shear strength properties of halloysite and allophane clays in Java, Indonesia. *Geotechnique* **27**, 125–36.

Part III

SOIL PROPERTIES AND
PROCESS RECONSTRUCTION

9

Scanning electron microscopy and the sedimentological characterisation of soils

W. B. Whalley

Some sedimentological aspects of soil description

A common reaction of a geomorphologist, sedimentologist or pedologist on first examining a soil is to determine its particle size distribution. However, grain size is only one component of a complex set of interrelationships which characterise the physical and chemical properties of a soil or sediment (Whalley 1981). Although the aims of the three types of investigator may differ, the constraints imposed by the soil and the techniques of analysis are often very similar.

The sedimentological nature of soil components may help determine pedological characteristics, for example with respect to permeability and *in situ* weathering. The sedimentology and pedology of soils may be considered as separate entities, although for some soils they come close together or indeed overlap. This can be the case with laterites and duricrusts in general, as well as loess, where physical and chemical properties are particularly closely related.

Following Griffiths (1967) it is possible to describe a sediment functionally (Whalley 1981):

$$D = f\ (C,\ G,F,O,P\ )$$

where, for particles 1 to x, D is a defining description of the properties of the sediment, C is the mineralogical composition of particles $(1...x)$, G is the grading, consisting of sizes of particles $(1...x)$, F is the form of each particle $(1...x)$, O is the orientation of each particle $(1...x)$ and P is the fabric and packing of the sediment.

Such an overall description is clearly complex and only a small portion may be required for any one purpose. The various descriptors cannot be completely independent: F and G both affect P; C has an influence on F and G, thereby on P. Additionally, the components may change in time; mineralogy changes during weathering, and packing may alter with compaction and consolidation. These descriptors are themselves composite; for instance, the concept of form is of particular interest in the present context. Thus, Whalley (1972) considered a relationship:

$$F = f\ (Sh,Sp,A,R,T\ )$$

where, for any particle, Sh is the shape, Sp is the sphericity, A is the angularity, R is the roundness and T is the surface texture.

This chapter is chiefly concerned with some selected aspects of describing particle form (F), in particular, the angularity, roundness and surface texture. Attention is given to the role of the scanning electron microscope (SEM) and the way in which this instrument can also be used to help characterise other sedimentological properties of soils.

Particle form

The description of sedimentological particle form has a long, complex and sometimes controversial history (see, for example, recent discussions by Barrett 1980 and Winkelmolen 1982). The categorisation presented by Whalley (1972) and given in the previous section will be equated in the present instance with 'form' as the overall concept of the 'external morphology of an individual grain' (this differs from the usage of, for example, Winkelmolen). Descriptions of form may be needed for a variety of purposes, e.g.:

(a)   to characterise the provenance of the clastic compenent, i.e. for environmental discrimination;
(b)   to help determine the mechanism of sediment transport;
(c)   to help ascertain past depositional (diagenetic) processes;
(d)   to help determine certain geotechnical aspects of a soil;
(e)   to act as one of several descriptors in a complex property, e.g. packing, permeability, porosity.

A similar list could be drawn up for size (grading) characteristics, and indeed grading curves are frequently used for all of these purposes. It should be noted that most of the form descriptors can be used for different size ranges. That components of form are less frequently used probably reflects the fact that they are much less easy to quantify than grading characteristics. However, certain techniques are now available which suggest that form should be used more frequently in many aspects of sedimentological investigation. Most soils can be treated as sediments but not all soils consist of transported grains as they may be derived from *in situ* rock weathering; however, this is relatively rare.

Of the form descriptors used most often, shape and sphericity tend to employ axial measurements, usually in three dimensions. In so far as the measurement

of long axes of particles over about 10 mm presents few problems, so the determination of these form components (Sh, SP) is not difficult, even if there is a
plethora of indices available (Orford 1981). For soils where the main components are sandy, use of axial ratios, at least in three dimensions, is more difficult.
There are various ways in which this may be tackled (e.g. Griffiths 1967), but
none is entirely satisfactory.

Angularity/roundness measures have necessitated comparison of pebble- or
sand-size clasts with some standard sets of charts (e.g. Krumbein 1941, Shepard &
Young 1961). Sometimes, as in the Powers (1953) chart, angularity/roundness
(A/R) is cross-classified with a measure of sphericity. The great problem with
A/R measurements is that they are on a continuum and chart comparisons give
(and then not reliably) nine discrete subdivisions at most.

Although still not in widespread use, the pioneering work of Ehrlich and co-
workers (e.g. Ehrlich *et al.* 1980) has shown how, in two dimensions, descriptions
derived from Fourier analysis can be used to characterise basic particle shape.
There are great advantages in using Fourier descriptor methods as replacements
for chart comparisons (Whalley 1978, Whalley & Orford 1982) as well as general
shape measures (Ehrlich *et al.* 1980, Clark 1981). Furthermore, such methods
can be employed in an automated fashion (Telford *et al.* in press).

Despite some problems with the use of Fourier descriptors (Clark 1981) this
technique has much promise. To some extent the Fourier method can be pushed
down in scale to encompass surface texture (T) where this employs microscopical
techniques, e.g. SEM surface texture analysis (to be discussed in more detail
below), and many provide a useful link between numerical characterisation
and the visual inspection of micrographs. This is subject to certain limitations,
however, such as tilt and mounting orientation (Whalley 1978, Whalley &
Orford 1982). Only a few workers have made such a link between Fourier
methods and SEM surface texture examination (Czarnecka & Gillott 1980,
Dowdeswell 1982). There has, as yet, been no really satisfactory amalgamation
of Fourier descriptors and SEM surface textures.

Scanning electron microscopy

The SEM has been used widely as a tool to help discriminate between sedimen-
tary/geomorphic transport and depositional processes (see, for example, the
review by Bull 1981). The remainder of this chapter will consider aspects of
SEM texture analysis, and other associated techniques which help in soil description and analysis, particularly in relation to the five purposes listed above.
In some cases only an indication of the potential can be given, as little attention
has yet been paid to certain uses. It should be appreciated that the use of SEM
will not necessarily give anything like a 'complete' answer on its own and other
techniques must be used as well. On the other hand, it can provide new insights
into problems where traditional, conventional techniques have been used with
limited success.

The SEM can be considered as a rather complicated and expensive magnifying glass! However, its high resolving power, large depth of field and ease of photography would make it a useful tool even in the absence of the analytical attachments now available. Technical aspects of the SEM can be found in a variety of texts (e.g. Goldstein & Yakowitz 1975, Smart & Tovey 1982) and it is not the purpose here to provide an introduction to the instrument. Good discussions of techniques for preparation and analysis of geological specimens can be found in Walker (1978) and in the very comprehensive survey by Smart and Tovey (1982). Sample preparation for soils can be very easy, as with quartz grains for example, or it may be complex and prone to artefact formation. Clays are particularly difficult to examine.

When an electron beam hits a sample, several interactions occur and these can be examined with appropriate detectors. Some electrons are reflected by the specimen and some low energy electrons are emitted from atomic shells near the surface. Both types are collected to give the 'normal', visible image of the SEM. X-rays of characteristic energy are emitted from atoms which interact with incident electrons and these can be collected as an energy spectrum from which the elements can be identified. Similarly, the high-energy reflected (or back-scattered) electrons which are nearly of beam energy, can be collected and displayed. As they give atomic number information their distribution can also be analysed.

A number of techniques related to the SEM also give useful surface (or near-surface) information. These are somewhat esoteric methods but some have been found to give useful analytical data in the microscopic examination of soils (Bisdom 1981).

Most notable of the analytical facilities available is the energy dispersive X-ray analyser (EDS or EDXRA). This can be used to identify elements rapidly (normally for atomic number, Z, greater than 10) in spots, lines or areas. It may be possible to recognise grain mineralogy on the SEM without the use of EDS but for fully quantitative work polished flat specimens should be used although such conditions can be relaxed under some circumstances. For discussion on this large and complicated topic see Goldstein and Yakowitz (1975) and Hren *et al.* (1979). In the context of grain surface texture analysis, the use of EDS (much quicker to use but not necessarily any better than wavelength dispersive spectroscopy) is mainly of value for checking mineralogy; to check that it really is a quartz grain under study or, for instance, that a cement is calcareous (Pye 1983).

Analysis of other signals emitted by the specimen when irradiated by an electron beam may include light as cathodoluminescence (Grant 1978, Smart & Tovey 1982) and back-scattered electrons (BSE). The use of BSE imaging is becoming popular for geological materials (Hall & Lloyd 1981, Smart & Tovey 1982, Pye & Krinsley 1983, amongst others) but its optimal use is not an easy matter. Jongerius and Bisdom (1980) have used BSE imaging to determine porosities in conjunction with an image analyser. Attention in this chapter, however, will be paid to direct imaging of the material to be investigated as available in any SEM, i.e. the 'normal' or reflective−emissive mode.

The use of light microscopy should not be decried just because an SEM is available. A binocular microscope is essential for preliminary inspection and as an aid to mounting, and determination of mineralogy with a petrographic microscope can save much time (and thus expense) in the electron microscope laboratory.

Surface texture analysis of particles with the SEM

Ever since basic shape and angularity differences between grains were seen under hand lenses and light microscopes, it was recognised that these characteristics could be used to determine environmental provenance. Work on polishing and glazing of grains by Cailleux (1952) and Kuenen and Perdok (1962) started to extend these primitive ideas to the actual surface texture but they were limited to the rather poor depth of field in the light microscope. Then, as now, the idea was that characteristic textures were formed by specific environments and that such textures could be used to identify those environments. Early work with transmission electron microscopes (TEMs) showed what sort of information was held by surface textures of grains from a variety of environments (Krinsley & Takahashi 1962). Analysis by this method was fruitful but tedious and the advent of the scanning electron microscope in the late 1960s allowed more samples and grains in each sample to be analysed, with much greater ease (Krinsley & Doornkamp 1973). However, the number of grains usually examined is often rather small and there are undoubtedly statistical problems in the correct interpretation of some results. The numbers of grains used vary widely (Bull 1978); Baker (1976) suggested that 30–50 grains should be sufficient, while Tovey and Wong (1978) considered 20 to be too few and 50 likely to be representative. Another article with useful information about mounting and sample selection is Tovey *et al.* (1978).

From the early days of using electron microscopy, both transmission and scanning, Krinsley has shown how textures of various types can be related to specific geological environments (Krinsley & Doornkamp 1973) and Bull (1981) has produced an up-to-date summary of the basic rationale and achievements of the use of the interpretation of sand grain surface textures. The early paper by Schneider (1970) is still worth reading in the context of the general problems of the method and Bull's (1981) paper.

Strictly, the environments identified are those that impart change to the nature of the grain surface; that is, a modification of the surface texture. Thus, a grain modified by a high-energy river but suddenly sedimented as an overbank deposit will show features characteristic of the former transport mechanism rather than the actual depositional environment. In effect, the technique looks at modification of particle form produced by a given transportive energy. The basic shape and sphericity may be altered, as for instance with grain breakage, but changes are more usually confined to angularity/roundness alteration and the production of new surface textures.

The textures visible on quartz grains provide an easy distinction between a number of environments; low and high energy littoral, aeolian and diagenetic types, for example. The distinctiveness of the texture is not always marked, for instance because of the variability of action of the transport mechanism. Figure 9.1a shows a clean surface of a grain which has undergone a certain amount of subglacial attrition (Whalley 1978). Figure 9.1b on the other hand shows the rather altered surface of a grus grain (a grain weathered from a granite outcrop but not transported). Figure 9.1c shows textural detail of a grain from a fluvio-glacial deposit. It is possible, but by no means certain, that the original texture of the grain in Figure 9.1c was similar to that in Figure 9.1a. In this case, the grains were from geographically distinct areas, but it is possible to trace changes of action over a sedimentological energy sequence (Wilson et al. 1981, Whalley 1979). In any study, it is always as well to sample the unaltered and untransported source material if at all possible to allow comparison with stages in surface textural change. For instance, Figure 9.1d shows a grain that has undergone only a little fluvial transport in a mountain stream and which still shows 'original' or unaltered surface textures typical of a grus.

Photographs of grains taken with the SEM allow a subjective characterisation of surface texture; the least easily quantifiable component of particle form. Some recent attention has been given to the edges and corners of grains and their modification in various stages of grain attrition (Whalley 1978, Whalley et al. 1982, Whalley & Orford 1982). This is a link between angularity/roundness determination and surface textures. Although abrasion of edges and corners can probably be monitored by use of Fourier descriptions, there is no published study that looks at this important link between the two methods of angularity/roundness characterisation. Such potential may not even be possible for most of the surface textures which have been recognised and collated in the atlas produced by Krinsley and Doornkamp (1973). A lot more work has been done on surface texture analysis and interpretation since the 1973 'atlas' (Bull 1981) mainly in the context of textural relationships with geological environments identified using established methods and techniques.

Not only can individual (transport) environments be recognised from surface textures but it is also possible to identify several stages in the sedimentary history. Just how well this can be done depends very much upon the sequence and the extent to which one texture replaces another. There are, of course, both temporal and energy factors at work here, and no hard and fast rules of application can be laid down. There is still a need for some experimentation so that replications of environments and the textures they produce can be assessed, preferably as a function of energy imparted, or in relation to surrogates such as transport distance. Some such replications do exist; for instance, there have been several studies of aeolian environments (Kaldi et al. 1978, Linde & Mycielska-Dowgiałło 1980, Whalley et al. 1982) but there is still much work to be done. Correlations between 'energy' and surface texture are difficult to produce because of experimental problems of measuring the energy transfer from one colliding particle with another and relating this to the overall energy of a stream or wind. Addi-

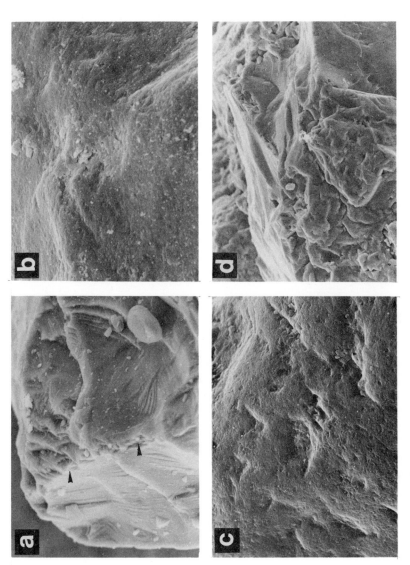

Figure 9.1 (a) A grain from a subglacial environment (Argentière glacier, France) showing an apparently unaltered surface with a little abrasion on one edge (between arrows). Picture width = 50 μm. (b) An untransported grain weathered out from a granite outcrop (grus) showing a surface altered by solution/deposition of silica (Powell Springs, Arizona, USA). Picture width = 65 μm. (c) A grain from a fluvioglacial environment (Chelford, Cheshire, UK) showing attrition produced by high-energy stream action. Although the surface looks slightly weathered (cf. Fig. 9.1b) the appearance is probably just the result of a smoothing process. Picture width = 75 μm. (d) A barely modified grain which has moved only a little way in fluvial transport from its source (Trassey River, Co. Down, UK). The clean or unweathered surface here contrasts strongly with the grus seen in Figure 9.1b. Picture width = 70 μm.

tionally, there is also the fact that the nature of the textures to be measured are so difficult to quantify.

The Krinsley and Doornkamp (1973) atlas has been the starting point for most investigations but the categories recognised by them have been catalogued by Margolis and Kennett (1971) and added to by, for example, Culver *et al.* (1983). The list of identified textures, upwards of 20, is somewhat bewildering and some practice is needed to recognise them, even from photographs. However, the success of the technique lies partly in the fact that the mind can 'hold' such images quite readily. With some experience, therefore, it is actually possible to take counts of features, marking their frequency on a table, directly at the SEM console without necessarily taking pictures. This saves time and money and ensures better statistical coverage of the sample, but some 'representative' photographs should always be taken. In the early stages of using the technique, however, it is best if plenty of photographs are taken so that a physical collection can be made for reference. The next section will examine some specific problems associated with the technique.

As well as what might be called a 'Krinsley' school of SEM surface texture recognition, there also exists a 'French' school, best exemplified by Le Ribault (e.g. Le Ribault 1978). There has never been much merging of ideas and texture classes between these two groups. This is also true of models concerning the way in which these features are produced.

Variability of textures and operators

One of the biggest problems in the interpretation of environments and transport processes from recognition of surface textures is the variability of the textures themselves. Texture variability is superimposed on the operator variance and limitations associated with recognition and classification of perceptional information (e.g. Miller 1956). These difficulties have combined to place important constraints upon the classification of discrete groups of angularity/roundness and also of surface textures. Because, as yet, there is no properly quantitative and objective measure of any surface texture, it is not possible to isolate either of these effects completely. Furthermore, any method which is used to classify textures will have to take account of both texture and operator variability.

It is possible to indicate some of the intersample variability with reference to selected samples. At an obvious level, differences in basic outline shape may be visible. An immediate first suggestion is that the sample might contain two (or more) mixed components of different origin (e.g. Whalley & Langway 1980). In Figures 9.2a and b (material from a subglacial till) the basic shape of the grains is similar but one grain (Fig. 9.2a) is relatively 'fresh' in appearance compared with the other (Fig. 9.2b) which has a more weathered look and 'plastered-on' fine particles (cf. Figs 9.1a & b). Notice, however, that even the grain in Figure 9.2a shows a little of the adhering debris which, depending upon the mounting angle of the grain, could present a greater or lesser proportion of the whole grain

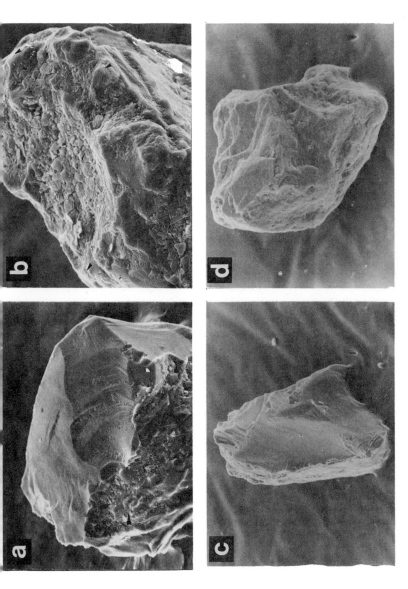

Figure 9.2 (a) A relatively fresh-looking grain from subglacial debris (Gruben glacier, Switzerland). Although there is some 'plastered-on' debris (arrow, centre left) there is no evidence visible of subglacial grinding producing edge attrition (cf. Fig. 9.1a). Picture width = 180 μm. (b) A grain from the same sample as Figure 9.2a but here showing much more adhering debris and evidence of more rounding on edges and corners (arrowed). Picture width = 85 μm. (c) A grain from a silt-derived soil (Newton Stewart, Scotland) with edge breakage features but a relatively clean surface. Picture width = 160 μm. (d) This is from the same sample as Figure 9.2c, but this grain shows much more surface 'weathering'. Picture width = 160 μm.

than the view actually shows. Figures 9.2c and d show a similar effect with differences in both overall shape and surface texture for two grains from the same loessic soil. It should be noted that apparent differences in outline shape occur if a non-uniform tilt is used on the specimens: examples given by Whalley and Orford (1982) show this effect. Similarly, Figures 9.1b and d contrast strongly in appearance although the grain shown in Figure 9.1d has been transported a little. Attritional surface textures, especially those due to breakage, can be added to a grain with a weathered surface as well as surface weathering being subsequent to attrition. This does, of course, provide some temporal information.

At the simplest level, this variability seen in the grains, whether or not it reflects mixing of populations, means that there is some considerable range in the textures encountered. It may be possible for simple 'eyeballing' methods to cope with this if photographs taken are arranged on a table and an assessment of the sample is made. Some overall judgement should then be made on that sample as to its possible origin and the variability of the textures seen.

Other samples can then be viewed in the same way and the reports of each sample compared. Preferably, trials such as this should be done 'blind' so that the investigator does not know the origin beforehand. Furthermore, each sample should be treated independently as far as possible. This was the method used by Smith and Whalley (1981) in their investigation of dune deposits and interridge fills. A more sophisticated approach is to use a chart or table with the surface textures and features listed so that presence or absence can be noted. This is the Margolis and Kennett (1971) method used, for example, by Bull (1978).

Tests to examine the operator variance of the technique have been run by Culver *et al.* (1983). They used an extension of the Margolis and Kennett (1971) method (for five operators and eight samples run blind) and found that, although there was considerable variability in the results, the actual discrimination was excellent. This has led Whalley (unpublished) to suggest that it may be possible to speed up the analysis (at least where a set of photographs is to be assessed) by looking at the overall impression with the brain acting as an image processor. This *'Gestalt'* idea is subjective but preliminary tests indicate that it may be extremely useful for the classification of sets of textures where it is difficult to give numerical values other than just 'presence' or 'absence'. Automatic image processing, although making great headway, is probably a long way off being able to recognise surface textures as accurately as a human brain.

A problem not often appreciated in discussions on surface textures is that the features (textures) listed by Margolis and Kennett (1971), for example, are by no means independent. Thus, there are many 'breakage' textures (e.g. conchoidal fractures, breakage blocks, semi-parallel step-like fractures) which are the result of brittle fracture and can be produced by a single mechanism. Gruses (weathered rock grains) frequently have such features, as do grains in moraines and a variety of deposits from glacial environments (Whalley & Krinsley 1974). It is not true to say that conchoidal breakage features and the like are positive indicators of glacial action, although some other features may be (Whalley 1978).

Surface textures and relative dating

One of the very first studies to use the surface texture of quartz grains to suggest that weathering of grain surfaces could be used as a time indicator was that of Andrews and Miller (1972). They were investigating areas of deglaciation in Baffin Island and were able to recognise zones of ice removal shown by surface texture differences. Unfortunately, no micrographs were presented. Douglas (1980) and Douglas and Platt (1977) have used a similar method to ascertain a time base for soil development. This would appear to be a useful technique but it needs careful cross checking or calibration and there have been no similar studies published. Some intriguing work by Kanaori et al. (1980) suggests that fault gouge material may be used as a relative dating method for the determination of fault movements, but again no similar work has been published which might indicate its general validity.

There seems to be no reason why, with suitable calibration, determination of relative ages of deposits could not be worked out from the surface textures. There are considerable problems to be investigated, however. First, the soil chemistry can play an important role in the development of certain textures (e.g. Wilson 1979) and this requires careful evaluation. Second, the source material should be examined, if possible, to check that the textures observed really are due to weathering of the grain at that specific locality and are not just inherited from a previous erosional cycle. Third, the within-sample variability needs to be investigated and, fourth, the textures need to be assessed with an appropriate method. This is certainly an area which needs more investigation before it can be used as a general tool.

Non-quartz grains and 'exotics' in soils

Nearly all environmental discrimination work using surface textures has been on quartz sands, although occasionally heavy minerals have been used (Setlow & Karpovich 1972, Folk 1975, Bull 1981). The surface textures, and thus the range of environments so recognised, have been rather less extensive than those with quartz. As yet, no studies have been published that compare quartz and heavy mineral textures from the same environments in a comprehensive manner, although some of the work just referred to provides a basis for such an examination.

There are minerals other than quartz and 'heavies' whose recognition could be useful in identifying modes of origin or provenance. Certain mineralogies have distinctive forms and, especially if an EDS system is available for easy checking of the elemental composition, important clues as to the origin of certain soil components can be gained. Such work with the SEM does not preclude, and is not really a substitute for, detailed work separating grains with differential flotation or by identification with other techniques such as X-ray diffraction. For instance, the presence of volcanic (pyroclastic) material can often be recog-

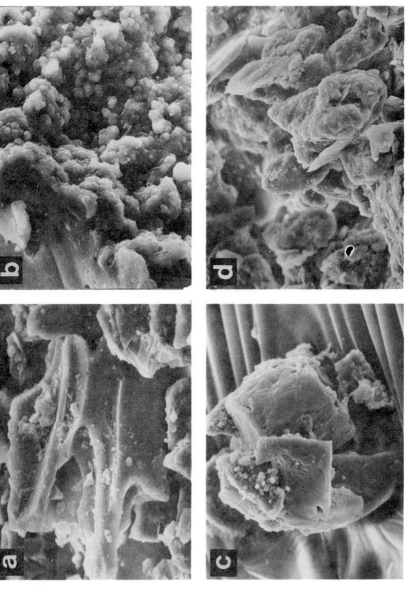

Figure 9.3 (a) A glassy shard from the pyroclastic deposits formed by the 1980 Mount St Helens eruption. Picture width = 40 μm. (b) Nodular siliceous concretions on a grain in a soil matrix from near a hot spring (Krisuvik, Iceland). Picture width = 60 μm. (c) An aggregate grain of quartz particles 'cemented' with clays (Tempe, Arizona, USA). Picture width = 60 μm. (d) Danubian terrace loess with aggregated clay, quartz and feldspar grains. Picture width = 75 μm.

nised easily by the form of glass shards. Figure 9.3a is an example from the Mount St Helens eruption. A basic atlas of material associated with pyroclastic deposits has been published by Heiken (1974) although more recent publications also include micrographs of this type of material (e.g. Kyle & Jezek 1978).

A further soil constituent which may be recognised easily is rapidly deposited silica. This may appear amorphous macroscopically but frequently has a characteristic globular appearance under the SEM (Fig. 9.3b). The presence of salts in soils can be determined with the SEM and EDS (Stoops et al. 1978, Vergouwen 1981). The complexity of minerals contained within soils means that an EDS system is an important tool, even if only for simple identification purposes. For example, although halite might be expected to show a distinctive cubic habit as a soil constituent, this is by no means always the case (Eswaran et al. 1980).

By far the most common soil particles other than quartz are likely to be feldspars and clay minerals. Feldspars have rarely been studied because of their relative ease of weathering. However, this may be a distinct advantage in some studies where the rather slow weathering of quartz may not give visible results over the time of interest, or be ambiguous. This problem is discussed in a subsequent section. Some recent work by Berner and Holdren (1977) and Curmi and Maurice (1981), although primarily concerned with the geochemistry of feldspar and mica weathering, may indicate ways of using such weathering textures as relative time indicators. However, problems need to be tackled similar to those discussed previously with respect to quartz grains and relative dating.

Clays have been studied a great deal by SEM and associated methods. It is possible to recognise certain clays easily from their appearance. Authigenic kaolinite 'books' provide an obvious example (see Blanche & Whitaker 1978, Waugh 1978 for good examples from sandstone rocks) but certain other clay minerals are also distinctive and provide useful clues as to provenance. Smith and Whalley (1982) found palygorskite 'rods' in a soil from north Africa after dissolution of surrounding dolomite and calcite. In this case, however, the rods are in fact composites of fine tubes of the clay mineral which can only be seen after ultrasonic treatment and examination under a transmission electron microscope. Palaeoenvironmental reconstructions were then possible, knowing something about the nature of palygorskite. The TEM, as with the light microscope, provides a useful complement to SEM work in some cases.

Other examples of clay minerals in soils examined with the SEM can be found in Gillott (1980), Keller (1978), Smart and Tovey (1981) and Sudo et al. (1981). It should be noted, however, that not all clays are easily recognisable by SEM examination and the usual methods of X-ray diffraction and thermal analysis should be employed to investigate fully the clays present (van Olphen & Fripiat 1979). Nevertheless, subsequent SEM examination may be very profitable for soil structural studies once the mineralogy is known and especially if an EDS attachment is available for rapid elemental analysis. A good example of such work is given by Kumai et al. (1978).

Aggregates and pores in soils and sediments

In many soils and sediments, the functional unit is frequently not an individual grain of sand, silt or clay but an aggregation of particles. This has long been recognised to be of importance in soil stability and erosion (Chepil 1957). Figure 9.3c, for example, shows a grain of airborne dust derived from soil of the Arizona desert. This grain is about 30 μm overall in diameter but consists of several quartz particles of fine silt size with some clay which possibly holds the grain together. The effects of aggregation may also be important in other areas. One significant problem concerns the way in which bonding is formed at a structural level in soils, as in quickclays, for example (Mitchell 1976, Lohnes & Demirel 1978, Smalley et al. 1980). Similarly, the microstructure of loess is especially significant as it has a bearing upon the geotechnical properties (Lutenegger 1981), origin (Pye 1983) and emplacement mode (Lee 1979). Figure 9.3d shows a loess from a terrace of the Danube which has aggregated silt and sand grains. In general, bonding of constituent grains can be by way of carbonates as well as by clays and other agents such as silica deposition. It is also possible to detect amorphous clay minerals (allophane) by electron microscopy as well as by other physical methods (van Olphen & Fripiat 1979) and they may play an important role in weathering and particle adhesion.

An aggregate may be the effective hydrodynamic unit of grain transport rather than the constituent particles. Thus breaking down a sediment or soil to give a grain size distribution may well be inappropriate for certain types of analysis. The examination of a sediment with SEM might help to ascertain whether it is important to consider the role of aggregates and use only a very gentle disaggregation to give the 'functional size range'. Alternatively, such examination may show whether it is more important to use, for example, ultrasonic methods to break all the linkages between particles. The reason for the size analysis must be looked at rather carefully for certain types of soil and sediment. This problem can be encountered with soils which have undergone a certain amount of compaction (e.g. those developed on overbank deposits or estuarine silts). In a geotechnical context, Davis (1967) and Sherwood (1967) have discussed problems of aggregation in Keuper marl. The SEM, used in conjuction with various preparation methods, allows a good view into the significance of aggregates in a soil.

Nearly all the studies involving the use of SEM to investigate pores and aggregates in soils have been for qualitative assessment. It is likely that the importance of SEM examination will be related to associated studies where quantitative data can be obtained easily. Porosity and permeability are two obvious examples; the methods employed for sandstones (Hancock 1978, Whalley et al. 1982) can also be used for soils (Bisdom & Thiel 1981). Pore size data can also be obtained and one hope is that such information can be related to micrographs of thin sections quantified by an image analyser (Bullock & Murphy 1980, Jungerius & Bisdom 1981). This is an area where the techniques are available and only software is needed to make some significant advances.

For further discussion on image analysers and electron microscopy see Smith (1982).

SEM and geotechnical characteristics of sands in soils

Geotechnical explanations of geomorphological phenomena are now becoming more common, making greater use of the properties of the materials which make up many landforms. In principle, nearly all such geotechnical explanations are based upon the strength relationships of the Mohr−Coulomb equation with respect to effective stresses. Thus, frictional and cohesive components of strength coupled to pore-water pressures are measured in terms of standard (geotechnical) tests. It has long been recognised, however, that (especially at low confining pressures) the mobilisation of friction and cohesion may show behaviour which is difficult to express in the usual ways. Thus liquefaction potential may be related to the surface texture characteristics of grains as discussed by Tovey (1978). Tovey used the SEM, although he did not quantify the frictional relationships. Tovey and Krinsley (1971), following a paper by Koerner (1970), made some suggestions regarding the use of the SEM to examine surface textures which were significant in providing the frictional component in soils. Frossard (1979) discussed dilatancy in soils and provided a mathematical analysis of aspects of grain form in this context using descriptors of shape and angularity/roundness, although not involving Fourier methods (Whalley 1982). The purely textural aspects of form and their contribution to liquefaction of sands and mobilisation of shear strength remain unquantified but the usefulness of the SEM for the investigation of 'locked sands' has been shown by Dusseault and Morgenstern (1979).

Considerable work has been done on the geotechnical characterisation of clays and clay fabrics in soils using the SEM. There are also significant geomorphological implications for this kind of approach (e.g. Derbyshire 1978). However, this increasingly important field lies on the fringes of the subject-matter of this chapter. There are, however, clearly common areas of interest in rapid means of data analysis, in particular, image analysis, and the means of quantitatively describing pores and pore geometries. Such work is likely to be of importance to soil investigations (e.g. Jongerius & Bisdom 1981) as well as geotechnical work. Mitchell (1976) provides an excellent introduction to the importance of electron microscopy applications to engineering soils and an up-to-date review on the subject is given by Smart and Tovey (1981, 1982). The paper by Sergeyev et al. (1980) should also be consulted (together with other papers in the same volume) for a discussion on the classification of soil microstructures and Tovey and Sokolov (1981) for quantitative methods of SEM soil fabric analysis. Progress in this area is rapid and there are strong links between geotechnical, sedimentological and pedological investigations with various means of quantifying micrographs likely to provide an increasingly important common ground.

Conclusions

The SEM is an increasingly important tool in the investigation of soils in the widest sense. It has shown its worth as a means of examining individual grains and structures simply because of the ease of obtaining high magnifications, and the large depth of field available. Much interest, however, lies in the future as EDS systems and a variety of imaging modes become more common. In addition, advances in both electronics and software will mean that the instrument will be easier to use and that image processing (even in 'real time') will extend its scope and usefulness to the geomorphological microscopist.

References

Andrews, J. T. and G. H. Miller 1972. Chemical weathering of tills and surficial deposits in east Baffin Island, N.W.T. In *International geography*, Vol. 1, W. P. Adams and F. H. Helleiner (eds), Paper 0102, 5−7. Toronto: University of Toronto Press.

Baker, H. W. 1976. Environmental sensitivity of submicroscopic surface textures on quartz sand grains − a statistical evaluation. *J. Sed. Petrol.* **46**, 871−80.

Barrett, P. J. 1980. The shape of rock particles, a critical review. *Sedimentology* **27**, 291−305.

Berner, R. A. and G. R. Holdren Jr 1977. Mechanism of feldspar weathering: some observational evidence. *Geology* **5**, 369−72.

Bisdom, E. B. A. (ed.) 1981. *Submicroscopy of soils and weathered rocks*. Wageningen: Centre for Agricultural Publishing and Documentation.

Bisdom, E. B. A. and F. Thiel 1981. Backscattered electron scanning images of porosities in thin sections of soils, weathered rocks and oil−gas reservoir rocks using SEM-EDXRA. In *Submicroscopy of soils and weathered rocks*, E. B. A. Bisdom (ed.), 191−206. Wageningen: Centre for Agricultural Publishing and Documentation.

Blanche, J. B. and J. H. McD. Whitaker 1978. Diagenesis of part of the Brent Formation (Middle Jurassic) of the northern North Sea Basin. *J. Geol Soc. Lond.* **135**, 73−82.

Bull, P. A. 1978. A statistical approach to scanning electron microscope analysis of cave sediments. In *Scanning electron microscopy in the study of sediments*, W. B. Whalley (ed.) 212−26. Norwich: Geo Abstracts.

Bull, P. A. 1981. Environmental reconstruction by electron microscopy. *Prog. Phys. Geog.* **5**, 368−97.

Bullock, P. and C. P. Murphy 1980. Towards the quantification of soil structure. *J. Microsc.* **120**, 317−28.

Cailleux, A. 1952. Morphoskopische Analyse der Geschiebe und Sandkörner und ihre Bedeutung für Paläoklimatologie. *Geol. Rundschau* **40**, 11−19.

Chepil, W. S. 1957. Sedimentary characteristics of dust storms, I Sorting of wind eroded material. *Am. J. Sci.* **255**, 12−22.

Clark, M. W. 1981. Quantitative shape analysis: a review. *Math. Geol.* **13**, 303−20.

Culver, S. J., P. A. Bull, S. Campbell, R. A. Shakesby and W. B. Whalley 1983. Environmental discrimination based on quartz grain surface textures: a statistical investigation. *Sedimentology* **30**, 129−36.

Curmi, P. and F. Maurice 1981. Microscopic characterization of the weathering in a granitic saprolite. In *Submicroscopy of soils and weathered rocks*, E. B. A. Bisdom (ed.), 249−70. Wageningen: Centre for Agricultural Publishing and Documentation.

Czarnecka, E. and J. E. Gillott 1980. Roughness of limestone and quartzite pebbles by the modified Fourier method. *J. Sed. Petrol.* **50**, 857−68.

Davis, A. G. 1967. On the mineralogy and phase equilibrium of Keuper marl. *Q. J. Engng. Geol.* **1**, 25−46.

Derbyshire, E. 1978. A pilot study of till microfabrics using the scanning electron microscope. In *Scanning electron microscopy in the study of sediments*, W. B. Whalley (ed.), 41−59. Norwich: Geo Abstracts.

Douglas, L. A. 1980. The use of soils in estimating the time of last movement of faults. *Soil Sci.* **129**, 345−52.

Douglas, L. A. and D. W. Platt 1977. Surface morphology of quartz and the age of soils. *Soil Sci. Soc. Am. J.* **41**, 641−5.

Dowdeswell, J. 1982. Scanning electron micrographs of quartz sand grains from cold environments examined using Fourier shape analysis. *J. Sed. Petrol.* **52**, 1315−22.

Dusseault, M. B. and N. R. Morgenstern 1979. Locked sands. *Q. J. Engng Geol.* **12**, 117−32.

Ehrlich, R., P. J. Brown, J. M. Yarus and D. T. Eppler 1980. Analysis of particle morphology data. In *Advanced particulate morphology*, J. K. Beddow and T. P. Meloy (eds), 101−19. Boca Raton: CRC Press.

Eswaran, H., G. Stoops and A. Abtahi 1980. SEM morphologies of halite (NaCl) in soils. *J. Microsc.* **120**, 343−52.

Folk, R. 1975. Glacial deposits identified by chattermark tracks in detrital garnets. *Geology* **8**, 473−5.

Frossard, E. 1979. Effect of sand grain shape on interparticle friction; indirect measurements by Rowe's stress dilatancy theory. *Geotechnique* **29**, 341−50.

Gillott, J. E. 1980. The use of the scanning electron microscope and Fourier methods in characterisation of microfabric and texture of sediments. *J. Microsc.* **120**, 261−77.

Goldstein, J. I. and H. Yakowitz (eds) 1975. *Practical scanning electron microscopy*. New York: Plenum.

Grant, P. 1978. The role of the scanning electron microscope in cathodoluminescence petrology. In *Scanning electron microscopy in the study of sediments*, W. B. Whalley (ed.), 1−11. Norwich: Geo Abstracts.

Griffiths, J. C. 1967. *Scientific method in analysis of sediments*. New York: McGraw-Hill.

Hall, M. G. and G. E. Lloyd 1981. The SEM examination of geological samples with a semiconductor back-scattered electron detector. *Am. Mineral.* **66**, 362−8.

Hancock, N. J. 1978. An application of scanning electron microscopy in pilot water injection studies for oilfield development. In *Scanning electron microscopy in the study of sediments*, W. B. Whalley (ed.), 61−70. Norwich: Geo Abstracts.

Heiken, G. 1974. An atlas of volcanic ash. *Smithsonian Contrib. Earth Sci.* **12**, 1−101.

Hren, J. J., J. I. Goldstein and D. C. Joy (eds) 1979. *Introduction to analytical electron microscopy*. New York: Plenum Press.

Jongerius, A. and E. B. A. Bisdom 1980. Porosity measurements using the Quantimet 720 on backscattered electron scanning images of thin sections of soils. In *Submicroscopy of soils and weathered rocks*, E. B. A. Bisdom (ed.), 207−16. Wageningen: Centre for Agricultural Publishing and Documentation.

Kaldi, J., D. H. Krinsley and D. Lawson 1978. Experimentally produced aeolian surface textures on quartz sand grains from various environments. In *Scanning electron microscopy in the study of sediments*, W. B. Whalley (ed.), 261−74. Norwich: Geo Abstracts.

Kanaori, Y., K. Miyakoshi, T. Kakuta and Y. Satake 1980. Dating fault activity by surface textures of quartz grains from fault gouges. *Engng Geol.* **16**, 243−62.

Keller, W. D. 1978. Classification of kaolins exemplified by their structures in scan electron micrographs. *Clays Clay Min.* **26**, 1−20.

Koerner, R. M. 1970. Effect of particle characteristics of soil strength. *Proc. Am. Soc. Civil Engrs. J. Soil. Mech. Found. Div.* **96**, 1221−34.

Krinsley, D. H. and J. C. Doornkamp 1973. *Atlas of quartz sand surface textures*. Cambridge: Cambridge University Press.

Krinsley, D. H. and T. Takahashi 1962. Application of electron microscopy to geology. *Trans N.Y. Acad. Sci.* **25**, 3–22.

Krumbein, W. C. 1941. Measurement and geological significance of shape and roundness of sedimentary particles. *J. Sed. Petrol.* **11**, 64–72.

Kuenen, Ph. H. and W. G. Perdok 1962. Experimental abrasion, 5. Frosting and defrosting of quartz grains. *J. Geol.* **70**, 648–58.

Kumai, M., D. M. Anderson and F. C. Ugolini 1978. Antarctic soil studies using a scanning electron microscope. In *Proceedings of the 3rd International Conference on permafrost*, Vol. 1, 107–12.

Kyle, P. R. and P. A. Jezek 1978. Compositions of three tephra layers from the Byrd Station Core, Antarctica. *J. Volc. Geothermal Res.* **4**, 225–32.

Lee, M. P. 1979. Loess from the Pleistocene of the Wirral Peninsula, Merseyside. *Proc. Geol Assoc.* **90**, 21–6.

Le Ribault, L. 1978. The exoscopy of quartz sand grains. In *Scanning electron microscopy in the study of sediments*, W. B. Whalley (ed.), 319–28. Norwich: Geo Abstracts.

Linde, K. and E. Mycielska-Dowgiałło 1980. Some experimentally produced microtextures in grain surfaces of quartz sands. *Geogr. Ann.* **62**, 171–84.

Lohnes, R. A. and T. Demirel 1978. SEM applications in soil mechanics. *Scan. Electron Microsc.* **1**, 643–54.

Lutenegger, A. J. 1981. Stability of loess in light of inactive particle theory. *Nature* **291**, 360

Margolis, S. V. and J. P. Kennett 1971. Cenozoic glacial history of Antarctica recorded in sub-Antarctic deep-sea cores. *Am. J. Sci.* **271**, 1–36.

Miller, G. A. 1956. The magic number 7, plus or minus 2: some limits on our capacity for processing information. *Psychol Rev.* **63**, 81–97.

Mitchell, J. K. 1976. *Fundamentals of soil behaviour*. New York: Wiley.

Orford, J. D. 1981. Particle form. In *Geomorphological techniques*, A. S. Goudie (ed.), 86–90. London: George Allen & Unwin.

Powers, M. C. 1953. A new roundness scale for sedimentary particles. *J. Sed. Petrol.* **23**, 117–19.

Pye, K. 1983. Grain surface textures and carbonate content of late Pleistocene loess from West Germany and Poland. *J. Sed. Petrol.* **53**, 973–80.

Pye, K. and D. Krinsley 1983. Mudrocks examined by backscattered electron microscopy. *Nature* **301**, 412–13.

Schneider, H. E. 1970. Problems of quartz grain morphoscopy. *Sedimentology* **14**, 325–35.

Sergeyev, Y. M., B. Grabowska-Olszewska, V. I. Osipov, V. N. Sokolov and Y. N. Kolomenski 1980. The classification of microstructures of clay soils. *J. Microsc.* **120**, 237–60.

Setlow, L. W. and R. P. Karpovich 1972. 'Glacial' microtextures on quartz and heavy mineral sand grains from the littoral environment. *J. Sed. Petrol.* **42**, 864–75.

Shepard, F. P. and R. Young 1961. Distinguishing between beach and dune sands. *J. Sed. Petrol.* **31**, 196–214.

Sherwood, P. T. 1967. Classification tests on Africa red clays and Keuper marl. *Q. J. Engng Geol.* **1**, 47–56.

Smalley, I. J., C. W. Ross and J. S. Whitton 1980. Clays from New Zealand support the inactive particle theory of soil sensitivity. *Nature* **288**, 576–7.

Smart, P. and N. K. Tovey 1981. *Electron microscopy of soils and sediments: examples*. Oxford: Clarendon Press.

Smart, P. and N. K. Tovey 1981. *Electron microscopy of soils and sediments: techniques*. Oxford: Clarendon Press.

Smith, B. J. and W. B. Whalley 1981. Late Quaternary drift deposits of north central Nigeria examined by scanning electron microscopy. *Catena* **8**, 345–67.

Smith, B. J. and W. B. Whalley, 1982. Observations on the composition and mineralogy of an Algerian duricrust complex. *Geoderma* **28**, 285–311.

Smith, K. C. A. 1982. On-line digital computer techniques in electron microscopy: general introduction. *J. Microsc.* **127**, 3–16.

Stoops, G., H. Eswaran and A. Abtahi 1978. Scanning electron microscopy of authigenic sulphate minerals in soils. In *Soil micromorphology*, M. Delgado (ed.), 1093–113. Proc. Fifth Int. Working Meeting Soil Micromorphology.

Sudo, T., S. Shimoda, H. Yotsumoto and S. Aita 1981. *Electron micrographs of clay minerals*. Amsterdam: Elsevier.

Telford, R. W., M. Lyons, J. D. Orford, W. B. Whalley and D. Q. M. Fay (in press). A low-cost, microcomputer-based image analysis system for characterisation of particle outline morphology. In *Clastic sediments*, J. R. Marshall (ed.), Stroudsburg: Dowden, Hutchinson & Ross.

Tovey, N. K. 1978. A scanning electron microscopy study of the liquefaction potential of sand grains. In *Scanning electron microscopy in the study of sediments*, W. B. Whalley (ed.), 83–94. Norwich: Geo Abstracts.

Tovey, N. K. and D. H. Krinsley 1971. Effect of particle characteristics on soil strength: a discussion. *Proc. Am. Soc. Civil Engrs Soil Mech. Found. Div.* **97**, 691–3.

Tovey, N. K. and V. N. Sokolov 1981. Quantitative SEM methods for soil fabric analysis. *Scan. Electron Microsc.* **1**, 537–54.

Tovey, N. K. and K. Y. Wong 1978. Preparation, selection and interpretation problems in scanning electron microscope studies of sediments. In *Scanning electron microscopy in the study of sediments*, W. B. Whalley (ed.), 181–99. Norwich: Geo Abstracts.

Tovey, N. K., N. Eyles and R. Turner 1978. Sand grain selection procedures for observation in the SEM. *Scan. Electron Microsc.* **1**, 393–400.

van Olphen, H. and J. J. Fripiat (eds) 1979. *Data handbook for clay mineral and other non-metallic minerals*. Oxford: Pergamon Press.

Vergouwen, L. 1981. Scanning electron microscopy applied on soils from the Konya Basin in Turkey and from Kenya. In *Submicroscopy of soils and weathered rocks*, E. B. A. Bisdom (ed.), 237–48. Wageningen: Centre for Agricultural Publishing and Documentation.

Walker, D. A. 1978. Preparation of geological samples for scanning electron microscopy. *Scan. Electron Microsc.* **1**, 185–92.

Waugh, B. 1978. Diagenesis in continental red beds as revealed by scanning electron microscopy. In *Scanning electron microscopy in the study of sediments*, W. B. Whalley (ed.), 329–46. Norwich: Geo Abstracts.

Whalley, W. B. 1972. The description of sedimentary particles and the concept of form. *J. Sed. Petrol.* **42**, 961–5.

Whalley, W. B. 1978. An SEM examination of quartz grains from sub-glacial and associated environments and some methods for their characterization. *Scan. Electron Microsc.* **1**, 353–60.

Whalley, W. B. 1979. Quartz silt production and grain surface textures from fluvial and glacial environments. *Scan. Electron Microsc.* **1**, 547–54.

Whalley, W. B. 1981. Materials properties. In *Geomorphological techniques*, A. S. Goudie (ed.), 11–38. London: George Allen & Unwin.

Whalley, W. B. 1982. Effect of sand grain shape on interparticle friction; indirect measurements by Rowe's stress dilatancy theory: comment on the paper by Frossard. *Geotechnique* **32**, 161–3.

Whalley, W. B. and D. H. Krinsley 1974. A scanning electron microscope study of surface of quartz sand grains from glacial environments. *Sedimentology* **21**, 87–105.

Whalley, W. B. and C. C. Langway Jr 1980. A scanning electron microscope examination of subglacial quartz grains from Camp Century Core, Greenland – a preliminary study. *J. Glaciol* 24, 125–31.

Whalley, W. B. and J. D. Orford 1982. Analysis of SEM images of sedimentary particle form by fractal dimension and Fourier analysis. Scan. Electron Microsc. **2**, 639–47.

Whalley, W. B., J. R. Marshall and B. J. Smith 1982. The origin of desert loess: some experimental observations. *Nature* **300**, 433–5.

Wilson, P. 1979. Experimental investigation of etch pit formation on quartz sand grains. *Geol. Mag.* **116**, 477–82.

Wilson, P., R. M. Bateman and J. A. Catt 1981. Petrography, origin and environment of deposition of the Shirdley Hill Sand of southwest Lancashire, England. *Proc. Geol Assoc.* **92**, 211–29.

Winkelmolen, A. M. 1982. Critical remarks on grain parameters, with special emphasis on shape. *Sedimentology* **29**, 255–65.

10

Soil particle size distribution and mineralogy as indicators of pedogenic and geomorphic history: examples from the loessial soils of England and Wales

J. A. Catt

Introduction

This chapter is concerned with particle size distribution and mineralogical composition as characteristics of soils in England and Wales resulting from geological processes of deposition and pedological processes of reorganisation and alteration. Both of these sets of processes are influenced by changing climatic and other environmental conditions during the Quaternary, so that the soil characteristics provide much useful evidence for landscape development.

Particle size distribution and, to a lesser extent, mineralogy are also important criteria in the system of soil classification developed for England and Wales by the Soil Survey (Avery 1980). The processes determining these two characteristics therefore partly account for the different units shown on official medium- and large-scale maps, an understanding of which is indispensable to reconstructions of landscape history. The emphasis especially on particle size distribution reflects three main factors:

(a) Particle size distribution and mineralogy are two of the most permanent, unchanging characteristics of soils; maps based on more transient characters, such as percentage organic matter, nitrogen content or soil structural features, might be of greater practical (e.g. agricultural) value initially, but would soon need revision.
(b) Particle size distribution and mineralogy influence several properties of agricultural and engineering importance, such as soil strength, stickiness, plasticity, the size, shape and degree of development of structural units

(peds), permeability and exchange capacity; these in turn determine ease of cultivation, trafficability, retention and release of water and plant nutrients, rates of leaching losses, and the incidence of some soil-borne crop pests and diseases.
(c)   Particle size distribution and mineralogy are the best indicators of soil parent materials and some soil-forming processes, such as clay illuviation and weathering.

Analytical methods

Particle size distribution can be assessed in the field by rubbing moistened soil between finger and thumb, and comparing it with samples of known clay, silt and sand contents. Experienced soil scientists can judge proportions of these three size fractions fairly accurately in horizons containing little or no organic matter (Hodgson *et al.* 1976), and this is usually adequate for interpretation of most agricultural and engineering properties of soil.

However, in organic horizons, including forest and agricultural topsoils, and where more detailed particle size analyses are required, laboratory methods are necessary. All of these depend upon efficient dispersion of the sample in water, so that the individual particles are separated from one another and not joined to form aggregates, which is the natural state for most soils. Dispersion is achieved by removal of cementing agents followed by prolonged (e.g. overnight) shaking in a dispersing agent such as a dilute alkaline sodium hexametaphosphate solution (Kilmer & Alexander 1949). The main cementing agents, organic matter and calcium carbonate, are usually removed by treatment of the air-dry soil sample with hydrogen peroxide and dilute hydrochloric acid respectively (Bascomb 1974). However, hydrochloric acid attacks several minerals other than calcium carbonate, and if the same sample is to be used for mineralogical studies after the particle size analysis, it should be decalcified instead with acetic acid buffered at pH 4−5. In samples where carbonates are important clastic components, analyses are done without prior decalcification (Bascomb 1974). Horizons cemented with iron oxides should be treated with dithionite-citrate (Mehra & Jackson 1960), and those containing soluble sulphates should be washed repeatedly with deionised water (Bascomb 1974). It is important not to oven-dry samples at temperatures $>30°C$ before particle size analysis, because some of the clay is baked and will not subsequently disperse. Because the amounts of the different size fractions are often eventually determined as oven-dry weights, this means that the moisture content of the air-dry sample must be determined on a separate subsample. Consequently, it is also important to homogenise the sample before analysis.

Ultrasonic treatment can also be used to aid dispersion (Edwards & Bremner 1967), but is often uneven in its effect, and should be used in several brief treatments for each sample with intervening stirring or shaking. It is important to remember that no pretreatment will completely disaggregate all soils into their ultimate constituent particles. Dispersion is usually partial, a function of the type of treatment and the length of time it is applied to the samples, and of the type and extent of bonding between particles. Some soils, containing mainly hard mineral particles (e.g. quartz) coarser than approximately 20 μm, are easily brought close to the ideal of 100% dispersion, whereas those containing fragments of partially compacted fine sedimentary rock (e.g. shales) may undergo slow, progressive disaggregation and possibly never reach 100% dispersion. For this reason it is always worth bearing in mind the purpose of the analysis; if one really wants to know the approximate size distribution of frost-shattered shale fragments, the sample should have little or no pretreatment, but if one wants to know how much clay and silt those fragments might yield in a strongly weathered soil the sample may have to be ground in an agate pestle and mortar and then given a series of prolonged ultrasonic treatments.

Sieves provide the best means of determining the proportions of coarser fractions (>62 μm). They are available with meshes equivalent to all appropriate quarter φ intervals (φ = −log₂ grain size in mm) coarser than 62 μm (= +4φ), and analysis at whole, half or quarter φ intervals allows the results to be summarised by standard methods (Folk & Ward 1957) and compared with those of most other workers. For analysis at quarter φ intervals, it is necessary to start with a sample large enough to provide approximately 200 g of sand (−1 to +4φ, (>2 mm) may be much greater, depending on the maximum size present (Milner 1962, Vol. 1, p. 171). Sieving is best done dry after removal of silt and clay (<62 μm) by wet-sieving through a nylon or stainless steel + 4φ mesh, though the wet sieve retains a little coarse silt which passes the same mesh size dry.

Sieves with meshes <62 μm are available, but do not work well, since the apertures become blocked and are difficult to clear; this applies even to fine sieves designed for use in ultrasonic baths. Fractions <62 μm are therefore usually analysed by one of several methods dependent upon the rates at which particles of different sizes settle through a column of water. The rates, as expressed by Stokes' Law, are in fact determined by the size, shape and specific gravity of the particles, and by the viscosity of the liquid medium, which in turn depends upon its temperature. In practice the size factor is isolated by holding the temperature of the column constant in a water bath or constant temperature room, by ignoring the effects of shape and by assuming a mean specific gravity of 2.65, i.e. that of quartz, which is usually the commonest mineral at least in silt fractions (2−62 μm). In soils containing large amounts of other minerals, different settling times can be calculated by substituting the relevant specific gravity factor in Stokes' Law.

The cheapest methods of size analysis in the subsieve range are the pipette sampling technique (see, for example, Bascomb 1974) and the hydrometer method of Bouyoucos (1927). In the former, samples of the dispersion are taken with a pipette inserted into the column to set depths and after set periods of time, both determined by Stokes' Law, so that all particles greater than certain limiting sizes will have fallen below the sampling depth. Tanner and Jackson (1947) give nomographs for settling times at a range of temperatures. The hydrometer method is slightly simpler, as the only operation involved after homogenising the dispersion is to measure its specific gravity at preset times dependent upon temperature. However, the results obtained are often less accurate, especially at the finer end of the size range.

Particle size analyses are possible at whole or half φ intervals by the pipette method, but quarter φ results are obtainable only if a very large column is used, as a small (e.g. 500 ml) column can be exhausted by removal of too many pipette samples. Alternatively, quarter φ results can be obtained by taking two staggered sets of pipette samples at half φ intervals from two columns of the same sample.

Another cheap apparatus for detailed analyses in the silt range is that proposed by Stairmand (1950), though this is not widely used as it requires special glassware. Other methods dependent upon settling rates include the sedimentation balance (Bostock 1952), Coulter counter (Pennington & Lewis 1979) and Sedigraph (Welch et al. 1979); the last two are rapid but very expensive. Laser diffractometers and image analysing computers are now increasingly used for sizing industrial products, but they are incapable of estimating total amounts of fine components such as clay (<2 μm). The pipette, hydrometer and Sedigraph methods enable total clay to be determined, but if the amounts of different clay sizes are required it is necessary to employ a centrifugation technique, such as that suggested by Bascomb (1974, p. 18). Amounts of fine clay (<0.2 μm) are useful for recognising argillic B horizons in soils (Avery 1980, p. 30), as illuvial clay is mainly of this size.

The mineralogical composition of most soils and clastic sediments varies with particle size, so it is usually necessary to separate size fractions before attempting mineralogical analysis. This has the additional advantage that most mineralogical techniques are best applied to particles of certain size ranges. Separation is best achieved by dispersion as for particle size analysis, followed by sieving for fractions >62 μm, repeated settling under gravity in aqueous suspension for 2−62 μm fractions, and centrifugation in water for clays <2 μm (Catt & Weir 1976, pp. 77−8). The choice of size fractions for study depends on the techniques available, the relative abundance of different sizes in the samples and the purpose for which they are being studied. For example, there is no point in laboriously separating clay from a sediment sample if none of the standard methods for clay mineral analysis is available, if there is very little clay in the sample, or if the intention is to find the source of its sand and silt fractions.

Sand and coarse silt (16 − 62 μm, +6 to +4φ) fractions are best analysed with a petrological microscope, different minerals being identified by crystal form and optical properties such as refractive index, birefringence, pleochroism, optical sign and the geometric relationships between optical and crystallographic directions (Hartshorne & Stuart 1950). It is often convenient to separate light minerals (quartz, felspars, muscovite, calcite, gypsum, flint, chert) from the usually wide range of rarer heavy minerals by flotation in bromoform (specific gravity 2.9). For sand fractions this can be achieved under gravity in a separating funnel, but coarse silts are best separated by centrifuge (Catt & Weir 1976, p. 79).

Clay fractions are composed mainly of platy layer silicate minerals, and are best analysed by X-ray diffractometry of aggregates oriented by drying on to glass slips (Brown & Brindley 1980). Clay minerals are too fine to study by optical microscope, but their shapes can be viewed by electron microscope (Gard 1971). Other methods useful for clay mineral analysis include infra-red spectroscopy (Farmer 1974), differential thermal analysis (Mackenzie 1957) and Mössbauer spectroscopy (Bancroft 1973).

Fine silt fractions (2–16 μm) are difficult to analyse mineralogically. They are usually mixtures of the layer silicates typical of clay fractions and granular minerals, such as quartz, normally abundant in coarse silt and sand fractions. The granular minerals often prevent the layer silicates from orienting well enough for successful diffractometry, and the individual particles are too small for ready identification by optical properties. Fortunately, fine silts are usually only minor components of soils and sediments, so it is rarely necessary to face these difficulties. Catt and Weir (1976) suggested some techniques to help with quantitative mineralogical analysis of fine silts, but they are time-consuming.

Factors controlling the particle size distribution and mineralogy of soils in England and Wales

Much the most important factor determining the particle size distribution and mineralogical constitution of soils in England and Wales is the original composition of their parent materials. These include igneous rocks, such as the granites of south-west England, and all the main Palaeozoic, Mesozoic and Tertiary sedimentary formations. Most of the soils derived from these pre-Quaternary materials (the 'solid' rocks of the British Geological Survey maps) show very small vertical changes in particle size distribution or mineralogy. Argillic horizons are enriched with no more than about 8% illuvial fine clay, which forms coats (argillans) around sand grains and linings to the walls of pores and fissures. The main mineralogical changes are:

(a)   dissolution of calcium carbonate (decalcification) in non-pyritic sediments to depths of 1.2 m or less;

(b)   oxidation of pyrite to iron oxide to depths of several metres, with associated decalcification and crystallisation of gypsum;

(c)   partial removal of interlayer potassium from clay mica forming interstratified mica-smectite to depths of 70 cm or less;

(d)   removal of sand- and silt-sized apatite, collophane, pyroxenes, biotite and glauconite in brown earths, and also of garnets, epidotes and amphiboles in podzols.

These effects of soil development are similar to those shown by soils in younger Quaternary sediments, such as Late Devensian loess (Weir et al. 1971) and tills (Madgett & Catt 1978), which can be attributed to pedogenesis during approximately the last 10 000 years. This suggests that earlier soils were extensively eroded from most outcrops of the 'solid' pre-Quaternary formations shortly before the close of the Devensian stage.

More strongly developed brown earths in England and Wales are restricted to level or very gently sloping interfluves, plateaux and higher terrace remnants. They are characterised by paleo-argillic B horizons (Avery 1980), which have reddish colours and up to 30% illuvial clay, mainly as disrupted argillans (papules) dissociated from the voids in which they originated and incorporated into strongly stress-reorganised soil matrices. Some also show more extensive mineral weathering, including depletion of amphiboles, garnets and epidotes in sand and silt fractions. Shown as various paleo-argillic subgroups on maps of the Soil Survey of England and Wales, they occur mainly in Clay-with-flints on Chalk interfluves south of Luton, in similar deposits capping the Upper Greensand plateau of east Devon, in the Angular Chert Drift on the Hythe Beds escarpment in west Kent and east Surrey, and on older glacial deposits in Lincolnshire, west Norfolk, Essex, Hertfordshire and north-west Warwickshire. Smaller areas occur on the Carboniferous Limestone in south Wales, Derbyshire and Staffordshire, on Jurassic rocks in south Wales and north-east Yorkshire, on higher terraces of the Thames, Kennet, Severn and other rivers in southern England, and on older gelifluction (head) deposits covering footslopes of Chalk, Carboniferous Limestone and other escarpments (Bullock & Murphy 1979, Catt 1979, 1983, Sturdy et al. 1979, Chartres 1980, Soil Survey of England and Wales 1983).

As the reddish colours of these soils are attributed to formation of haematite by weathering in a warm and seasonally dry climate (Schwertmann & Taylor 1977), a process that did not occur in Britain during the Holocene (Catt 1979), they are regarded as relict interglacial soils. This implies that the level surfaces on which they occur escaped most of the periglacial erosion that affected other areas. In some paleo-argillic brown earths (e.g. Bullock & Murphy 1979) there is strong micromorphological evidence for two or more episodes of clay illuviation and reddening (rubefaction) separated by phases of disturbance, possibly by cryoturbation, so the implied periods of land-surface stability do not necessarily date only from the last interglacial (Ipswichian); some of the soils are thought to cover periods since the Cromerian. There is little evidence in relict soils in England and Wales for earlier (e.g. Late Tertiary) soil development (Catt 1983).

Although soil-forming processes such as mineral weathering and clay illuviation account for some small changes in mineralogy and particle size distribution within English and Welsh soil profiles, many soil types exhibit larger differences between horizons, which cannot reasonably be attributed to such processes. These result from original inhomogeneity of the soil parent materials, which may arise in two main ways:

(a) the profile has developed in a single 'solid' or Quaternary formation composed of thin beds of lithologically different sediment (e.g. interbedded sands and clays), and there has been insufficient disturbance of the soil by frost, tree-fall, animal burrowing or deep cultivation to homogenise the layers below 25−40 cm depth; this type of profile occurs on surfaces eroded down to fresh rock or unconsolidated sediment during or since the Devensian, and also on surfaces built up by Devensian or Holocene deposition;

(b)   the profile has developed in two or more formations of different ages, the thin upper deposit or deposits usually resulting from processes of terrestrial Quaternary sedimentation, such as glacial, fluvial, aeolian or slope deposition; in this type of profile there is often evidence for soil development before as well as after deposition of the younger parent materials.

Evidence for loess in the soils of England and Wales

Small isolated deposits of loess (aeolian silt, usually of periglacial origin) have been known in England for many years; they were reported in County Durham (Trechmann 1920), near Portsmouth (Palmer & Cooke 1923) and at Pegwell Bay in east Kent (Pitcher *et al.* 1954), and a partly loessial origin for many of the deposits shown as 'brickearth', 'head brickearth', 'river brickearth' or 'loam' on British Geological Survey maps was inferred by Dines *et al.* (1954) and others. The deposit at Warren House Gill on the Durham coast was buried beneath thick glacial deposits, but most of the remainder occur at the surface, and are shown on maps of the Soil Survey of England and Wales as Hamble, Hook and Park Gate series, depending on the extent of gleying (Hodgson 1967).

The minimum thickness of loess for recognition of these three series is 90 cm. Profile characteristics, including the extents of clay translocation and mineral weathering, indicate soil development during the Holocene, as there are no paleo-argillic (interglacial) features. This accords with widespread stratigraphic, mineralogical and archaeological evidence for a Late Devensian age for the loess (Catt 1978), and with a thermoluminescence date of 14 800 years for a sample from Pegwell Bay (Wintle 1981). The soils are decalcified to 70−90 cm depth, and various forms of white secondary carbonate concentrations occur below. The decalcified soil horizons are also weathered brown where ungleyed, and are much looser and softer than the calcareous material; consequently they do not fit the strict definitions of loess laid down by Russell (1944) and others, that the deposit is 'unstratified, homogeneous, porous, calcareous silt ... yellowish or buff, tends to split along vertical joints, maintains steep faces, and ordinarily contains concretions, and snail shells.' Nevertheless, there can be no doubt that the decalcified soil horizons are derived from the typical calcareous loess; apart from small differences of clay content and heavy mineral composition, which can reasonably be attributed to Holocene pedogenesis, they closely resemble the calcareous loess in particle size distribution and mineralogy.

Figure 10.1 shows in black the distribution of Hamble, Hook and Park Gate series, taken from the 1 : 250 000 soil map (Soil Survey of England and Wales 1983), which is the only complete soil map of the country based on a uniform coverage of soil observations. They occupy 0.42% of the country, and occur mainly over Chalk and Thanet Beds in north and east Kent, on the low ('25 ft') raised beach in West Sussex and south-east Hampshire, and on low terrace gravels of the Rivers Thames and Lea.

The existence of thinner, completely decalcified loess deposits in other parts of the country was first postulated by ecologists investigating anomalous plant communities that indicate base-deficient soils on base-rich rocks, such as Chalk (Perrin 1956), serpentine (Coombe et al. 1956) and Carboniferous Limestone (Pigott 1962). The soils were found to contain large amounts of silt and minerals not present in the underlying rocks. Subsequently, routine mapping and particle size analyses by the Soil Survey of England and Wales showed that the silt contents of soils on many other substrata also increase upwards through the profile. Of itself, this does not confirm the occurrence of loess, as many other deposits may be silt-rich. However, in many situations the additional silt has been shown to contain the same mineral assemblage as the undoubted loess deposits in north-east Kent and West Sussex. This is true of silty surface or subsurface horizons over Clay-with-flints and Coombe Deposits on the Chilterns (Avery et al. 1959, 1969, 1972) and South Downs (Hodgson et al. 1967), over Clay-with-flints-and-cherts, Budleigh Salterton Pebble Beds, Haldon Gravels, Devonian Limestones and the Dartmoor Granite in east Devon (Harrod et al. 1973), over pre-Devensian tills and glacial gravels in north Norfolk (Catt et al. 1971), over the Chalk of eastern Yorkshire and Lincolnshire (Catt et al. 1974), and over flinty valley gravels on the Dorset Downs (Catt et al. 1982).

The only differences in the silty additions to soils from Yorkshire to Sussex and east Devon are a progressive westward increase in amounts of flaky heavy minerals, such as chlorite and biotite, and a progressive westward decrease in the modal particle size of the silt (Catt 1978). Both of these changes can be attributed to the winnowing effect of easterly winds transporting the loess across the country from a source in the North Sea basin. The mineralogical similarity of the loess to the Late Devensian Skipsea Till of eastern England supports both the inferred age of the loess and its derivation from glacial detritus accumulating in the North Sea basin when the sea level was depressed eustatically. Catt et al. (1974) suggested that the loess was deposited before the ice reached its limit, and therefore occurs at the surface only beyond the till margin, though it also extends locally beneath the till in eastern Yorkshire and possibly as far north as Trechmann's site at Warren House Gill, Co. Durham.

The loess overlying serpentine rocks on the Lizard Peninsula, originally described by Coombe et al. (1956), is mineralogically different from that in other parts of southern England and coarser than the deposits in east Devon. This suggests that it was derived from a different source, possibly glacial detritus in the Irish Sea Basin (Catt & Staines 1982). Other silty soils in north western England and north Wales have also been described as containing loess derived from the Devensian Irish Sea glacier (Lee 1979, Lee & Vincent 1981, Vincent & Lee 1981, 1982), but without supporting mineralogical evidence.

Within most of the extensive loess areas in England, such as north Kent,

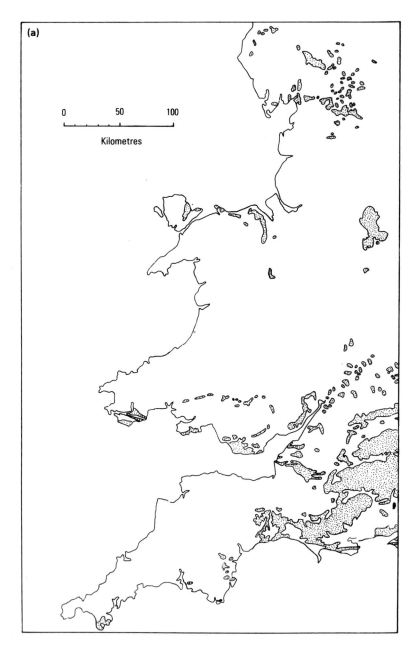

Figure 10.1 Surface distribution of loess in England and Wales, based on Soil Survey 1 : 250 000 map 1983 (loess > 1 m thick, black; loess > 30 cm thick and often partly mixed with subjacent deposits, stippled).

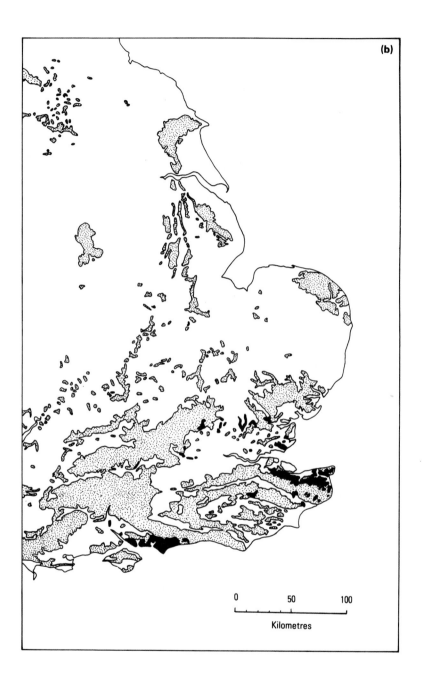

0   50   100

Kilometres

West Sussex, north Norfolk and east Yorkshire, the deposit can be detected on almost all facets of the land surface. It is usually thickest on broad interfluves and low terraces close to the valley bottoms, but is also spread thinly on many slopes, usually as a minor component of shallow soils such as rendzinas on limestones. Using particle size analyses at whole or half φ intervals and mineralogical analyses of coarse silt (16−62 μm) fractions, it is often possible to identify loess forming as little as 5% of the shallow, valley side soils (e.g. Catt & Staines 1982). The only soils with no evidence of loess are those on steep, rock-dominated slopes subject to recent erosion, and those composed of recent peat in wet valley bottoms or upland hollows. Almost everywhere else the soil contains silt of remarkably uniform mineralogical composition. Only fairly recent wind deposition can account for this.

The Holocene alluvial deposits of most English rivers are very silty, suggesting incorporation of loess eroded from the valley sides and headwater catchments (Burrin 1981). There is also similar evidence for widespread incorporation of loess into Devensian gelifluction ('head') deposits (Mottershead 1971, Catt 1982) and Holocene colluvial deposits resulting from Neolithic and later soil erosion. Secondary reworking of loess by these processes, possibly assisted by wind erosion, could account for its absence from soils in many parts of the country.

Figure 10.1 shows stippled the areas of soils thought to contain loess equivalent to a fairly continuous cover at least 30 cm thick (but <1 m), though this is often mixed in varying proportions with the subjacent deposits. The areas are taken from the 1 : 250 000 soil map (Soil Survey of England and Wales 1983) and include the associations listed in Table 10.1. The total area is 12.69% of the country. In southern England the loessial soils occur on a wide range of substrata, including the Lizard Serpentine, Devonian Limestones, Carboniferous Limestone, Lias, the Weald and other Mesozoic clays, London Clay, river terrace gravels, and pre-Devensian tills, glaciofluvial gravels and gelifluction deposits. However, north of the Severn−Wash line they are virtually restricted to calcareous substrata, such as the Carboniferous Limestone, Chalk and Jurassic

Table 10.1 Soil associations of England and Wales derived wholly or partly from loess (after Soil Survey of England and Wales 1983).

Soil association	Map symbol and subgroup	Main substrata
Wetton 2	311d	Carboniferous Limestone
Crwbin	313c	Carboniferous Limestone
Icknield	341	Chalk
Upton 1	342a	Chalk
Andover 1	343h	Chalk
Andover 2	343i	Chalk
Panholes	511c	Chalk

Soil association	Map symbol and subgroup	Main substrata
Blewbury	511d	Chalk
Coombe 1	511f	Chalk
Coombe 2	511g	Chalk
Badsey 1	511h	river terrace gravels
Badsey 2	511i	river terrace gravels
Aswarby	512a	Jurassic Limestone
Milton	512f	river terrace gravels
Malham 1	541o	Carboniferous Limestone
Malham 2	541p	Carboniferous Limestone
Waltham	541q	Carboniferous Limestone
Wick 2	541s	glaciofluvial gravels
Wick 3	541t	glaciofluvial gravels
East Keswick 3	541z	Palaeozoic limestones
Ston Easton	571a	Jurassic and Carboniferous Limestones
Frilsham	571j	Chalk
Charity 1	571l	Clay-with-flints
Charity 2	571m	chalky gelifluction deposits
Efford 1	571s	marine and river terrace gravels
Efford 2	571t	glaciofluvial gravels
Sutton 1	571u	river terrace gravels
Sutton 2	571v	river terrace gravels
Ludford	571x	glaciofluvial gravels
Hamble 1	571y	loess
Hamble 2	571z	loess
Rowton	571A	glaciofluvial gravels
Bignor	572k	Cretaceous sandstone
Ratsborough	572r	Eocene clay
Waterstock	573a	river terrace deposits
Nordrach	581a	Carboniferous Limestone
Carstens	581d	Clay-with-flints
Marlow	581e	Clay-with-flints
Batcombe	582a	Clay-with-flints
Hornbeam 1	582b	Clayey glaciofluvial gravels
Hornbeam 2	582c	Clay-with-flints
Hornbeam 3	582d	pre-Devensian tills
Tendring	582e	glaciofluvial gravels
Wickham 1	711e	Cretaceous clays
Wickham 2	711f	Jurassic and Cretaceous clays
Wickham 4	711h	Tertiary clays
Wickham 5	711i	Cretaceous clays and sandstones
Gresham	711v	pre-Devensian tills
Croft Pascoe	711w	ultrabasic igneous rocks
Dunkeswell	714a	Clay-with-flints-and-cherts
Oak 2	714c	pre-Devensian tills
Essendon	714d	pre-Devensian head deposits
Kelmscot	832	river terrace gravels
Shabbington	841d	river terrace sands
Park Gate	841e	loess

limestones, with a few isolated patches on Mesozoic clay formations. Catt *et al.* (1974) attributed this to mixing of the loess with frost-shattered limestone to form silty and very calcareous head deposits, in which Early Holocene pedogenesis created subsoil horizons cemented with secondary carbonate that were able to resist subsequent erosion. On non-calcareous substrata no cementation occurred, and the loess was easily removed by the various erosion processes discussed above.

Three major differences between the distribution of loess shown in Figure 10.1 and that given in earlier maps, such as Catt (1977, Fig. 16.1) and Catt (1978, Fig. 1), require explanation. First, the large area along the Thames valley through London is omitted because urban areas were unsurveyed for the 1 : 250 000 soil map. Second, the areas of loess-containing soils on Dartmoor, Bodmin Moor and part of the Bunter (Sherwood) Sandstone outcrop near Ollerton, Nottinghamshire, are omitted because the more recent surveying has shown that they are subordinate in these areas to loess-free soils. Likewise, there are many areas not stippled in Figure 10.1 where isolated patches of loess are known to occur, but are too small to show. Third, the later surveying has extended the areas of predominantly silty soils, especially in the Weald, Lincolnshire, north Wales, northern Pennines and areas adjacent to the Severn estuary. The loessial origin of the silt in these and some other areas has yet to be confirmed by detailed particle size and mineralogical studies, and is meanwhile assumed on the evidence that the subjacent strata are generally silt-deficient.

Importance of loess in soil and landscape development histories

Almost all the soils in England and Wales that contain paleo-argillic B horizons also have overlying A and E horizons rich in loess. The loessial horizons usually do not show paleo-argillic features, thus supporting the suggestion that the paleo-argillic horizons originated in the Ipswichian and earlier interglacials. The boundary between the paleo-argillic horizons and those rich in loess is often gradational, reflecting disturbance and mixing of near-surface horizons by several possible processes, but locally is sharper with structures suggesting frost disturbance, such as wedges and involutions.

In several deep excavations for 'brickearth' on the Chilterns, Avery *et al.* (1982) found horizons with paleo-argillic features derived mainly from loess. Their coarse silt fractions differed mineralogically from Devensian loess, and were explained as weathered mixtures of silt from Reading Beds or Clay-with-flints and loess similar to either the Anglian Barham loess of Rose and Allen (1977) or the Wolstonian loess at Northfleet, Kent (Burchell 1933). Interpretation of the pedogenic history of these sites was confounded because the brickearths occurred in dolines and were derived from surrounding land surfaces partly by mass movement (gelifluction), so some of the paleo-argillic features could have been transported intact from soils in Clay-with-flints. Nevertheless, the ability to distinguish loesses of different cold stages from their mineralogical composition raises hope of using coarse silt mineralogy of loessial horizons in undisturbed profiles to relate the paleo-argillic features in these and other horizons to soil-forming episodes in different interglacials. At present too little is known about the mineralogical differences within and between British loesses older than the Late Devensian for this technique to be applied extensively, but as more older loesses dated by thermoluminescence or other methods become available for study, it could prove useful in retracing the history of individual profiles. This could be of considerable geomorphological significance, because the inferred time for the start of soil development provides a minimum age for the landscape facet or facets on which that and similar soils occur.

Conclusions

Routine soil mapping and particle size analyses, supported by detailed particle size and mineralogical studies of selected samples, have shown that thin loess deposits are much more widely distributed in England and Wales than was previously supposed. Almost all the loess in profiles on the present land surface was deposited in the Late Devensian south of the margin of the last ice sheet. Older loesses can be distinguished mineralogically from the Late Devensian loess, but occur only locally and in buried situations or deep enclosed hollows such as dolines. In many profiles, the Devensian loess overlies paleo-argillic horizons resulting from pedogenesis during the Ipswichian and earlier interglacials. The precise history of the paleo-argillic soils, and the age of the land surfaces on which they occur, may be clarified in future by micromorphological examination of the paleo-argillic horizons and by mineralogical comparisons of associated loessial horizons with loesses of known ages. In areas, either with or without a thin cover of Devensian loess, where the soils are otherwise developed in earlier Quaternary or pre-Quaternary sediments and show as little decalcification, clay translocation and mineral weathering as those formed in thicker Devensian sediments, a Devensian episode of widespread surface erosion can usually be inferred. However, it is possible that some such areas were not eroded at this time, because they have, for example, poorly drained soils in which there was little or no decalcification, clay translocation or weathering even during interglacial periods.

The areas with soils containing detectable amounts of Late Devensian loess either have suffered little erosion during the Late Devensian and Holocene, or have been built up by deposition of gelifluction, colluvial or alluvial sediments derived partly from the loess. Most of the loess of this age was blown westwards across southern England from a source area of glacial outwash in the North Sea basin. Apart from gradual westward changes in modal size and content of flaky minerals, which can be attributed to the winnowing effect of the easterly periglacial winds, the Devensian loess is remarkably uniform over large areas, and in favourable circumstances can be detected in amounts as small as 5% of a soil horizon from the characteristic particle size distribution and mineralogical composition. Anomalous modal sizes and mineral suites in western Cornwall probably result from a different source area, possibly outwash in the Irish Sea basin, but the extent to which this contributed to thin deposits in Wales and north-west England remains uncertain.

The uniformity of Late Devensian loess composition in eastern and southern England suggests that it originally covered even larger areas than it does now, possibly extending over almost the whole land area south of the Devensian ice margin. Many areas with little or no remaining cover, especially those adjacent to regions of almost continuous loess-containing soils, must therefore have undergone extensive erosion by Late Devensian gelifluction or Late Holocene fluvial and colluvial activity following Neolithic and later deforestation. Because it is so easily eroded, a thin loess cover is an especially sensitive indicator of land-surface stability over the period since it was deposited. Particle size and mineralogical analyses as methods of identifying loess are therefore techniques of considerable value to the geomorphologist.

Acknowledgement

I thank E. M. Thomson, Soil Survey of England and Wales, who helped with the preparation of Figure 10.1.

References

Avery, B. W. 1980. *Soil classification for England and Wales (Higher Categories)*. Soil Survey Tech. Monogr, No. 14.

Avery, B. W., P. Bullock, J. A. Catt, A. C. D. Newman, J. H. Rayner and A. H. Weir 1972. The soil of Barnfield. *Rothamsted Exp. Stn Rep. for 1971*, Pt. 2, 5–37. Harpenden: Rothamsted Experimental Station.

Avery, B. W., P. Bullock, J. A. Catt, J. H. Rayner and A. H. Weir 1982. Composition and origin of some brickearths on the Chiltern Hills, England. *Catena* 9, 153–74.

Avery, B. W., P. Bullock, A. H. Weir, J. A. Catt, E. C. Omerod and A. E. Johnston 1969. The soils of Broadbalk. *Rothamsted Exp. Stn Rep. for 1968*, Pt. 2, 63–115. Harpenden: Rothamsted Experimental Station.

Avery, B. W., I. Stephen, G. Brown and D. H. Yaalon 1959. The origin and development of brown earths on Clay-with-flints and Coombe Deposits. *J. Soil Sci.* 10, 177–95.

Bancroft, G. M. 1973. *Mossbauer spectroscopy: an introduction for inorganic chemists and geochemists*. Maidenhead: McGraw-Hill.

Bascomb, C. L. 1974. Physical and chemical analyses of ‹2 mm samples. In *Soil Survey laboratory methods*, B. W. Avery & C. L. Bascomb (eds) 14–41. Soil Survey Tech. Monogr. No. 6. Harpenden: Rothamsted Experimental Station.

Bostock, W. 1952. A sedimentation balance for particle size analysis in the sub-sieve range. *J. Scient. Instrum.* 29, 209–11.

Bouyoucos, G. J. 1927. The hydrometer as a new method for the mechanical analysis of soil. *Soil Sci.* 23, 343–53.

Brown, G. and G. W. Brindley (eds) 1980. *X-ray diffraction procedures for clay mineral identification*. London: Mineralogical Society.

Bullock, P. and C. P. Murphy 1979. Evolution of a paleo-argillic brown earth (Paleudalf) from Oxfordshire, England. *Geoderma* **22**, 225–52.

Burchell, J. P. T. 1933. The Northfleet 50-foot submergence later than the Coombe Rock of post-early Mousterian times. *Archaeologia* **83**, 67–92.

Burrin, P. J. 1981. Loess in the Weald. *Proc. Geol Assoc.* **92**, 87–92.

Catt, J. A. 1977. Loess and coversands. In *British Quaternary studies recent advances.* F. W. Shotton (ed.) 221–9. Oxford: Clarendon Press.

Catt, J. A. 1978. The contribution of loess to soils in lowland Britain. In *The effect of man on the landscape: the lowland zone*, S. Limbrey & J. G. Evans (eds), 12–20. Council for Brit. Arch. Res. Rep. No. 21.

Catt, J. A. 1979. Soils and Quaternary geology in Britain. *J. Soil Sci.* **30**, 607–42.

Catt, J. A. 1982. The Quaternary deposits of the Yorkshire Wolds. *N. Eng. Soils Disc. Grp Proc.* **18**, 61–7.

Catt, J. A. 1983. Cenozoic pedogenesis and landform development in south-east England. In *Residual deposits: surface related weathering processes and materials*, R. C. L. Wilson (ed.), 251–8. Oxford: Blackwell.

Catt, J. A. and S. J. Staines 1982. Loess in Cornwall. *Proc. Ussher Soc.* **5**, 368–75.

Catt, J. A. and A. H. Weir 1976. The study of archaeologically important sediments by petrographic techniques. In *Geoarchaeology: earth science and the past*, D. A. Davidson & M. L. Shackley (eds), 65–91. London: Duckworth.

Catt, J. A., W. M. Corbett, C. A. H. Hodge, P. A. Madgett, W. Tatler and A. H. Weir 1971. Loess in the soils of north Norfolk. *J. Soil Sci.* **22**, 444–52.

Catt, J. A., M. Green and A. J. Arnold 1982. Naleds in a Wessex downland valley. *Dorset Nat. Hist. Arch. Soc. Proc.* **104**, 69–75.

Catt, J. A., A. H. Weir and P. A. Madgett 1974. The loess of eastern Yorkshire and Lincolnshire. *Proc. Yorks. Geol Soc.* **40**, 23–39.

Chartres, C. J. 1980. A Quaternary soil sequence in the Kennet valley, central southern England. *Geoderma* **23**, 125–46.

Coombe, D. E., L. C. Frost, M. Le Bas and W. Watters 1956. The nature and origin of the soils over the Cornish serpentine. *J. Ecol.* **44**, 605–15.

Dines, H. G., S. C. A. Holmes and J. A. Robbie 1954. *Geology of the country around Chatham.* Mem. Geol. Surv. UK.

Edwards, A. P. and J. M. Bremner 1967. Dispersion of soil particles by sonic vibration. *J. Soil Sci.* **18**, 47–63.

Farmer, V. C. (ed.) 1974. *The infrared spectra of minerals.* London: Mineralogical Society.

Folk, R. L. and W. C. Ward 1957. Brazos River bar: a study in the significance of grain-size parameters. *J. Sed. Petrol.* **27**, 3–26.

Gard, J. A. (ed.) 1971. *The electron-optical investigation of clays.* London: Mineralogical Society.

Harrod, T. R., J. A. Catt and A. H. Weir 1973. Loess in Devon. *Proc. Ussher Soc.* **2**, 554–64.

Hartshorne, N. H. and A. Stuart 1950. *Crystals and the polarising microscope.* London: Edward Arnold.

Hodgson, J. M. 1967. *Soils of the West Sussex Coastal Plain.* Soil Survey of Great Britain, Bull. 3.

Hodgson, J. M., J. A. Catt and A. H. Weir 1967. The origin and development of Clay-with-flints and associated soil horizons on the South Downs. *J. Soil Sci.* **18**, 85–102.

Hodgson, J. M., J. M. Hollis, R. J. A. Jones and R. C. Palmer 1976. A comparison of field estimates and laboratory analyses of the silt and clay contents of some West Midlands soils. *J. Soil Sci.* **27**, 411–19.

Kilmer, V. J. and L. T. Alexander 1949. Methods of making mechanical analyses of soils. *Soil Sci.* **68**, 15–24.

Lee, M. P. 1979. Loess from the Pleistocene of the Wirral Peninsula, Merseyside. *Proc. Geol. Assoc.* **90**, 21–6.

Lee, M. P. and P. J. Vincent 1981. The first recognition of loess from North Wales. *Manchester Geographer* (NS) **2**, (2), 45−53.

Mackenzie, R. C. (ed.) 1957. *The differential thermal investigation of clays.* London: Mineralogical Society.

Madgett, P. A. and J. A. Catt 1978. Petrography, stratigraphy and weathering of Late Pleistocene tills in east Yorkshire, Lincolnshire and north Norfolk. *Proc. Yorks. Geol. Soc.* **42**, 55−108.

Mehra, O. P. and M. L. Jackson 1960. Iron oxide removal from soils and clays by a dithionite-citrate system buffered with sodium bicarbonate. *Clays Clay Min.* 7, 317−27.

Milner, H. B. 1962. *Sedimentary petrography,* Vol. 1. London: George Allen & Unwin.

Mottershead, D. N. 1971. Coastal head deposits between Start Point and Hope Cove, Devon. *Field Stud.* **3**, 433−53.

Palmer, L. S. and J. H. Cooke 1923. The Pleistocene deposits of the Portsmouth district and their relation to man. *Proc. Geol. Assoc.* **34**, 253−82.

Pennington, K. L. and G. C. Lewis 1979. A comparison of electronic and pipette methods for mechanical analysis of soils. *Soil Sci.* **128**, 280−4.

Perrin, R. M. S. 1956. The nature of 'Chalk Heath' soils. *Nature Lond.* **178**, 31−2.

Pigott, C. D. 1962. Soil formation and development on the Carboniferous Limestone of Derbyshire. 1. Parent materials. *J. Ecol.* **50**, 145−56.

Pitcher, W. S., D. J. Shearman and D. C. Pugh 1954. The loess of Pegwell Bay and its associated frost soils. *Geol. Mag.* **91**, 308−14.

Rose, J. and P. Allen 1977. Middle Pleistocene stratigraphy in south-east Suffolk. *J. Geol. Soc.* **133**, 83−102.

Russell, R. J. 1944. Lower Mississippi Valley loess. *Bull. Geol. Soc. Am.* **55**, 1−40.

Schwertmann, U. and R. M. Taylor 1977. Iron oxides. In *Minerals in soil environments*, J. B. Dixon, S. B. Weed, J. A. Kittrick, M. H. Milford & J. L. White (eds), 145−80. Madison: Soil Science Society of America.

Soil Survey of England and Wales 1983. *Soil Map of England and Wales Scale 1 : 250 000.* Harpenden: Soil Survey of England and Wales.

Stairmand, C. J. 1950. A new sedimentation apparatus for particle size analysis in the sub-sieve range. In *Symposium on particle size analysis*, 128−34. London: Institute of Chemical Engineers and Society of Chemistry and Industry.

Sturdy, R. G., R. H. Allen, P. Bullock, J. A. Catt and S. Greenfield 1979. Paleosols developed on chalky boulder clay in Essex. *J. Soil Sci.* **30**, 117−37.

Tanner, C. B. and M. L. Jackson 1947. Nomographs of sedimentation times for soil particles under gravity or centrifugal acceleration. *Proc. Soil Sci. Soc. Am.* **12**, 60−5.

Trechmann, C. T. 1920. On a deposit of interglacial loess, and some transported preglacial fresh-water clays on the Durham coast. *Q. J. Geol. Soc. Lond.* **75**, 173−201.

Vincent, P. J. and M. P. Lee 1981. Some observations on the loess around Morecambe Bay, north-west England. *Proc. Yorks. Geol. Soc.* **43**, 281−94.

Vincent, P. J. and M. P. Lee 1982. Snow patches on Farleton Fell, south-east Cumbria. *Geogr. J.* **148**, 337−42.

Weir, A. H., J. A. Catt and P. A. Madgett 1971. Postglacial soil formation in the loess of Pegwell Bay, Kent (England). *Geoderma* **5**, 131−49.

Welch, N. H., P. B. Allen and D. J. Galindo 1979. Particle size analysis by pipette and Sedigraph. *J. Environ. Qual.* **8**, 544−6.

Wintle, A. G. 1981. Thermoluminescence dating of Late Devensian loesses in southern England. *Nature Lond.* **289**, 479−80.

11

Geomorphological applications of soil micromorphology with particular reference to periglacial sediments and processes

C. Harris

Introduction

Micromorphological analysis may prove of considerable value in the future interpretation of periglacial soils and sediments. Such interpretation depends on identifying diagnostic features, indicating such factors as the presence or absence of permafrost, the thickness of the active layer, ice segregation and the operation of processes of periglacial mass-wasting. While much remains to be learned concerning the reorganisation of soil materials by periglacial processes, certain micromorphological features have been recognised as characteristic of soils in periglacial environments. In this review these features are described, possible modes of formation are outlined, and their use in palaeoenvironmental reconstruction is discussed.

Soil fabric was defined by Brewer (1964) as the physical constitution of a soil material as expressed by the spatial arrangement of the solid particles and associated voids. Major micromorphological components include: (a) plasma, comprising fine-grained, often colloidal, material liable to translocation; (b) skeletal grains, consisting of larger grains which are relatively stable and not easily translocated; and (c) voids or pore spaces. Plasma, skeletal grains and voids may all be incorporated in the S-matrix, which is material within the simplest peds (aggregates), or apedal soil materials in which pedological features occur (Brewer & Pawluk 1975)

Micromorphological features of periglacial soils

The following four categories of micromorphological phenomena have been recognised in periglacial soils: platy or lenticular structure (sometimes associated with horizontal planar voids), grain coatings (cutans), reoriented skeletal grains and vesicular voids.

Platy or lenticular structure

Ice segregation during soil freezing results from the migration of soil water towards the freezing plane from unfrozen soil below (Arakawa 1966, Williams 1968). This migration of water may cause dehydration of the soil immediately below the freezing plane. The development of heaving pressures may also lead to compaction of the unfrozen soil. Dense platy peds containing skeletal grains and plasma, separated by large planar voids, have been ascribed to compaction during soil freezing in a variety of environments (e.g. McMillan & Mitchell 1953, FitzPatrick 1956, Crampton 1965, Dumanski & St Arnaud 1966, O'Brien et al. 1979, Van Vliet-Lanoë & Langohr 1981), the resulting fabrics being referred to as isoband fabric. FitzPatrick (1956) described the excavation and subsequent thaw of permafrost soils in Spitzbergen, the resulting thawed sediment showing a platy structure inherited from the former presence of ice lenses. He went on to suggest that the laminar structure of indurated layers in Scottish soils reflected the former presence of permafrost.

Bunting (1983) refers to laterally extensive voids in soils from Arctic Canada as elongated metavughs. These mark the former sites of ice lenses, and as a result of compaction of the intervening soil matrix during ice lense growth, often show thin stress coatings of fine material along their walls. Saturation following thaw, however, may lead to fines on the upper surfaces settling through the water-filled voids, forming concentrations on the lower void surfaces.

Platy, foliated or laminar structures are reported in thin sections from seasonally frozen soils above permafrost (Morozova 1965, Bunting & Federoff 1974, Pawluk & Brewer 1975) and from seasonally frozen soils where permafrost is absent (Fedorova & Yarilova 1972). Van-Vliet-Lanoë (1976, 1980, 1982) and Van Vliet-Lanoë and Langohr (1981) described platy structures in the active layer which become more blocky below the permafrost table, and O'Brien et al. (1979) showed a similar transition from platy to blocky structure with depth in Elephant Island, South Shetland Islands.

In addition to macro and micro platy peds defined largely by the presence of planar voids, the S-matrix may also appear concentrated into denser horizontal bands or lenses when observed in vertical thin sections (Figs. 11.1a & b). Van Vliet-Lanoë (1976, 1982) observed such fabrics in the indurated layers or 'fragipans' (Muir 1949) present in silty soils in Belgium. Coarser skeletal grains were concentrated at the base of each band and finer-grained plasma formed an irregular coating which often showed a flecked or laminated extinction pattern parallel to the banding. Banded, or plectic, fabrics (Brewer & Pawluk 1975) are encountered in the active layers above permafrost in sub-Arctic and Arctic regions (e.g. Morozova 1965, Brewer & Pawluk 1975, Bunting 1983, Hughes et al. 1983) and in boreal podzolic soils (e.g. Dumanski & St Arnaud 1966, Björkhem & Jongerius 1974, McKeague et al. 1974). They may also be found in chernozemic soils where soil freezing occurs (Coen et al. 1966). Suggested mechanisms in-

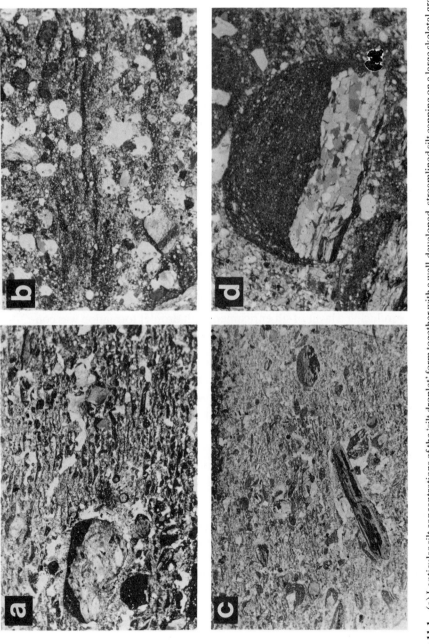

Figure 11.1 (a) Lenticular silt concentrations of the 'silt droplet' form together with a well-developed, streamlined silt capping on a large skeletal grain: 50 cm depth, turf-banked gelifluction lobe on a silty sand till, Okstindan, Norway. Plain light. Frame length 17 mm. (b) Silt bands with vesicular voids (white), 10 cm depth, fine-grained centre of sorted circle, Gråsubre, Jotunheimen, Norway. This site, at 1850 m, is probably underlain by permafrost. Plain light. Frame length 14 mm. (c) Soliflucted till, 25 cm depth, Okstindan, Norway. Grains with silt cappings have been rotated, smooth, rounded silt aggregates are present and there is a weakly developed 'silt droplet' or banded structure. Plain light. Frame length 17 mm. (d) Skeletal grain with capping of silt and fine sand. Note flecked extinction pattern in the capping parallel to its smooth surface. 10 cm depth, gelifluction lobe, Okstindan, Norway. Partially crossed polarisers. Frame length 4.5 mm.

volved in the formation of plectic fabrics include compaction during ice segregation (FitzPatrick 1956, Crampton 1965, Morozova 1965, Bunting 1983), dehydration during ice segregation (FitzPatrick 1956, Morozova 1965, Pissart 1970, Van Vliet-Lanoë 1982) and increasing electrical charge associated with flocculating ions in the soil solution during freezing (Morozova 1965, Van Vliet-Lanoë 1982).

Interception of downward migrating plasma by platy aggregates is generally considered to produce the increase in density and fining upwards often observed in individual bands (Bunting 1983). Downward migration of fines may result from frost-sorting processes (Romans et al. 1966, Bertouille 1972, Van Vliet-Lanoë 1976, Fox and Protz 1981), rapid eluviation of silts in tills immediately following deposition (Romans et al. 1980), and hydration of colloids during thaw, which leads to deflocculation and subsequent downward washing (Morozova 1966, Van Vliet-Lanoë 1976, 1982).

In upland Britain, Romans et al. (1966) and Romans and Robertson (1974) described soil horizons with silt accumulation in the form of lenticular zones of silt enrichment, referred to as silt droplets. Laboratory experimentation suggested that seasonal soil freezing accompanied by ice segregation produced an ice-rich surface layer above a dry, porous subsoil. Thawing of the ice-rich layer released silty meltwater onto the desiccated subsoil, producing silt droplets. Variations in soil moisture status at the onset of freezing and in the severity of winter cooling led to variation in the depth of silt droplet formation from year to year, so producing a layer of silt enrichment. Silt droplet formation was considered, therefore, to depend not only on soil freezing, but also on soil moisture conditions. Romans et al. (1980) considered Iceland to be marginal for the development of silt droplet structures, owing to high precipitation and therefore soil moisture contents that are too high for a desiccated subsoil to form. On micromorphological evidence Romans et al. (1980) suggested that in the Brecon Beacons, south Wales, the pre-Boreal climate might have resembled that of present-day Iceland.

Similar lenticular or banded silt concentrations were observed in vertical thin sections of soils developed on till in Norway (Ellis 1980a, 1980b, 1983, Harris & Ellis 1980) and interpreted as marking the former locations of ice lenses. Examples are shown in Figure 11.1a

Disturbance of platy matrix concentrations by frost creep and gelifluction leads to aggregates becoming broken and rounded (Fig. 11.1c), and often becoming enclosed by coatings of plasma (Harris & Ellis 1980, Van Vliet-Lanoë 1982). Cryoturbation may also produce granular, degraded aggregates (Romans et al. 1966, Romans & Robertson 1974, Van Vliet-Lanoë 1982, Ellis 1983).

Skeletal grain coatings

In their early work on indurated horizons in Scottish soils, Glentworth (1944, 1954) and FitzPatrick (1956) reported plasma coatings on the surfaces of skeletal grains and larger clasts. FitzPatrick (1956) observed that coatings in soils from Spitzbergen developed following melt of ice sheaths around clasts. Although grain coatings of silt and clay have been described in soils developed over permafrost in northern Alaska by Tedrow (1965) and Kubiena (1972), they are reported to be rare in soils of the Canadian High Arctic (Bunting & Federoff 1974) and of the Canadian sub-Arctic and southern Arctic (Brewer & Pawluk 1975). Bunting and Federoff considered that where coatings were observed, which was mainly in sites with patterned ground features, they resulted from thixotropic behaviour of the soil mass during thaw. Björkhem and Jongerius (1974) proposed a similar explanation for coatings on sand grains in podzolic soils in Sweden, the thixotropic behaviour dating from an earlier period of greater climatic severity. Brewer and Pawluk (1975) observed coatings in thin sections at only one site in the Canadian forest−tundra transition, on the Peel Plateau. This skelsepic matrichlamydic fabric (matrix coatings on skeletal grain surfaces with flecked extinction pattern parallel to the surface) was considered to result from drying and wetting associated with freezing and thawing, which caused the finer matrix to be plastered on to the surfaces of framework members. Silt cappings were also weakly developed in soils on Elephant Island (O'Brien *et al.* 1979), again over permafrost.

Coatings on skeletal grains are widely reported in seasonally frozen soils where permafrost is absent (e.g. Federova & Yarilova 1972, Björkhem & Jongerius 1974, McKeague *et al.* 1974, Ellis 1980a, 1980b, 1983 Harris & Ellis 1980, Van Vliet-Lanoë 1982), although Dumanski (1969) stressed that the coating of plasmic material on skeletal grains may also occur in non-periglacial soils, owing to wetting and drying.

Coatings may incorporate clay, silt and, depending on overall texture, fine sand. Ellis (1980a) described a marked increase in the $20-60$ μm fraction with depth in Okstindan, northern Norway, associated with grain coatings in the profile. Flecking in the extinction pattern parallel to the surface of the coating (Fig. 11.1d) which is commonly observed, results from the presence of oriented silt grains.

Van Vliet-Lanoë (1982), Ellis (1980a, 1980b, 1983) and Harris and Ellis (1980) argued that illuviation of fines is responsible for accumulations on the upper surfaces of sand and gravel grains. Harris and Ellis (1980) showed that on a till-covered slope in Norway, drainage was impeded by a still-frozen subsoil during spring and early summer, but that subsequent clearance of ground ice led to rapid vertical drainage. In permafrost soils, particularly where gradients are low and the active layer thin, such free vertical drainage is prevented by the permafrost table so that illuviation of fines and the generation of grain coatings by this mechanism is unlikely to occur.

The effects of gelifluction on grain coatings were described by Harris and Ellis (1980) and Harris (1981a, 1981b), in Okstindan, northern Norway. Here coatings were smooth-surfaced and streamlined. Below the depth of active soil displacement coatings were restricted to the upper surfaces of sand grains, and were often found in association with lenticular or banded silt concentrations (Fig. 11.1a). In the gelifluction layer, however, rotation of grains had apparently taken place and coatings were observed on all surfaces (Fig. 11.1c). Plasma may have been plastered onto the surfaces of skeletal grains during gelifluction to produce streamlined coatings of oriented silt.

Reorientation of skeletal grains

At least two processes may operate in periglacial soils to reorient elongate skeletal grains: frost sorting and gelifluction. Corte (1966) showed that downward penetration of a freezing plane leads to coarser material moving upwards towards the surface and fines migrating downwards in front of the freezing plane. Stones become vertically oriented during this sorting process (Fig. 11.2a). Fox and Protz (1981) and Hughes et al. (1983) reported similar vertical orientation of elongate sand grains in turbic cryosols from the Mackenzie River Valley, Canada. Fines concentrated beneath skeletal grains were considered to occupy voids left by melting ice which developed beneath the grains during each frost-sorting cycle. Benedict (1969) showed that frost-sorting near the surface in a degraded earth hummock on the Niwot Ridge, Colorado, led to a strong vertical orientation of elongate sand grains. Within the two earth hummocks investigated, sand grain orientations in the vertical plane closely reflected the disturbance of soil horizons by cryoturbation and frost heave.

Fox and Protz (1981), Hughes et al. (1983), Fedorova and Yarilova (1972) and Koniscev et al. (1973) described skeletal grains organised into circular or elliptical arrangements, sometimes associated with aggregates of fines. These circular patterns were observed in both vertical and horizontal thin sections. Such fabrics were termed orbiculic by Fox and Protz (1981), and were considered to result from cryoturbation. In Arctic Canada, Bunting (1983) observed similar ovoid fabrics in areas with wet soils, such as late snowbank sites or solifluction deposits. He suggested that relatively large particles, peds or rock fragments are rotated, and fines are generated by the mutual abrasion and/or compression of such entities, forming belts or bands of fine material.

It is widely recognised that reorientation of stones takes place during gelifluction, leading to strongly developed orientations parallel to the direction of flow and dips parallel or slightly imbricate to the ground surface (Harris 1981b). Benedict (1969, 1970) and Harris and Ellis (1980) have shown similar orientation of sand grains in active solifluction lobes. Generally the downslope-preferred orientation observed in horizontally cut thin sections is weaker and more multimodal than is commonly reported for larger stones, although the tendency for grains exposed in vertical thin sections to dip parallel to the ground surface is strong (Fig. 11.2b). Evaluation of skeletal grain orientations observed in thin section is, however, clearly complicated by the influence of grain shape on a two-dimensional view.

Vesicular voids

Large, smooth-walled, bubble-like pores or vesicles have been reported from many arctic and alpine soils, particularly those in areas with patterned ground. Vesicles have been described from Greenland (Ugolini 1966), the Canadian

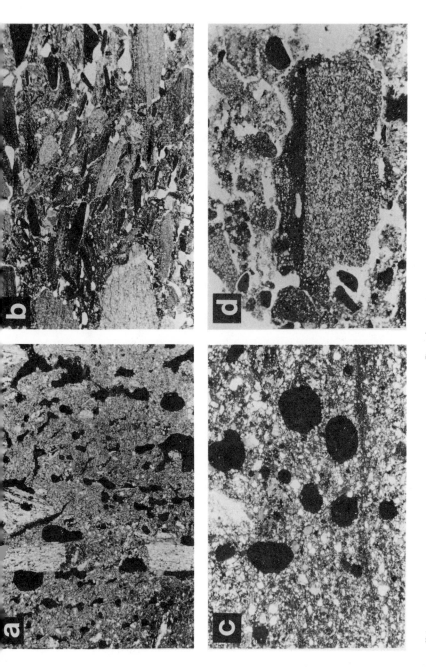

Figure 11.2 (a) Large vesicular voids from 5 cm below the surface, small non-sorted circle, Alaska – Yukon boundary, Sixtymile Highway, Klondike Plateau. Note the vertical alignment of larger rock fragments resulting from frost-sorting. Permafrost is present at this site. Crossed polarisers. Frame length 14 mm. (b) Platy skeletal grains of Devonian slate, in solifluction deposit from a coastal exposure at Wembury, south Devon. Note imbricate dips of many fragments (surface slope is 5° to the left) and relatively clean surfaces of rock fragments. Plain light. Frame length 24 mm. (c) Vesicles or bubble pores (black), 10 cm depth, fine-grained centre of sorted circle, Gråsubre, Jotunheimen, Norway. Crossed polarisers. Frame length 7 mm. (d) Capping of silt and fine sand on clast of Old Red Sandstone (Devonian) in a solifluction deposit, Brecon Beacons, Wales. Plain light. Frame length 18 mm.

High Arctic (Bunting & Jackson 1970, Bunting & Federoff 1974, Bunting 1977, 1983), Spitzbergen (FitzPatrick 1956, Chandler 1972), the South Shetlands (O'Brien *et al.* 1979), Norway (Harris 1977, 1983, Harris & Ellis 1980, Ellis 1983,) and southeastern Iceland (Romans *et al.* 1980). They range in diameter up to 2 or 3 mm, and are often abundant in thin sections (Figs 11.1b, 11.2a & c). Soils in which vesicles develop are often wet, at least during the thaw period.

FitzPatrick (1956) proposed that vesicular voids form as a result of expulsion of air during freezing of wet soils, while Bunting (1977, 1983) suggested drying as the cause of vesicles in Devon Island, North West Territories, Canada. Wetting of dry soils may also produce vesicles (Romanov 1974), and Bunting (1983) and Bunting and Federoff (1974) showed that thixotropic flow of wet soils might also generate air bubbles. Chaplow (1974) found that bubble voids developed in borehole samples of wet sand owing to compaction and consequent liquefaction. Harris (1983) performed a series of simple laboratory experiments which supported the conclusion that vesicles observed in fine sediments from the centres of sorted circles and from soliflucted till resulted from thixotropic behaviour of soils during thaw consolidation. He showed that vesicles may easily be formed simply by puddling a wet soil sample, and that freezing need not be directly responsible for their formation. The presence of vesicles in periglacial soils may therefore indicate liquefaction during thaw consolidation, which is likely to be an important process in the development of many patterned ground features.

Micromorphology and environmental reconstruction

Soil induration as an indicator of former permafrost
Glentworth (1944, 1954), Glentworth and Dion (1949) and FitzPatrick (1956, 1971) have described a compact silty layer of variable thickness in many freely drained Scottish soils, generally 40−60 cm below the surface. This indurated layer has a sharply defined upper surface, diffuse lower boundary and a platy to massive structure, the platy structure being best developed towards the top. Within the indurated layer, pore space is greatly reduced and skeletal grains show a capping of fines. FitzPatrick (1956) considered that compaction and the platy structure resulted from ice segregation during a gradual rise in the permafrost table following deglaciation, and that the depth of the indurated layer corresponded closely to the depth of the active layer.

Stewart (1961), in central Wales, and Crampton (1965), in south Wales, described similar indurated horizons where, unlike the Scottish examples, there was evidence for clay illuviation from above. Compaction, a platy structure which becomes massive with depth, and the presence of an indurated layer in large solifluction terraces, led Crampton to conclude that, as in the Scottish soils, induration resulted from permafrost conditions.

The presence of an indurated layer (fragipan) has been utilised by Van Vliet-Lanoë and Langohr (1981) to identify areas of Weichselian permafrost in the loess deposits of Belgium and northern France. They showed that fragipans are encountered at around 50 cm below the surface in better-drained sites and 30 cm below the surface in poorly drained areas. The depth of the fragipan surface is considered to represent the former depth of the permafrost table.

Silt droplet structures as indicators of former permafrost

Romans et al. (1966) described silt droplet structures in alpine soils in Scotland which incorporated broken fragments of an earlier generation of droplets (silt granules). They showed that the first period of droplet formation probably occurred during the Late Devensian (late Pollen Zone II and Zone III times) and the second period occurred during the Early Flandrian (Pollen Zone IV). Romans et al. (1966) and Romans and Robertson (1974) considered that droplets formed only in freely drained sites where a dehydrated zone developed annually in the active layer between a freezing front advancing downwards from the surface and underlying permafrost.

Romans and Robertson (1974) used the depth of silt droplets and their geographical distribution to reconstruct the extent of discontinuous permafrost in Britain at the beginning of the Flandrian. They concluded that permafrost was present above 595 m in Wales, above 550 m in the Lake District, and above 390 m in the Scottish Highlands. It should be noted, however, that lenticular silt concentrations apparently very similar to silt droplets have been observed in periglacial areas lacking permafrost (e.g. Harris & Ellis 1980). This raises the possibility that permafrost may not be essential for these features to develop.

Micromorphology of fossil periglacial slope sediments

Harris (1981a) described silt and fine sand coating the upper surfaces of skeletal grains in solifluction deposits in the Brecon Beacons (parent material Old Red Sandstone) and the south Wales coalfield (parent material Carboniferous sandstone and shale). Although both coatings and grains were somewhat larger than those observed in Okstindan, Norway (cf. Figs. 11.1d & 11.2d), Harris suggested that downward translocation of fines was responsible for the development of coatings in both cases. It is unlikely, therefore, that drainage was severely impeded by a shallow permafrost table during accumulation of the solifluction deposits in south Wales. In contrast, thin sections of similar solifluction deposits at Wembury (bedrock Devonian slates) in south Devon and Bude (bedrock Carboniferous sandstones and shales) in north Cornwall revealed little evidence for downward translocation of fines, and no grain coatings (Fig. 11.2b). At Wembury, severe impedence of drainage and in consequence abundant surface runoff is indicated by layers of surface wash material consisting largely of silt and clay, which are interbedded with the solifluction deposits (Harris 1981b). Since the south Wales sites described by Harris (1981a) were ice-covered during the Devensian maximum, while the sites in south-west England were not, it would seem likely that in south-west England solifluction occurred under severe periglacial conditions with a shallow permafrost table and restricted subsurface drainage, possibly during the period of full glaciation

further north, while in the south Wales uplands solifluction followed deglaciation and took place under less severe climatic conditions.

Micromorphological analysis in non-periglacial geomorphology: some examples

Many sediments may be usefully described through optical microscopy, particularly those modified by post-depositional reworking. An example of this
type of study is the effect of surface wash on loess. Micromorphological criteria
for surface wash have been identified by Jungerius and Mücher (1972), Kwaad
and Mücher (1977, 1979), Mücher *et al.* (1972), and Mücher (1974) and micromorphological analysis in the interpretation of loess stratigraphy is illustrated
by the work of Mücher and Vreeken (1981). Laboratory studies of crusting and
erosion by rainsplash and surface wash have also employed micromorphological
analysis to describe changes in aggregate and pore sizes (e.g. Mücher & De
Ploey 1977, Mücher *et al.* 1981, Imeson & Jungerius 1976, Farres 1978).

Micromorphological descripition of glacial tills may provide valuable evidence on modes of deposition (e.g. Holmes 1941, Sitler 1968), ice flow directions, from sand grain orientation (e.g. Sitler & Chapman 1955, Harrison 1957,
Ostry & Deane 1963, Penny & Catt 1967) and textural properties (e.g. Huddart
1971). However, possibly owing to difficulty in preparing thin sections of fine-
grained sediments with low permeabilities, electron microscopy has recently
been more widely used than optical microscopy in the analysis of till fabrics (e.g.
Derbyshire *et al.* 1976, McGown & Derbyshire 1977).

A final example of wider geomorphological applications of soil microscopy is
provided by the use of palaeosols as stratigraphic markers. Micromorphological
study of the palaeosol often facilitates both assessment of the nature and degree
of soil development prior to burial, and the possibility of truncation of the profile
by surface erosion during or prior to burial (e.g. Griffey & Ellis 1979, Ellis &
Matthews 1984).

Conclusions

Micromorphological description of periglacial soils and sediments provides a
potentially valuable tool for the geomorphologist as well as the pedologist.
However, much remains to be learned of the precise mechanisms involved in
the formation of many micromorphological phenomena. In this respect, laboratory experimentation would appear to offer the best means of progress. Environmental reconstruction based on micromorphological analysis must also
take account of the possibility that similar micromorphological features may be
produced by different processes, such as the effects of freezing, thawing, wetting
and drying.

In addition to the interpretation of fossil periglacial sediments, micromorphological investigation may also be of value in modern process studies. Micro-fabrics could, for instance, provide information on patterns of soil displacement by gelifluction or cryoturbation, and on the effectiveness of frost-sorting processes. The modes of formation of many patterned ground phenomena may therefore be profitably investigated by micromorphological analysis, as Bunting (1983) has recently demonstrated in the case of earth hummocks. Post-depositional modification of glacial sediments by periglacial processes might similarly be usefully explored through micromorphological studies. Since such processes often operate in only a relatively thin active layer, their role in modifying superficial sediments and generating small-scale morphological features must be of interest not only to the geomorphologist, but also to the pedologist. Equally, many soil-forming processes are also of considerable importance geomorphologically. The potential for co-operation between soil scientists and geomorphologists in this field is clearly great, and techniques developed in one discipline are likely to yield valuable information to the other.

References

Arakawa, K. 1966. Theoretical studies of ice segregation in soil. *J. Glaciol.* **6**, 255–60.

Benedict, J. B. 1969. Microfabric of patterned ground. *Arctic Alpine Res.* **1**, 45–8.

Benedict, J. B. 1970. Downslope soil movement in a Colorado Alpine region: rates, processes and climatic significance. *Arctic Alpine Res.* **2**, 165–226.

Bertouille, H. 1972. Effect du gel sur les sols fins. *Rév. Géomorph. Dynamique* **21**, 71–84.

Björkhem, V. and A. Jongerious 1974. Micromorphological observations in some podzolised soils from central Sweden. In *Soil microscopy*, G. K. Rutherford (ed.), 320–32. Kingston: Limestone Press.

Brewer, R. 1964. *Fabric and mineral analysis of soils*. New York: Wiley.

Brewer, R. and S. Pawluk 1975. Investigations of some soils developed in hummocks of the Canadian sub-Arctic and southern Arctic regions, I. Morphology and micromorphology. *Can. J. Soil. Sci.* **55**, 301–19.

Bunting, B. T. 1977. The occurrence of vesicular structures in arctic and subarctic soils. *Z. Geomorph.* **21**, 87–95.

Bunting, B. T. 1983. High Arctic soils through the microscope: prospect and retrospect. *Ann. Assoc. Am. Geogs.* **73**, 609–16.

Bunting, B. T. and N. Fedoroff 1974. Micromorphological aspects of soil development in the Canadian High Arctic. In *Soil microscopy*, G. K. Rutherford (ed.), 350–65. Kingston: Limestone Press.

Bunting, B. T. and R. M. Jackson 1970. Studies of patterned ground in south-west Devon Island, N.W.T. *Geogr. Ann.* **52A**, 194–208.

Chandler, R. J. 1972. Periglacial mudslides in Vestspitzbergen and their bearing on the origin of fossil 'solifluction' shears in low angled clay slopes. *Q. J. Eng. Geol.* **5**, 225–41.

Chaplow, R. 1974. The significance of bubble structures in borehole samples of fine sand. *Géotechnique* **24**, 223–41.

Coen, G. M., S. Pawluk and W. Odynsky 1966. The origins of bands in sandy soils of the stony plain area. *Can. J. Soil. Sci.* **46**, 245–54.

Corte, A. 1966. Particle sorting by repeated freezing and thawing. *Biul. Peryglac.* **15**, 176–240.

Crampton, C. B. 1965. An indurated horizon in soils of south Wales. *J. Soil Sci.* **16**, 230–41.

Derbyshire, E., A. McGown and A. Radwan 1976. Total fabric of some till landforms. *Earth Surf. Proc.* **1**, 17–26.

Dumanski, J. 1969. Micromorphology as a tool of Quaternary research. In *Pedology and Quaternary research*, S. Pawluk (ed.), 39–52. Edmonton: University of Alberta Press.

Dumanski, J. and R. J. St Arnaud 1966. A micromorphological study of eluviated horizons. *Can. J. Soil Sci.* **46**, 287–92.

Ellis, S. 1980a. An investigation of weathering in some arctic–alpine soils on the northeast flank of Oksskolten, north Norway. *J. Soil Sci.* **32**, 371–85.

Ellis, S. 1980b. Physical and chemical characteristics of a podzolic soil formed in Neoglacial till, Okstindan, northern Norway. *Arctic Alpine Res.* **12**, 65–72.

Ellis, S. 1983. Micromorphological aspects of arctic–alpine pedogenesis in the Okstindan Mountains, Norway. *Catena* **10**, 133–48.

Ellis, S. and J. A. Matthews 1984. Pedogenic implications of a ^{14}C-dated paleopodzolic soil at Haugabreen, southern Norway. *Arctic Alpine Res.* **16**, 77–91.

Farres, P. 1978. The role of time and aggregate size in the crusting process. *Earth Surf. Proc.* **3**, 243–54.

Fedorova, N. N. and T. Yarilova 1972. Micromorphology and genesis of prolonged seasonally frozen soils in western Siberia. *Geoderma* **7**, 1–13.

FitzPatrick, E. A. 1956. An indurated soil horizon formed by permafrost. *J. Soil Sci.* **7**, 248–54.

FitzPatrick, E. A. 1971. *Pedology*. London: Oliver & Boyd.

Fox, C. A. and R. Protz 1981. Definition of fabric distributions to characterize the rearrangement of soil particles in turbic cryosols. *Can. J. Soil Sci.* **61**, 29–34.

Glentworth, R. 1944. Studies on the soils developed on basic igneous rocks in central Aberdeenshire. *Trans R. Soc. Edin.* **61**, 149–70.

Glentworth, R. 1954. The soils of the country round Banff. Huntly and Turriff. *Memoirs of the soil survey of Great Britain:* Scotland. Edinburgh: HMSO.

Glentworth, R. and H. G. Dion 1949. The association or hydraulic sequence in certain soils of the podzolic zone of north-east Scotland. *J. Soil Sci.* **1**, 35–49.

Griffey, N. J. and S. Ellis 1979. Three *in situ* paleosols buried beneath Neoglacial moraine ridges, Okstindan and Jotunheimen, Norway. *Arctic Alpine Res.* **11**, 203–14.

Harris, C. 1977. Engineering properties, groundwater conditions, and the nature of soil movement on a solifluction slope in north Norway. *Q. J. Eng. Geol.* **10**, 27–43.

Harris, C. 1981a. Microstructures in solifluction sediments from south Wales and north Norway. *Biul. Peryglac.* **28**, 221–6.

Harris, C. 1981b. *Periglacial mass wasting: a review of research*. BGRG Res. Monograph 4. Norwich: Geo Abstracts.

Harris, C. 1983. Vesicles in thin sections of periglacial soils from north and south Norway. In *Proceedings of the 4th International Conference on Permafrost*, Fairbanks, Alaska, 445–9.

Harris, C. and S. Ellis 1980. Micromorphology of soils in soliflucted materials, Okstindan, northern Norway. *Geoderma* **23**, 11–29.

Harrison, P. W. 1957. A clay till fabric: its character and origin. *J. Geol.* **65**, 275–308.

Holmes, C. D. 1941. Till fabric. *Bull. Geol Soc. Am.* **52**, 1299–354.

Huddart, D. 1971. Textural distinction of Main Glaciation and Scottish Readvance tills in the Cumberland Lowland. *Geol. Mag.* **108**, 317–24.

Hughes, O. L., R. D. van Everdingen and C. Tarnocai 1983. Regional setting physiography and geology. In *Guidebook to permafrost and related features of the northern Yukon Territory and Mackenzie Delta, Canada*, H. M. French and J. A. Heginbottom (eds), 5–34. Guidebook 3, 4th Int. Conf. Permafrost, Fairbanks, Alaska.

Imeson, A. C. and P. D. Jungerius 1976. Aggregate stability and colluviation in the Luxembourg Ardennes, an experimental and micromorphological study. *Earth Surf. Proc.* **1**, 259–71.

Jungerius, P. D. and H. J. Mücher 1972. The micromorphology of fossil soils in the Cyprus Hills,

Alberta, Canada. In *Soil micromorphology*, S. Kowalinski (ed.) 617–27. Proc. 3rd Working Meeting on Soil Micromorphology, Wroclaw, Poland.

Koniscev, V. N., M. A. Faustova and V. V. Rogov 1973. Cryogenic processes as reflected in ground microstructures *Biul. Peryglac.* **22**, 213–19.

Kubiena, W. L. 1972. On the micromorphology of soils of the Arctic of North Alaska. In *Soil micromorphology*, S. Kowalinski (ed.), 235–43. Proc. 3rd Int. Working Meeting on Soil Micro-scopy, Warsaw.

Kwaad, F. J. P. M. and H. J. Mücher 1977. The evolution of soils and slope deposits in the Luxembourg Ardennes near Wiltz. *Geoderma* **17**, 1–37.

Kwaad, F. J. P. M. and H. J. Mücher 1979. The formation and evolution of colluvium on arable land in northern Luxembourg. *Geoderma* **22**, 173–92.

McGown, A. and E. Derbyshire 1977. Genetic influences on the properties of tills. *Q. J. Engng Geol.* **10**, 389–410.

McKeague, J. A., C. J. Acton and J. Dumanski 1974. Studies of soil micromorphology in Canada In *Soil microscopy*, G. K. Rutherford (ed.), 84–100. Kingston: Limestone Press.

McMillan, N. J. and J. Mitchell 1953. A microscopic study of platy and concretionary structures in certain Saskatchewan soils. *Can. J. Agric. Sci.* **33**, 179–83.

Morozova, T. D. 1965. Micromorphological characteristics of pale yellow permafrost soils in central Yakutia in relation to cryogenesis. *Soviet Soil Sci.* **7**, 1333–42.

Mücher, H. J. 1974. Micromorphology of slope desposits; the necessity of a classification. In *Soil microscopy*, G. K. Rutherford (ed.), 553–66. Kingston: Limestone Press.

Mücher, H. J. and J. De Ploey 1977. Experimental and micromorphological investigation of erosion and redeposition of loess by water. *Earth Surf. Proc.* **2**, 117–24.

Mücher, H. J. and W. J. Vreeken 1981. (Re) Deposition of loess in southern Limbourg, the Netherlands, 2. Micromorphology of the lower silt loam complex and comparison with deposits produced under laboratory conditions. *Earth Surf. Proc. Landforms* **6**, 355–63.

Mücher, H. J., J. de Ploey and J. Savat. Response of loess materials to simulated translocation by water: micromorphological observations. *Earth Surf. Proc. Landforms* **6**, 331–6.

Mücher, H. J., T. Carballas, D. Guitián, P. D. Jungerius, S. B. Kroonenberg and M. C. Villar 1972. Micromorphological analysis of effects of alternating phases of landscape stability on two soil profiles in Galicia, N.W. Spain. *Geoderma* **8**, 241–66.

Muir, A. 1949. *Report on a visit to the United States and Canada.* Rep. Agric. Res. Council of Gt Britain, No. 11828.

O'Brien, R. M. G., J. C. C. Romans and L. Robertson 1979. Three soil profiles from Elephant Island, South Shetland Islands. *Brit. Antarctic Surv. Bull.* **48**, 1–12.

Ostry, R. C. and R. E. Deane 1963. Microfabric analysis of till. *Bull. Geol Soc. Am.* **74**, 165–8.

Pawluk, S. and R. Brewer 1975. Micromorphological and analytical characteristics of some soils from Devon and King Christian Islands. NWT *Can. J. Earth Sci.* **55**, 349–61.

Penny, L. F. and J. A. Catt 1967. Stone orientations and other structural features of tills in East Yorkshire. *Geol. Mag.* **104**, 344–60.

Pissart, A. 1970. Les phénomènes physiques essentiels liés au gel, les structures periglacaires qui en résultent, et leur signification climatique. *Ann. Soc. Géol. Belg.* **93**, 7–49.

Romanov, Y. E. A. 1974. Effect of trapped gas on some processes in soils. *Sov. Soil Sci.* **4**, 222–8.

Romans, J. C. C. and L. Robertson 1974. Some aspects of the genesis of alpine and upland soils in the British Isles. In *Soil microscopy*, G. K. Rutherford (ed.), 498–510. Kinston: Limestone Press.

Romans, J. C. C., L. Robertson and D. L. Dent 1980. The micromorphology of young soils from south-east Iceland. *Geog. Ann.* **62A**, 93–103.

Romans, J. C. C., J. Stevens and L. Robertson 1966. Alpine soils of northwest Scotland. *J. Soil Sci.* **17**, 184–99.

Sitler, R. F. 1968. Glacial till in oriented thin section. *23rd Int. Geol. Congr.* **8**, 283–95.

Sitler, R. F. and C. H. Chapman 1955. Microfabrics of till from Ohio and Pennsylvania. *J. Sed. Petrol.* **25**, 262–9.

Stewart, V. L. 1961. *A permafrost horizon in the soils of Cardiganshire.* Welsh Soils Disc. Group Rep. No. 2, 19–22.

Tedrow, J. C. F. 1965. Concerning genesis of the buried organic matter in tundra soil. *Soil Sci. Soc. Am. Proc.* **29**, 89–90.

Ugolini, F. C. 1966. Soils of the Mesters Vig. district, northeast Greenland. *Med. Grønland* Vol. 176.

Van Vliet-Lanoë, B. 1976. Traces de ségrégation de glace en lentilles associées aux sols et phénomènes periglacaires fossiles. *Biul. Peryglac.* **26**, 41–55.

Van Vliet-Lanoë, B. 1980. Corrélations entre fragipan et permagel. Applications aux sols lessivés loessiques. *C. R. Groupe de Travail 'Regionalization du Periglacaire'* **5**, 9–22. Strasbourg.

Van Vliet-Lanoë, B. 1982. Structures et microstructures associées à la formation de glace de ségrégation: leurs consequences. In *R. J. E. Brown Memorial Volume, Proceedings of the 4th Canadian Permafrost Conference* Calgary, 116–22.

Van Vliet-Lanoë, B. and R. Langohr 1981. Correlation between fragipans and permafrost with special reference to Weischsel silty deposits in Belgium and northern France. *Catena* **8**, 137–54.

Williams, P. J. 1968. Ice distribution in permafrost profiles. *Can. J. Earth Sci.* **5**, 1381–6.

12

The mineralogy and weathering history of Scottish soils

M. J. Wilson

Introduction

In general, Scottish soils are relatively young, being developed mainly upon glacial drift (comprising till, fluvioglacial and solifluction deposits) associated with the Devensian period. The mineralogical composition of the tills and solifluction deposits often compares quite closely with that of the directly subjacent or closely associated rocks. Indeed, the Soil Survey of Scotland makes extensive use of this relationship in that its soil maps are, to a large extent, based upon the 'soil association' which consists essentially of a group of soils developed from common parent material. Each association consists of a number of 'series', ideally members of a genetic soil group, which often form a hydrological sequence (catena) related to topography. The principal genetic soil types formed in Scotland are somewhat limited in number in that they are typically related to podzols, brown forest soils or gleys; other soil types do occur but to a lesser extent. On the other hand, the types of parent materials on which Scottish soils have developed are extremely varied. These materials may be derived from granite, gabbro, basalt/andesite, mica-schist and other metamorphic types, Lower Palaeozoic greywackes and shales, Old Red Sandstone sediments, Carboniferous sediments, fluvioglacial sands and gravels, or estuarine silts and clays. Many other types of parent material also occur and many soils are developed on tills of mixed parentage, but the above materials form the most widespread soil associations.

Given the relative youth of Scottish soils, it might be anticipated that there has been insufficient time for pedogenesis to have exercised a marked influence on their mineralogy. Indeed, the sand and silt fractions separated from some soils may contain abundant quantities of fresh feldspar, amphibole and biotite, minerals usually considered to be relatively susceptible to weathering. Nevertheless, careful study of soil clay fractions (<2 μm in diameter) reveals that they often contain minerals that could not have been directly inherited and whose origin, therefore, may not be immediately obvious. These clay minerals could have been formed either during modern (Holocene) pedogenetic weathering or, alternatively, could have been inherited from previous weathering episodes, presumably of Preglacial or interglacial age. In this paper the possible modes of origin of the clay minerals in Scottish soils are outlined and examples of the clay mineralogy of the main Scottish soil associations given. An attempt is then made to synthesise this information so that mineralogical criteria can be developed which may be of general use in the interpretation of the weathering history of Scottish soils.

Origin of the clay minerals in Scottish soils

Direct inheritance

This factor is predominant because of the relative youth of Scottish soils and, in general, it is prudent to assume that *all* the clay minerals in a soil have originated in this way unless there is strong evidence to the contrary. The overwhelming influence of inheritance can be easily shown by comparing clay fractions from C horizons, which usually consist of fresh glacial till, with clay fractions separated from the underlying or nearby rock formations from which the till derives. Frequently such clay fractions are indistinguishable. Where soils are developed on tills derived from sedimentary rocks, as in the soils derived from Old Red Sandstone and Carboniferous material, the influence of inheritance is, of course, obvious. It is sometimes not recognised, however, that igneous rocks can be also an important source of hydrous layer silicates. For example, soils formed from basaltic material may contain abundant smectite and it is often inferred that this is because the base-rich soils provide an environment favourable for the pedogenic crystallisation of this mineral. In Scotland, however, it has been found that many basaltic lavas have been hydrothermally altered and contain abundant smectite, infilling vesicles, pseudomorphing ferromagnesian minerals or being widely disseminated throughout the body of the rock. This smectite may, therefore, simply be inherited into overlying soils (Wilson 1976). Even apparently fresh granite can yield a significant clay fraction. Thus, the constituent feldspars may be intensely sericitised and the effects of mild hydrothermal activity, which may not be obvious from the appearance of the rock, may have resulted in the formation of small amounts of kaolinite and smectite. This has been observed in the Bennachie granite near Aberdeen, where nontronite and kaolinite occur locally in small pockets (Wilson, unpublished). An even more unlikely source of clay minerals was found in the high-grade Dalradian meta-limestones of north-east Scotland which consistently contain small amounts of saponite and interstratified expansible minerals (Wilson & Bain 1970).

These examples illustrate the dangers of making assumptions about igneous and metamorphic rocks *not* containing clay minerals, and indicate that the role of inheritance in soil clay formation must be determined carefully before conclusions are reached with regard to effects of subsequent weathering or pedogenesis.

Preglacial and interglacial weathering

Although the effects of Preglacial and interglacial weathering episodes can rarely be distinguished unambiguously from each other, the weight of evidence suggests that pre-Pleistocene weathering is more significant (FitzPatrick 1963). This point is brought out by a consideration of the geomorphological evolution of Scotland. It is now generally recognised that the Scottish landscape can only be explained in terms of the interplay of a number of factors dating at least from Tertiary times and possibly earlier (George 1965). Many of the dominating features of the Scottish landscape are recognisably superimposed upon earlier primary landforms and, in general terms, upland Scotland can be considered as a dissected plateau that was gently undulating. This plateau surface was, at one time, thought to have been a reflection of the Cretaceous sea floor but this concept has now been generally abandoned in favour of one relating planation to periods of stability during Palaeogene or Neogene times when most, if not all, of Scotland was an area of positive relief.

During much of Tertiary times the climate was hot and humid and Scotland was, therefore, subjected to a period of intense weathering. This resulted in the formation of a deep weathering cover, most of which was stripped away following pulses of uplift and possible tilting in the Pliocene, and several periods of glaciation during the Pleistocene. Nevertheless, remnant patches of this cover are still to be found in many parts of Scotland, particularly the north-east corner which, to a large extent, escaped deep glacial erosion. In their most completely developed form, these remnants show profound weathering where the primary minerals are altered to kaolinite and, occasionally, gibbsite (Hall 1983). Such weathering obviously influences soil mineralogy when soils are directly developed on this material but there could be an effect on a wider scale than is generally appreciated. Thus, that part of the ancient weathering cover removed by glacial erosion must, to some extent, have been incorporated into the glacial tills which form the immediate parent material of most Scottish soils. In general, the influence of Preglacial weathering must be suspected where deeply weathered rock is common, as in the north-east of Scotland. The occurrence of kaolinite and gibbsite in soils of highly variable drainage status, and where inheritance can be excluded, is certainly strong circumstantial evidence for the minerals to be attributed to pre-Holocene weathering episodes, although it may be difficult to say whether this weathering is Preglacial or interglacial in age.

Recent work by Hall (1983) has, however, provided a wealth of evidence that may enable some distinctions to be made in this respect. He distinguished two types of weathered saprolite, namely gruss and clayey gruss on the basis of their granulometry, geochemistry and clay mineralogy. Gruss normally has a low clay content and a heterogeneous clay mineral composition whereas clayey gruss has an elevated clay content, a kaolinite−illite clay mineral assemblage and may show rubefaction. It was concluded that the balance of evidence indicated that gruss-type weathering occurred under a humid temperate climate in the Late Pliocene and Early Pleistocene, while the clayey gruss weathering occurred under a hot humid climate probably during the Miocene with possible continued formation in the Early Pliocene.

Holocene pedogenesis

In general, the influence of Holocene pedogenesis upon the clay mineralogy of Scottish soils is relatively minor although some effects do occur, as can be shown by careful comparison of C horizon clay fractions with those separated from A and B horizons from profiles developed upon uniform parent material. These changes often relate to the transformation of pre-existing layer silicates. An example is the conversion of biotite to vermiculite which, at its simplest level, involves the replacement of interlamellar K^+ ions by more hydratable ions such as Ca^{2+} or Mg^{2+}. This reaction can occur rapidly in soils, although concomitant oxidation reactions are almost always involved, and may result in a vermiculite-dominated clay fraction (Wilson 1970). Vermiculitisation of dioctahedral illite also occurs particularly in A horizons of acid soils and in these circumstances both di- and trioctahedral minerals may contain abundant interlayer aluminium, often in a non-exchangeable form. Complete breakdown of trioctahedral expansible minerals such as vermiculite and saponite may occur where the soil pH falls below a value of 5 and, under more acid conditions, chlorite behaves similarly (Bain 1977). The synthesis of new clay minerals does not seem to have occurred to any great extent during Holocene pedogenesis, although the formation of smectite under poorly drained conditions does undoubtedly occur (Wilson & Berrow 1978). Surprisingly, however, the formation of smectite in the A horizons of podzols which has been widely reported from Canada (Brydon et al. 1968) and Scandinavia (Gjems 1967) has yet to be shown in Scotland. There is certainly no evidence that kaolinite is forming at the present time in Scottish soils.

Clay mineralogy of some major soil associations

Illustrating the relationship between mineralogy and weathering history in Scottish soils is facilitated by emphasising those associations developed upon a uniform type of parent material with a limited ability for contributing inherited clay, particularly of the kaolinite group. In general, these associations derive ultimately from igneous and metamorphic rocks, although the provisos mentioned previously must always be borne in mind. The clay mineralogy of all the major Scottish soil associations has recently been reviewed (Wilson et al. 1985) and the following information represents a selection of this work.

Soils derived from granitic material

These soils, which are usually mapped as Countesswells Association, occur widely throughout Scotland and occupy 5.8% of the total land area. Their clay fractions are usually dominated by micaceous minerals and their weathering products, often in the form of interstratified mica—vermiculite and aluminium-interlayered vermiculite. Kaolinite is common in minor amounts, and in eastern Scotland both poorly ordered kaolinite and dehydrated halloysite occur (Wilson & Tait 1977), the latter in the form of irregularly scrolled laths. The clay mineralogy of these soils can be interpreted mainly in terms of Holocene pedogenesis on previously inherited fine-grained micaceous material, corresponding · essentially to comminuted sericite, muscovite and biotite. As biotite weathers much more rapidly than the aluminous micas, it is almost certainly responsible for the vermiculitic clays. The common, and sometimes abundant, occurrence of kaolinite and halloysite, however, is believed to be largely a relic of Preglacial or interglacial weathering. This is suggested by the abundance of these minerals in the granitic soils of eastern Scotland, where deep weathering is a conspicuous feature, and by the fact that they occur in soils of widely varying drainage status, ranging from freely drained brown forest soils to very poorly drained gleys.

Soils derived from gabbro

These soils are mapped as Insch Association and occupy 0.7% of Scotland. Their clay fractions are usually dominated by trioctahedral vermiculite, which may or may not be aluminium-interlayered. In poorly drained soils, however, this material behaves like a smectite in that it expands after glycerol solvation. Kaolin minerals occur in considerable amounts, the most common type being halloysite, which becomes progressively more hydrated with depth in poorly drained soils (Wilson & Tait 1977). The vermiculitic nature of the clay fractions can usually be attributed to the decomposition of biotite (a common constituent of these particular gabbros) as was first shown in the classic work of Walker (1949, 1950). It was later shown that macroscopic vermiculites can be aluminium-interlayered, that these vermiculites may also contain discrete zones of kaolinite and gibbsite (Wilson 1966) and that the decomposition of biotite may play a key role in determining the nature of the clay fraction (Wilson 1967). The smectite in the poorly drained Insch soils − the so-called 'cardenite' (MacEwan 1954) − is also considered to be a product of biotite weathering at a higher pH value (approximately 6) where oxidation results in a decrease of surface charge (Ismail 1969). The kaolin minerals are again thought to be related to Preglacial or interglacial weathering, as is indicated by the extensive occurrence of deeply decomposed rock in the Insch valley as well as the universal occurrence of kaolinite/halloysite in soils of all drainage classes.

Soils derived from schist

These soils are mapped under various associations including Strichen, Foudland and Arkaig. They extend over huge areas of northern, eastern and western Scotland and represent 27.4% of the Scottish land area. As expected, the clay fractions of these soils are dominated by micaceous minerals and their weathering products. In eastern Scotland the clay mineralogy is rather similar to that of the granite-derived soils with di- and trioctahedral mica, interstratified mica—vermiculite and vermiculite being conspicuous. The extent of vermiculitisation depends very largely on whether the parent rock contains biotite in addition to the more resistant muscovite. A surprising feature of these soils in eastern Scotland is that they contain large amounts of kaolin minerals, consisting of poorly ordered kaolinite, halloysite or a mixture of both. Tubular halloysite is particularly well developed in the Foudland Association soil clays. Much lesser amounts of kaolin seem to occur in the Arkaig Association soils (Wilson & Tait 1977) which cover larger areas of the Highlands, but nevertheless it is usually present. The clay mineral composition of these soils can once again be attributed to the combined effects of Holocene pedogenesis and Preglacial and interglacial weathering on inherited micaceous material.

Soils derived from Lower Palaeozoic sedimentary rocks

These soils are mapped mainly as Ettrick Association and are developed on drifts derived from weakly metamorphosed greywackes and shales. The Ettrick Association is the most extensive association south of the Highland Boundary Fault and occupies 9.3% of Scotland. The clay mineralogy of the C horizons of these soils is almost invariably dominated by trioctahedral chlorite and dioctahedral illite, both undoubtedly derived from the parent greywackes and shales. In the surface horizon illite tends to be vermiculitised, whereas chlorite appears to decompose completely without leaving any well-defined weathering product (cf. Bain 1977). Minor amounts of kaolinite occur widely in all the soil types of the Ettrick Association. In the montane soils formed on the summits of the Merrick and Kells Hills, the clay fractions may contain up to 30% gibbsite (Wilson & Bown 1976). It is not inconceivable that these gibbsitic soils could have formed during recent pedogenesis, in view of their extremely well-drained nature, although micromorphological evidence and the finding of haematitic pedorelicts suggest formation on a previously weathered mantle. Preglacial or interglacial weathering is also thought to be responsible for the small amounts of kaolinite referred to above.

Soils derived from other parent materials

In the soils representative of many of the other major associations, the large amounts of inherited clay make it difficult to discern the relationship between mineralogy and weathering history. Nevertheless, once the nature of the inherited material is clear, it becomes possible to make some deductions relating to weathering history. For example, the clay fractions of the soils derived from basalt and andesitic material (Darleith and Sourhope Associations) contain trioctahedral interstratified minerals and smectite. These minerals weather out in the surface horizons of acid soils with pH <5, leading to X-ray amorphous products, but the trioctahedral minerals persist even into the surface horizons where the soil pH is >6, as in some gleys (Wilson 1973). These lava-derived soils often contain small amounts of illite and kaolinite but the origin of these minerals is not clear.

The soils derived from Old Red Sandstone sediments have been mapped under a large number of association names and are differentiated according to the part of the Old Red Sandstone sequence from which the parent tills derive. As it happens, these distinctions correspond quite closely with marked clay mineralogical variations in the rocks themselves. Thus in the associations formed from Lower Old Red Sandstone material (Balrownie, Forfar, Glenalmond/ Maybole, Laurencekirk and Stonehaven Associations) the soil clays contain large amounts of some interstratified minerals almost identical with those occurring in the sedimentary rocks (Wilson 1971, 1973). The Middle Old Red Sandstone-derived soils of northern Scotland (Thurso and Canisbay Associations) contain a chlorite—illite assemblage identical to that of the calcareous flagstones which form the main parent rock. Again, the clay fractions of the soils derived from the Upper Old Red Sandstone rocks (Dinnet, Elgin, Hobkirk, Lauder and Kippen/Largs Associations) are often dominated by kaolinite and illite which originate to a great extent from diagenetic kaolinite and illite in the parent rocks. The illite often has an unusual lath-like form and the kaolinite occurs in well-shaped hexagonal crystals, both morphologies being unmistakable under the electron microscope. The possible effects of interglacial or Preglacial weathering are difficult to establish with certainty in the Old Red Sandstone-derived soils, but at least the effects of Holocene pedogenesis are clear. These effects are similar to those described previously and include vermiculitisation of dioctahedral illite with concomitant precipitation of hydroxy-aluminium in exposed interlamellar spaces, and degradation of trioctahedral expansible minerals.

Finally, the clay material contained in the many associations developed upon fluvioglacial and raised beach sands and gravels may be mentioned. The clay fractions separated from these soils are of some interest because, although small in amount, these fractions are frequently dominated by amorphous material, which is certainly the result of Holocene pedogenesis. In podzols, this amorphous material has been shown to consist largely of imogolite and allophane (Farmer *et al.* 1980), a finding that has led to a new theory of podzolisation, where aluminium is regarded as having been transported in the form of hydroxy-aluminium orthosilicate sols rather than as part of the traditionally recognised organic complex.

Discussion

The relationship between mineralogy (essentially clay mineralogy) and weathering history of Scottish soils can be elucidated only by first separately analysing the influence of inheritance, which as demonstrated above, tends to be overwhelming. Nevertheless, once the contribution of inheritance has been set aside, the effects of Preglacial weathering and Holocene pedogenesis can be brought more sharply into focus. These effects can be better understood by taking into account the many factors involved in the geological history of Scotland that affected the nature of the prevailing processes and products of weathering from Tertiary times onwards. In the following discussion, the conclusions of Hall (1983) in relation to the timing of the clayey gruss and gruss type weathering are accepted as a reasonable working basis and weathering history is considered within the context of the following periods.

Eocene to Pliocene

During this time, Scotland was an area of positive relief and was subjected to periods of uplift and concomitant denudation. The main phase of uplift occurred from the Middle Palaeocene to the Early Eocene after which, following a period of subsidence in the Middle Eocene, there was little tectonic activity until the Middle Oligocene. Renewed uplift occurred at this time, sometimes at a local level, until a period of stability was estabished in the Middle Miocene. Thereafter, it is probable that pulsed uplift occurred through the Pliocene and perhaps into the Pleistocene in response to denudational unloading. The prevalent climate during Tertiary times was somewhat variable but the evidence suggests that the Eocene was a period of extraordinary warmth and high rainfall, with latosols developing throughout western Europe (Millot 1970), and that in the Middle to Late Miocene conditions were warm temperate to tropical. The combination of tectonic stability and a hot, humid climate undoubtedly favoured profound rock decomposition under the 'geochemical' type of weathering regime described by Duchaufour (1979). Geochemical weathering occurs under the influence of water, more or less charged with carbon dioxide, and in the virtual absence of organic matter. It is best developed in the soils and weathered rocks of humid tropical regions and is characterised by the rapid crystallisation of clay minerals like kaolinite, and of simple oxides and hydroxides like haematite and gibbsite. These weathering products are exactly those found in the profoundly decomposed rocks of north-east Scotland, the clayey gruss type described by Hall (1983). Here, the feldspars can be almost completely kaolinised and a fine-grained haematitic pigment, presumably originating from the weathering of ferromagnesian minerals, may be widely disseminated. Gibbsite is usually a minor constituent, and of local occurrence.

Pliocene to Holocene

Tectonic activity during this period was characterised by pulsed uplift with possible tilting in the Highlands. During the Late Miocene and Early Pliocene, temperatures dropped sharply and a humid temperate climate became established. Following a brief warm period in the Middle Pliocene, the climate became even cooler and in the Early Pleistocene deteriorated into cold-temperate cycles, which with increasing frequency developed into the onset of regional glaciation. Hall (1983) has presented evidence to support the view that this period saw the stripping of the previously formed kaolinitic weathering cover, which was then replaced by a granular gruss type of weathering. Characteristically, these grusses have low clay contents and show minor alteration of primary rock-forming minerals, but despite this they often contain kaolinite and sometimes gibbsite. This has been interpreted by Hall (1983) as a reflection of the free-draining nature of this material under the temperate climate prevailing in the Pliocene and Early Pleistocene, although another interpretation could be that the granular gruss is simply the basal part of a deep profile formed during Eocene to Miocene times that has been mostly stripped away during Pleistocene glaciation. The nature of the weathering products that may have formed in the Pliocene and Early Pleistocene, and in the interglacials, is indeed problematical, although it is tempting to attribute the formation of halloysite to this period. The fact that halloysite forms rapidly during the weathering and pedogenesis of different types of parent material in temperate climates (e.g. in New Zealand, Birrell *et al.* 1955 and Japan, Aomine & Wada 1962) and yet is not found to any great extent in the geological record, suggests that the mineral is thermodynamically metastable and probably eventually converts to kaolinite. It seems unlikely, therefore, that the halloysite in Scottish soils is of Palaeogene origin, or indeed that it developed during Holocene pedogenesis, and in these circumstances formation in the Pliocene–Pleistocene periods seems most probable.

Holocene

Since the last Devensian glaciation, Scottish soils have generally been subjected to the 'biochemical' type of weathering described by Duchaufour (1979). This is strongly influenced by the presence of organic matter and living organisms and is characterised by extremely slow crystallisation processes and poorly ordered materials and organo-mineral complexes. Biochemical weathering predominates in the soils of cool temperate climates and is largely controlled by the effects of organic acids, involving simple acidity, chelation or a combination of both. In these circumstances transformation reactions of pre-existing layer silicates assume an important role, exemplified by the vermiculitisation of micaceous minerals in the A horizons of Scottish soils. The concurrent precipitation of hydrous aluminium species in the exposed interlamellar spaces of the expanded layer silicates is also probably related to the chelating effects of organic acids, as aluminium would normally be quite insoluble in an aqueous environment in the common soil pH range of $4 - 6$. The influence of organic acids in this respect has been fully supported by the experiments of Robert *et al.* (1979). Biochemical weathering also accounts for the complete decomposition of vulnerable clay minerals like saponite and chlorite. Furthermore, formation of the X-ray amorphous minerals imogolite and allophane in the illuvial (Bs) horizons of podzols (Farmer *et al.* 1980) is likely to have involved the participation of chelating organic acids, at least in the initial mobilisation of aluminium from primary minerals. Finally, it may be noted that the characteristic iron oxide minerals associated with Holocene pedogenesis are goethite, often in microcrystalline form, ferrihydrite and, in gleyed soils, lepidocrocite. The occurrence of haematite is usually associated with inheritance (from Old Red Sandstone sediments, for example) or with Preglacial or interglacial weathering.

Conclusions and geomorphological implications

The clay mineralogy of many Scottish soils can be interpreted as representing the products of several different episodes of weathering. The earliest such episode occurred in Miocene or earlier times, under a hot, humid, subtropical climate and resulted in the establishment of deep weathering profiles containing abundant kaolinite, occasional gibbsite and widely disseminated haematite. These minerals originated by neoformation following the decomposition of feldspars and ferromagnesian minerals. The marked cooling that occurred in the Pliocene and Early Pleistocene culminated in a humid temperate climate, and although the above clay minerals may have continued to form, albeit on a much reduced scale, it is suggested that halloysite was the characteristic weathering product of this period. Halloysite requires careful differentiation from kaolinite by X-ray diffraction and may be more widespread than previously realised. It is probable that the Pliocene witnessed a massive denudation of the kaolinitic weathering cover, but removal was far from complete at the onset of regional glaciation in the Pleistocene. The effects of Preglacial weathering on the composition of glacial tills can be readily discerned, not only in north-east Scotland where outcrops of deeply decomposed rock are common, but also in many areas where deep weathering is not conspicuous. These effects are indicated by the widespread occurrence of minor amounts of kaolinite in circumstances where the mineral cannot have been inherited from parent rocks and where it is most unlikely to have formed under the Holocene pedogenic regime. With the waning of the last glaciation and the establishment of a cool temperate climate, weathering conditions became strongly influenced by organic acids generated from decomposing organic matter. This type of weathering led to the transformation of layer silicates, as in the conversion of mica to vermiculite, and in the breakdown of other silicates to yield X-ray amorphous products. Goethite, ferrihydrite and, occasionally, lepidocrocite are the typical iron oxide minerals formed in these conditions.

It is clear from the above observations that there may be many geomorphological problems in Scotland where study of the mineralogy of soils, glacial tills or Preglacially weathered materials could yield information of considerable, if not crucial, importance. In general, mineralogical data have not been widely used in this context although the work of Hall (1983) may be cited as a notable exception. Hall reconstructed the denudational history and the evolution of landforms in north-east Scotland by characterising the distribution and the physical, chemical and mineralogical nature of the Preglacially weathered rock of the area. In particular, this information was used to make inferences about the formation of high- and low-level erosion surfaces, the development of Preglacial meso-scale landforms such as basins and valleys, and the delimitation of patterns of glacial erosion. Undoubtedly, a similar approach could be used in other areas of Scotland where glacial scouring has not been severe. Such areas may be more widespread than is generally appreciated and may even include montane areas where intense scouring might be anticipated to be the general rule. For example, the finding of pedogenic kaolinite and haematite in some alpine podzol soils developed from chlorite–mica schist on the Ben Lawers Massif in Tayside (Stevens & Wilson 1970) suggests that these soils formed before the last glaciation, during which time the mountains may have been nunataks. Again, the gibbsitic soils developed from hornfelsic rocks on the summit areas of the Merrick and Kells Hills in the Southern Uplands were interpreted by Wilson and Bown (1976) as being developed upon Preglacially weathered material which had been preserved because glacial erosion was of the ice sheet type, being characterised by slow ice movement over broad, non-constricting surfaces.

Similar gibbsitic soils are developed upon weathered material on the Cairngorm plateau (Wilson, unpublished, Hall 1983). From geomorphological evidence, the plateau itself is thought to be part of an erosion surface of Eocene to Miocene age and as this was a period characterised by intense subtropical weathering, it is possible that the gibbsitic material represents residual fragments of a previously extensive deeply weathered regolith dating from this time. Although this suggestion is only a tentative one, it does at least illustrate the point that in Scotland there may well be scope for more detailed collaborative studies between geomorphologists and soil mineralogists. Such collaboration would be of mutual benefit and would inevitably lead to a deeper understanding of the weathering history of Scottish soils. It is very much hoped that the information and ideas presented in this chapter will help to stimulate this kind of joint approach.

References

Aomine, S. and K. Wada 1962. Differential weathering of volcanic ash and pumice, resulting in formation of hydrated halloysite. *Am. Min.* **47**, 1024–48.

Bain, D. C. 1977. The weathering of ferruginous chlorite in a podzol from Argyllshire, Scotland. *Geoderma* **17**, 193–208.

Birrell, K. S., M. Fieldes and K. I. Williamson 1955. Unusual forms of halloysite. *Am. Min.* **40**, 122–4.

Brydon, J. E., H. Kodama and G. J. Ross 1968. Mineralogy and weathering of clays in orthic podzols and other podzolic soils in Canada. In *Transactions of the 9th International Congress on Soil Science*, Vol. III, 41–51.

Duchaufour, P. 1979. Introduction generale. In *Alteration des roches cristallines en milieu superficiel*, 87–9. Versailles: L'Association Francaise pour l'Etude du Sol.

Farmer, V. C., J. D. Russell and M. L. Berrow 1980. Imogolite and proto-imogolite allophane in spodic horizons: evidence for a mobile aluminium silicate complex in podzol formation. *J. Soil Sci.* **31**, 673–84.

FitzPatrick, E. A. 1963. Deeply weathered rock in Scotland, its occurrence, age and contribution to the soils. *J. Soil Sci.* **14**, 33–43.

George, T. N. 1965. The geological growth of Scotland. In *The geology of Scotland*, G. Y. Craig (ed.), 1–49. Edinburgh: Oliver & Boyd.

Gjems, D. 1967. Studies on clay minerals and clay mineral formation in soil profiles in Scandinavia. *Medd. Norske Skogfors Vers.* **21**, 303–415.

Hall, A. H. 1983. *Weathering and land form evolution in north-east Scotland.* Unpublished Ph.D. thesis, University of St Andrews.

Ismail, G. T. 1969. Role of ferrous iron oxidation in the alteration of biotite and its effect on the type of clay minerals formed in soils of arid and humid regions. *Am. Min.* **54**, 1460–6.

MacEwan, D. M. C. 1954. 'Cardenite', a trioctahedral montmorillonoid derived from biotite. *Clay Min. Bull.* **2**, 120–6.

Millot, G. 1970. *The geology of clays* (translation by W. R. Farrand and H. Pacquet). London: Chapman & Hall.

Robert, M., M. K. Razzaghe, M. A. Vincente and G. Veneau 1979. Role du facteur biochimique dans l'alteration des mineraux silicates. In *Alteration des roches cristallines en milieu superficiel*, 153–74. Versailles: L'Association Francaise pour l'Etude du Sol.

Stevens, J. H. and M. J. Wilson 1970. Alpine podzol soils on the Ben Lawers Massif, Perthshire. *J. Soil Sci.* **21**, 85–95.

Walker, G. F. 1949. The decomposition of biotite in the soil. *Min. Mag.* **28**, 693–703.

Walker, G. F. 1950. Trioctahedral minerals in the soil clays of north-east Scotland. *Min. Mag.* **29**, 72–84.

Wilson, M. J. 1966. The weathering of biotite in some Aberdeenshire soils. *Min. Mag.* **35**, 1080–93.

Wilson, M. J. 1967. The clay mineralogy of some soils derived from a biotite-rich quartz-gabbro in the Strathdon area, Aberdeenshire. *Clay Min.* **7**, 91–100.

Wilson, M. J. 1970. A study of weathering in a soil derived from a biotite-hornblende rock. I. Weathering of biotite. *Clay Min.* **8**, 291–303.

Wilson, M. J. 1971. Clay mineralogy of the Old Red Sandstone (Devonian) of Scotland. *J. Sed. Petrol* **41**, 995–1007.

Wilson, M. J. 1973. Clay minerals in soils derived from Lower Old Red Sandstone till: effects of inheritance and pedogenesis. *J. Soil Sci.* **24**, 26–41.

Wilson, M. J. 1976. Exchange properties and mineralogy of some soils derived from lavas of Lower Old Red Sandstone (Devonian) age. II. Mineralogy, *Geoderma* **15**, 289–304.

Wilson, M. J. and D. C. Bain 1970. The clay mineralogy of the Scottish Dalradian meta-limestones. *Contr. Min. Petrol* **26**, 285–95.

Wilson, M. J. and C. J. Bown 1976. The pedogenesis of some gibbsitic soils from the southern Uplands of Scotland. *J. Soil Sci.* **27**, 513–22.

Wilson, M. J. and J. M. Tait 1977. Halloysite in some soils from north east Scotland. *Clay Min.* **12**, 59–66.

Wilson, M. J. and M. L. Berrow 1978. The mineralogy and heavy metal content on some serpentinite soils in north-east Scotland. *Chemie Erde* **37**, 181–205.

Wilson, M. J., D. C. Bain and D. M. L. Duthie 1985. The soil clays of Great Britain. II. Scotland. *Clay Min.* **19**, 709–35.

13

Geomorphological linkages between soils and sediments: the role of magnetic measurements

J. A. Dearing, B. A. Maher and F. Oldfield

Introduction

The ubiquitous nature of iron oxides in soils makes their study significant from a number of viewpoints. From a pedological outlook, iron oxides possess two exceptional sets of properties. First, the iron present within a soil is very responsive to its chemical and physical environment, displaying a tendency to transform to different types of oxide and hydroxide in sympathy with edaphic conditions. Second, although the new oxide or hydroxide is often only metastable with respect to subsequent environmental changes, very slow reaction rates will tend to preserve the new formation even in somewhat disequilibrium conditions (Schwertmann & Taylor 1977). Recent advances in geophysical theory and instrumentation may now add to the already acknowledged usefulness of soil iron oxide characterisation, by focusing on those oxides that can be identified by their inherent magnetic structure and by their response to artificially induced external magnetic fields. Mineral magnetic studies open up the possibilities of (a) describing soil iron distributions and their causative processes, and (b) tracing their movements within and between environmental systems. The advantages of mineral magnetic techniques are their ease, safety and rapidity of application, their non-destructive nature and relative economy, and, most importantly, their power to detect a number of iron oxides even in concentrations well below the levels of detection by other methods, such as X-ray diffraction, differential thermal analysis, infra-red spectroscopy and differential chemical extractions.

Mineral magnetism: theory and measurement

The physical basis of the magnetism of natural materials lies in the electronic structure of their constituent atoms, where the electrons move in orbital circulation around the central nucleus. The movement of the electrically charged electron generates an electric current which results in a magnetic moment. A second magnetic moment is generated by the spin movement of the electrons (that is, the orbiting electrons are themselves spinning), and the product of the two moments confers a magnetic dipole on the atoms. Thus, all substances exhibit some degree of magnetic behaviour which is determined by the interactions between the electron spins and the nature of their alignment when the substance is placed in an external magnetic field (Table 13.1).

A further factor involved in the nature of the magnetic response of samples is the size of their magnetic grains: where the grain is large (>0.05 μm), it is energetically favourable for the atoms of the grain to group themselves into regions of similar direction of magnetisation. These regions are termed domains. As the grain size decreases, a threshold is reached where the grain will act as one stable single domain (<0.05 μm); a further reduction in grain size (<0.02 μm) results in the grain exhibiting a superparamagnetic response, with thermal randomisation acting to flip the magnetisation direction continually from one axis to another. Figure 13.1 illustrates relationships between the magnetic parameters and mineral grain size.

In the context of mineral magnetic studies, it is those minerals spanning the solid solution series between magnetite (Fe_3O_4) and maghemite (γFe_2O_3) which are most important. These minerals are ferrimagnetic (Table 13.2) and thus will often dominate much of the magnetic information gained from bulk sample measurements, even when present in very small concentrations. Non-ferrimagnetic minerals contributing to a magnetic signal are normally either

Table 13.1 Types of magnetic behaviour.

Magnetic state	Nature of alignment	Resulting force
paramagnetic	electron and orbital spins have a small magnetic moment, which tends to align *with* the applied field	very weak, not capable of holding a remanence
diamagnetic	electron and orbital spins balance out but induced magnetic moments align in opposite direction to the applied field.	very weak, not capable of holding a remanence
antiferromagnetic	electron spins alternate, atom by atom, thus cancelling out each generated moment	weak, capable of remanence acquisition
ferrimagnetic	two of every three atoms line up in one direction, the third oppositely	strong, capable of remanence acquisition

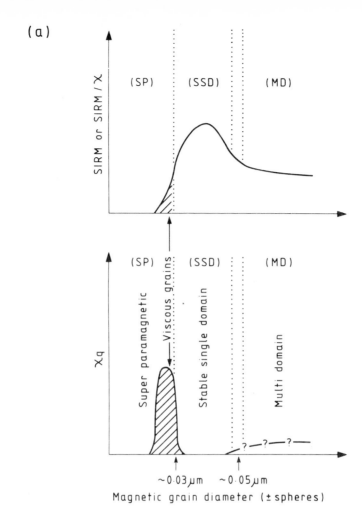

(a)

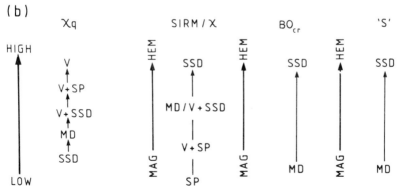

(b)

Figure 13.1 (a) The relationships between SIRM (SIRM/χ) and χ$_q$, and mineral grain size, showing approximate ranges of multi-domain, stable single domain and viscous grains. (b) A summary of the general influences on χ$_q$, SIRM/χ, (Bo)cr, and S values of mineral grain size (MD, SSD, SP and V; see (a) and mineral type (ferrimagnetic 'magnetite' (MAG) and antiferro-magnetic 'haematite' (HEM).

Table 13.2 Soil iron oxides and oxyhydroxides.

	Chemical formula	Magnetic state
goethite	αFeOOH	antiferromagnetic
haematite	αFe$_2$O$_3$	antiferromagnetic
lepidocrocite	γFeOOH	paramagnetic
maghemite	γFe$_2$O$_3$	ferrimagnetic
magnetite	Fe$_3$O$_4$	ferrimagnetic
ferrihydrite	5Fe$_2$O$_3 \cdot$9H$_2$O	paramagnetic

antiferromagnetic or paramagnetic (Table 13.1), and although these will only dominate the magnetic properties of a sample when they exist in extremely high proportions, their presence is often still indicated through the use of a combination of mineral magnetic measurements.

Table 13.3 provides a brief introduction to mineral magnetic parameters and their interpretation.

Magnetic mineral assemblages in soil

The factors controlling soil magnetic mineral assemblages are essentially the same as for other soil properties, with parent material, climate, organisms, relief and drainage, and time all playing a role. The distribution of magnetic minerals in a soil profile is governed by a number of processes, and it is convenient to distinguish between those which effect an input into the soil system and those which transform iron in a way that alters the magnetic properties. The end products of these two sets of processes are referred to as primary and secondary minerals respectively.

Weathering
The influence of weathering as a major control on soil magnetic characteristics depends primarily on the type of substrate available. A basalt or andesite bedrock may contain up to 10% primary magnetite (or more likely titanomagnetite) together with a large quantity of Fe in other minerals. In contrast, a sedimentary substrate, such as Triassic sandstone, will contain much less Fe and that in the form of haematitic coatings around silica grains. In the case of basalt, the effect of weathering is twofold: first, resistant grains of primary ferrimagnetic magnetite are released into the silt and sand-sized fractions of the soil and, second, chemical weathering effects the release of ferrous iron (Fe^{2+}) from the iron-bearing silicates. This ferrous iron may then either be slowly oxidised *in situ*, or be mobilised and redistributed before being oxidised and precipitated. The newly precipitated ferric oxide may initially be amorphous or cryptocrystalline, but ageing and topotactic reactions involving internal rearrangement act to

Table 13.3 Magnetic parameters and instrumentation.

K χ	*Magnetic susceptibility*: the ratio of magnetisation induced to intensity of the magnetising field. This is measured *within* a small magnetic field, and is reversible (i.e. no remanence is induced). Can be measured on a volume (K) or mass specific (χ) basis. Roughly proportional to the concentration of ferrimagnetic minerals within a sample. Units: K (dimensionless), χ (m^3 kg^{-1}).
	Instrumentation: 20 cm search loop, ferrite probe, core loop (K); single sample susceptibility sensor (χ).
χ_q	*Quadrature* (or frequency-dependent) *susceptibility* : the variation of susceptibility with frequency. This parameter indicates the presence of viscous grains lying at the stable single domain/superparamagnetic boundary, and their delayed response to the magnetising field. Expressed as a % of total low frequency susceptibility (χ_q%), or specific quadrative susceptibility (χ_q). Units: χ_q (m^3 kg^{-1}).
	Instrumentation: dual frequency susceptibility sensor.
SIRM	*Saturation isothermal remanent magnetisation*: the highest volume of magnetic remanence that can be produced in a sample by application of a very high field (usually >0.85T). SIRM relates to both mineral type and concentration. Units: A m^2kg^{-1}.
	Instrumentation: pulse magnetiser, fluxgate magnetometer.
SIRM/χ	The ratio of these two parameters can be diagnostic of either mineralogy type (e.g. a low, theoretically zero, ratio indicates the presence of paramagnetic minerals), or, where samples have similar mineral types and concentrations, the dominant magnetic grain size. Units: Am^{-1}.
(Bo)cr IRM$_{-x}$/SIRM, 'S'	*Demagnetisation parameters*: obtained by applying one or more reversed magnetic fields to a previously saturated sample. The reverse field strength (T) required to return a magnetised sample from its SIRM to zero is termed the coercivity of remanence (Bo)cr.
	The loss of magnetisation at other selected backfields can be expressed as a ratio, IRM$_{-x}$/SIRM (giving a result between +1 to −1); the ratio obtained using IRM$_{-0.1T}$ (a backfield which discriminates between ferrimagnetic and antiferromagnetic mineral types) has been termed the 'S' value.
	Instrumentation: Pulse magnetiser, fluxgate magnetometer.

produce antiferromagnetic oxides, such as goethite ($\alpha FeOOH$) or haematite (αFe_2O_3); the former tending to occur under moist, cool conditions, and the latter predominating in warmer, drier regimes (Schwertmann & Taylor 1977).

Fire and fermentation

The formation of ferrimagnetic minerals within the upper layers of a soil, even in lithologies devoid of primary ferrimagnetic minerals, is a widely observed

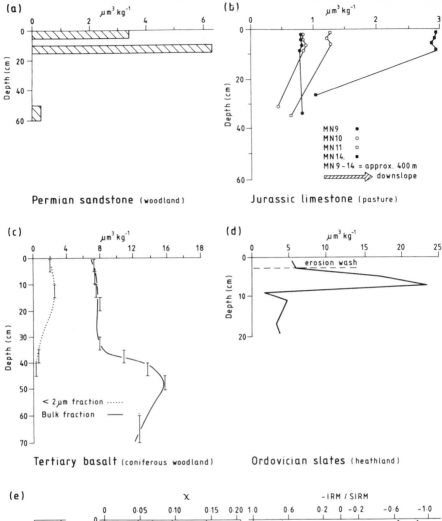

Permian sandstone (woodland)

Jurassic limestone (pasture)

Tertiary basalt (coniferous woodland)

Ordovician slates (heathland)

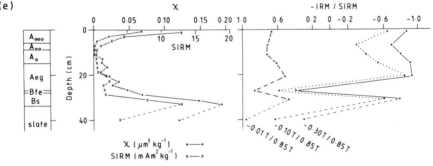

Iron - pan podzol

Figure 13.2 Susceptibility profiles of a range of soil types: (a) Enhancement on sandstone derived soil, probably caused by the 'fermentation' mechanism (after Oldfield *et al.* 1979); (b) A topo-sequence of enhanced rendzina soils showing redistribution effects by hillslope processes (after Dearing, 1979); (c) Bulk sample profile in a highly ferrimagnetic basaltic soil, with apparent enhancement in the clay fraction (after Dearing, 1979); (d) Enhanced topsoil caused by natural firing, underlying eroded subsoil material (after Rummery *et al.* 1979); (e) Relationship between χ, SIRM and S values (at three demagnetisation fields) and horizons of iron pan podzol developed in slate (after Yates 1983).

phenomenon in temperate zones (Germany, Neumeister & Peschel 1968; France, Le Borgne 1955; UK, Tite & Mullins 1971, Oldfield *et al.* 1979). These formations of secondary minerals confer a rise, or 'enhancement', in the topsoil bulk values of χ and SIRM (Fig. 13.2 a–c). Mechanisms for enhancement are described by Mullins (1977), who recognises two types. First, natural or man-induced firing produces high soil temperatures and a reducing atmosphere; under these conditions, non- ferrimagnetic oxides and oxyhydroxides, present in the soil as weathering end products, are reduced to ferrimagnetic magnetite, which may reoxidise to ferrimagnetic maghemite as air enters the soil on cooling. (Fig. 13.2d) In poorly drained soils, maghemite may be formed directly through the dehydration of paramagnetic lepidocrocite (γFeOOH). Secondly, it is most probable that a 'fermentation' mechanism exists, consisting of oscillations between oxidation and reduction within the soil microenvironment. The specific processes involved are as yet ill-defined but microbial action is implicated, and it seems likely that the end product is a non-stoichiometric ferrimagnetic mineral within the solid solution series between magnetite and maghemite (cf. Longworth *et al.* 1979). The decomposition of organic matter may constitute a further contribution to surface enhancement, releasing up to 3% of soil Fe. Soil microfauna are intimately associated with this process, and both direct and indirect accretion of iron oxides by bacterial action has been noted (Aristovskaya 1974). In some soils, atmospheric fall-out from industrial processes or volcanic activity may contribute to topsoil enhancement, thus complicating any interpretation of soil magnetism in terms of pedological control alone. Fall-out products usually include significant quantities of magnetite which can be assumed on the basis of historical studies of sediment cores (Oldfield *et al.* 1980, Scoullos *et al.* 1979) to be persistent and long-lived. Industrially derived magnetic spherules, formed through the release of iron impurities or the action of ablation on the industrial structures, appear to be characterised by the presence of both magnetite and haematite, often with trace element associations, and by a relatively large magnetic grain size.

Podzolisation and gleying

Both these processes appear to be inimical to the neoformation of secondary ferrimagnets, and may be destructive to primary ferrimagnets and antiferromagnets. The loss of magnetic iron oxides from the Aeg horizon in a gleyed podzol (Fig. 13.2e) is clearly shown by the χ and SIRM curves. The response to demagnetisation fields (Fig. 13.2e) distinguishes between the different magnetic mineral assemblages; a magnetically 'soft' response in the organic, eluviated (Aeg) and spodic (Bfe and Bs) horizons indicates the presence of various concentrations of ferrimagnets, together with some paramagnetic lepidocrocite or ferrihydrite, as reflected in the χ and SIRM curves, and an extremely resistant or magnetically 'hard' response in the Bfe indurated iron pan typifies the antiferromagnetic minerals, in this case probably goethite. A number of studies have noted the association between soil waterlogging and greatly reduced χ values (Rummery *et al.* 1979, Vadyunina *et al.* 1974, Oldfield *et al.* 1979). Dis-

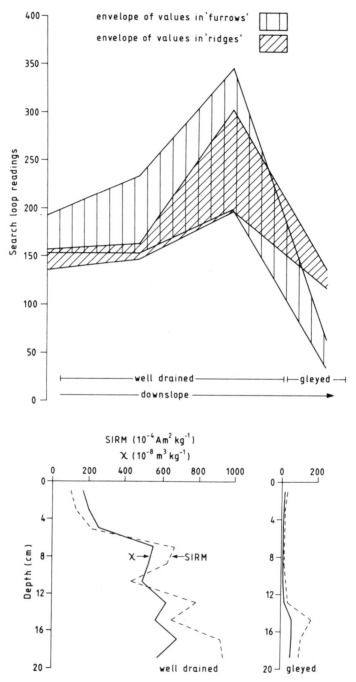

Figure 13.3 Redistribution of ferrimagnetic minerals on a 'ridge and furrow' slope, as observed by scanning loop (K) measurements (not calibrated) in the field, and the effects of drainage on bulk χ and SIRM values of soil profiles sampled from the upper and lower slope.

solution of magnetic grains seems to be linked with biochemical effects of gleying, and is most effective on small grains (<0.05 μm) (B. A. Maher unpublished data). Oxidation subsequent to dissolution, along old root channels for example, can lead to the neoformation of paramagnetic oxides (Schwertmann & Kampf in press). Figure 13.3 shows the down-profile variations in χ for gleyed and well-drained sites at Ardnamurchan, West Scotland; the dominant weathering input of titanomagnetite is clearly defined with χ increasing towards the base, but within the gleyed profiles χ values are an order of magnitude less.

Application to geomorphological studies

The preceding sections illustrate the magnetic characterisation of soil horizons and substrates. The opportunities for characterising soil-forming processes and mapping soil types are undoubtedly large, and have yet to be fully explored; further, it seems likely that mineral magnetic studies of palaeosols (Poutiers 1975, Harvey *et al.* 1981) may provide insight into past weathering and climatic regimes, and aid correlation of equivalent horizons in discontinuous palaeosol sequences. But at present, the most detailed examinations of the application of mineral magnetic studies to geomorphological problems have been in the fields of sediment source tracing and erosional processes. The following sections review the scope of these applications and the problems to be encountered, and provide a generalised interpretative model for soil−sediment linkages based on mineral magnetic studies.

Hillslope processes
Soil creep and soil wash will alter the spatial pattern of topsoil depth and particle size distributions. On a slope where magnetic enhancement is widespread, surface-scanning sensors can be used to produce magnetic toposequences (cf. Fig. 13.2b) and maps to show zones of erosion and deposition. Knowledge of the relationship between K (or χ if single samples are analysed) and particle size can be used to assess the role of specific detachment and transportation processes. Even where primary ferrimagnetic minerals effectively mask enhancement, the movements of certain particle size fractions containing relatively high concentrations of the primary minerals can be identified. The previously cited Ardnamurchan study (Fig. 13.3) on a series of 'ridge and furrow' features running parallel to the slope produced results illustrating the gross downslope accumulation of ferrimagnet-rich material, and the microtopographical movement of similar material from ridge to furrow. An exciting prospect is that surface horizon enhancement on slopes, expressed as a cascading model, might be used to test theoretical models of hillslope processes.

Sediment source tracing: topsoil/subsoil contrasts
Discussion of hydrological processes frequently makes the distinction between surface and channel processes, and yet there remains no single effective tech-

nique to distinguish between the sources of eroded sediment in systems where both processes contribute to the sediment load. In catchments where magnetic enhancement is widespread and fairly uniform, mineral magnetic studies of suspended stream sediment or lake sediment may provide a means of examining the relative importance of different erosional processes and their timing. Studies undertaken in the Jackmoor Brook catchment, Devon (Oldfield *et al.* 1979, Walling *et al.* 1979), showed that the majority of soils, mainly brown earths developed in Permian substrates, exhibited enhanced values of χ and SIRM (cf. Fig. 13.2a), and that additional magnetic parameters, (Bo)cr, SIRM/χ and IRM$_{-0.1T}$/SIRM effectively 'fingerprinted' topsoil and subsoil material. Mineral magnetic measurements made on suspended stream sediments trapped through a number of storm events revealed the timing of sediment detachment and transport from different sources in relation to total suspended sediment concentrations and stream discharge (Fig. 13.4).

Similar mineral magnetic studies of lake, reservoir and estuarine sediment cores can be used to extend the 'monitoring' timescale of $10^{-2} - 10^{1}$ years into the 'geomorphological' timescale of $10^{1} - 10^{3}$ years (Dearing *et al.* 1982, Foster *et al.* 1985). In the Moroccan catchment of Lake Roumi, Flower *et al.* (1984) found enhanced topsoils developed in deeply weathered Miocene sediments and by analysing a ^{210}Pb-dated 1.6 m sediment core were able to identify the timing of accelerated topsoil erosion linked to the expansion of olive plantations. Microscopic examination of ferrimagnetic fractions in both soils and lake sediment, in combination with the use of the grain size-dependent parameter SIRM/χ, confirmed both the widespread presence of fine-grained ferrimagnetic minerals of pedogenic origin, and the use of bulk sediment χ measurements to infer influx of topsoil material (Figure 13.5). Similar studies by Dearing (1979) in the Lac d'Annecy lake catchment, Haute Savoie, France, demonstrated linkages between high sediment χ and eroded rendzina topsoil; Mössbauer effect studies (Longworth *et al.* 1979) confirmed the presence of a high proportion of ultra-fine superparamagnetic grains in enhanced topsoil (cf. Fig. 13.2b). A problem may arise when attempting to identify the presence of secondary ferrimagnetic minerals in sediments which might also include small but significant proportions of coarser multidomain primary minerals. Enhancement processes probably produce both superparamagnetic and stable single domain minerals and, as Figure 13.1 shows, these secondary minerals do not display, as a group, a unique set of responses to demagnetisation field parameters such as 'S' and (Bo)cr or to SIRM/χ. A drop in value of any of these parameters can be the result of high concentrations of either superparamagnetic grains (i.e. a topsoil source) or multidomain grains (i.e. a non-enhanced soil source). Although superparamagnetic grains possess a higher χ than either stable single domain or multidomain minerals, a peak in bulk χ may be due to a greater influx of multidomain minerals, concentrated in a specific silt or sand-sized fraction (see below), rather than an influx of topsoil material. In such a situation, it might be necessary to exclude the majority of multidomain grains from the analysis, either by microsieving to 2 μm or by simple dispersion and decanting methods, before studying

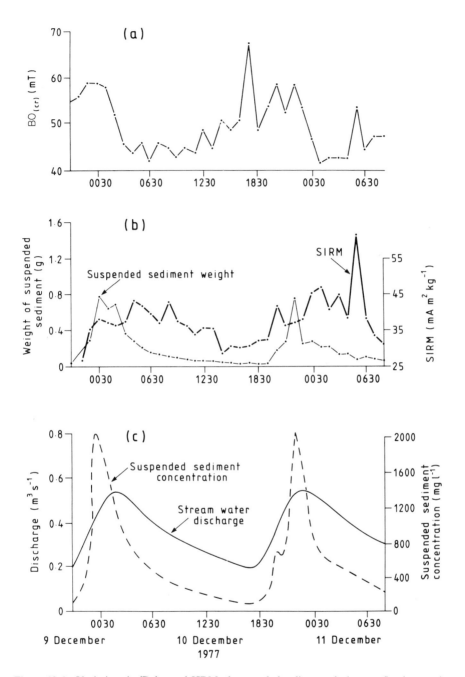

Figure 13.4 Variations in (Bo)cr and SIRM of suspended sediments during two flood events in the Jackmoor Brook, Devon (soil χ shown in Fig. 13.2a). Note the increased levels of SIRM (b) and reduced levels of (Bo)cr (a) on the falling limbs of the hydrograph (c), consistent with rising discharge levels causing initial erosion of subsoil (low SIRM, high (Bo)cr), before topsoil (high SIRM, low (Bo)cr) reaches the stream by surface runoff processes (after Walling *et al.* 1979).

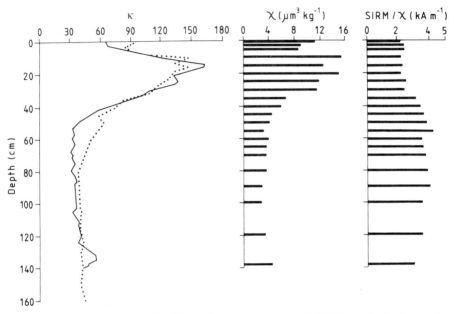

Figure 13.5 Volume susceptibility (K) profiles (two cores), χ and SIRM/χ profiles for lake sediment from Lake Roumi, Morocco. Susceptibility begins to rise at c. A.D. 1935 (50 cm), largely as a consequence of an increased influx of very fine secondary ferrimagnetic minerals (low SIRM/χ) caused by cultivation (after Flower et al. 1984).

the mineral magnetic properties of the clay fraction. Preferable to this would be to measure χ_q, the parameter affected by viscous grains in secondary mineral formations (Fig. 13.1). Oldfield (1983) described how χ and χ_q were used to distinguish between topsoil and industrial sources of sediments deposited in the Elefsis Gulf, Greece.

Sediment source tracing: particle size relationships
Sediments derived from soils for which the concentration of secondary minerals is low relative to that of primary minerals, are likely to possess mineral magnetic properties controlled by particle-size variations. Fig. 13.6 shows the variations in χ values for different fractions of lake sediment sampled in Loch Lomond (Thompson & Morton 1979) and three Icelandic lakes (Bradshaw unpublished data), and of bedload material sampled from streams in Blekinge, South Sweden (Björck et al. 1982). The results indicate that χ values of different fractions of a bulk sample can vary by up to a factor of 3 or 4. Consequently, small changes in the overall particle size distribution of a sediment sample can produce large changes in bulk χ values. The precise shape of a χ/particle size curve will depend upon the mineralogy of the local rock type or deposit and the degree to which attritional processes have concentrated the ferrimagnetic minerals in different fractions. In the most degraded deposits, primary magnetite will be concentrated in a relatively narrow range of fractions corresponding to the mineral

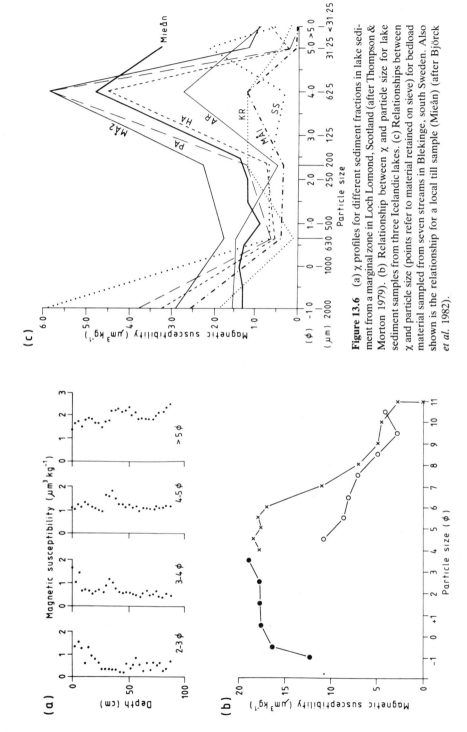

Figure 13.6 (a) χ profiles for different sediment fractions in lake sediment from a marginal zone in Loch Lomond, Scotland (after Thompson & Morton 1979). (b) Relationship between χ and particle size for lake sediment samples from three Icelandic lakes. (c) Relationships between χ and particle size (points refer to material retained on sieve) for bedload material sampled from seven streams in Blekinge, south Sweden. Also shown is the relationship for a local till sample (Mieån) (after Björck et al. 1982).

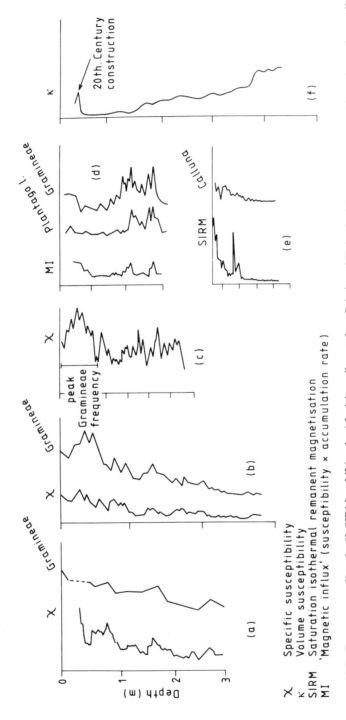

Figure 13.7 Downcore profiles of χ, K, SIRM, or MI (see key) for lake sediments from Britain and Finland, with pollen analytical evidence for catchment disturbance: (a) Llyn Geironydd (after Turner 1979); (b) Lough Neagh (after Thompson et al. 1975); (c) Loch Lomond (after Dickson et al. 1978; (d) Loch Davan (after Edwards 1978); (e) Lough Fea; (f) Lake Vuokonjarvi (after Stober & Thompson 1977). Note the high degree of correlation between magnetic and 'erosional indicator' peaks.

χ Specific susceptibility
K   Volume susceptibility
SIRM   Saturation isothermal remanent magnetisation
MI   'Magnetic influx' (susceptibility × accumulation rate)

dimensions. In certain situations, the curves may contain the legacy of past weathering and transporting processes, in which case they can be used to 'finger-print' potential sediment sources and to infer source—sediment linkages. Björck *et al.* (1982) were able, by this method, to identify the specific till unit from which stream bedload originated (Fig. 13.6c), and to confirm regional lithostrati-graphic divisions of glacial deposits.

In lake sediments, bulk χ and SIRM can be surrogate measures of particle size distribution, highly sensitive to subtle changes in sediment composition in a particle size range that frequently only encompasses the clays and silts. Figures 13.6b and c indicate that when secondary ferrimagnetic minerals are virtually absent, bulk sediment χ should reflect the extent to which relatively coarser particles in the medium silt to fine sand range contribute to sediment composi-tion. There is much circumstantial evidence to suggest that peak values of the concentration parameters χ, and SIRM in fine sediments sampled from central and deep water zones reflect a coarsening of particle size attributable to catch-ment disturbances, as inferred from pollen analytical evidence (Fig. 13.7). The precise hydrological implications of these observed particle size variations are far from clear. Dearing and Flower's (1982) results from trapped sediments in Lough Neagh (cf. Fig. 13.7) showed a significant relationship ($r=0.72$, $p= 0.01$) between χ of monthly trapped sediment and rainfall over a year's monitoring period, indicating that higher stream discharge levels increased the amount (rate of transport) of silt reaching the lake bed. The silt encompasses the modal grain size ($10-40$ μm) of ferrimagnetic titanomagnetite derived from basaltic soils. Thus it seems likely that in lake catchments where soil enhancement is minimal, mineral magnetic properties of cores may provide a rapid qualitative means of assessing hydrological changes induced by climate or man.

Some recent studies (Banerjee *et al.* 1981, King *et al.* 1982) have used mineral magnetic measurements to infer changes in the *grain size* of ferrimagnetic minerals. Grain size is rarely synonymous with particle size and cannot form the basis for an interpretation of hydrological processes (cf. Banerjee *et al.* 1981), which by and large act on specific fractions of minerogenic material.

Sediment source tracing: mineralogical comparisons
In catchments where the three-dimensional variability in mineral magnetism is great, a switch in sediment source may override a particle size control on sedi-ment properties. Consequently, several attempts have been made to construct soil—sediment linkages by comparing, usually in scatter diagrams, values of χ, SIRM and demagnetisation parameters for a large number of samples; the assumption being that different sediment sources have unique assemblages of magnetic minerals. Oldfield (1983) showed how an 'envelope' comparison of S/SIRM plots helped to distinguish between topsoil and eroding cliff sources for deposited sediment in the Rhode River estuary, Maryland (Fig. 13.8). Dating of central cores indicated that a shift to topsoil erosion occurred in the mid-19th century, a period correlating with pollen analytical evidence for an intensifica-tion of agriculture. Successful results such as these often depend on the catch-

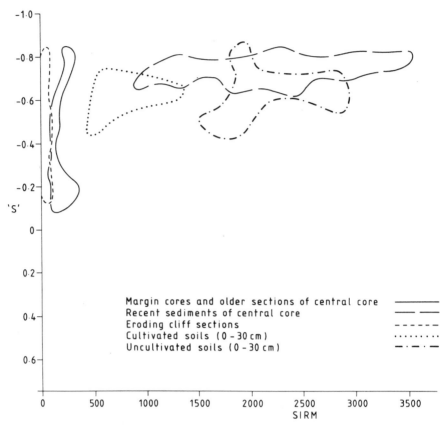

Figure 13.8 Magnetic characterisation of Rhode River, USA, sediments and their potential sources on the basis of S (IRM$_{-0.1T}$/SIRM) and SIRM (after Oldfield 1983).

ment displaying a range of lithologies or magnetic soil horizonations, and failure to identify the suitability of a catchment for this approach can produce fortuitous or misleading soil–sediment linkages. For instance, Stober and Thompson (1979) compared plots of SIRM/χ from stream sediments and soils in several Finnish catchments (Fig. 13.9). The lack of overlap between the 'envelopes', especially between the lake sediments and the other sets, indicated that the soils did not constitute the major source of eroded material. However, the uniform nature of the soils and underlying deposits in all the catchments suggests that the dominant control on lake sediment magnetic properties is particle size variation within the <32 μm fractions rather than the magnetic mineralogy.

There is some evidence to suggest that in oxidising lake sediments, precipitated paramagnetic iron oxides (e.g. ferrihydrite, lepidocrocite) reflect the occurrence and intensity of iron mobilisation in the catchment soils. The oxides are characterised by relatively high χ and extremely low (theoretically zero) SIRM values (cf. Bloemendal 1982, Edwards 1978) but their identification is problematical if they coexist with significant concentrations of ferrimagnetic minerals (primary

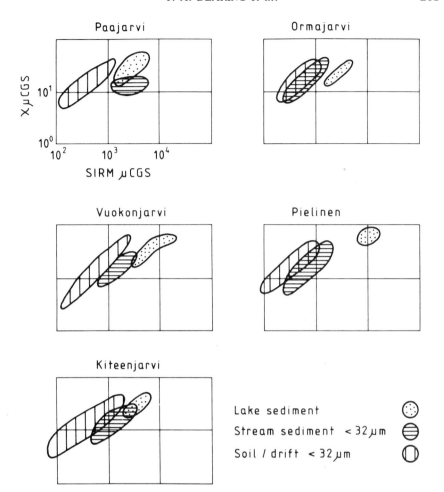

Figure 13.9 Magnetic characterisation of lake sediment, stream sediments and soils from five Finnish lake catchments on the basis of χ and SIRM (note CGS units) (after Stober & Thompson 1979).

or secondary). Hence, optimum sites for observing past leaching phases are normally upland, with poorly enhanced catchment soils and weakly ferrimagnetic substrates.

An interpretative model for susceptibility-based linkages

The recent development of relatively economic and portable magnetometers and susceptibility meters now means that mineral magnetic studies can be undertaken without recourse to established geophysical facilities. With this in mind, Figure 13.10 has been compiled to provide a simple model for the interpretation of susceptibility-based linkages between soils and sediments. It provides a review

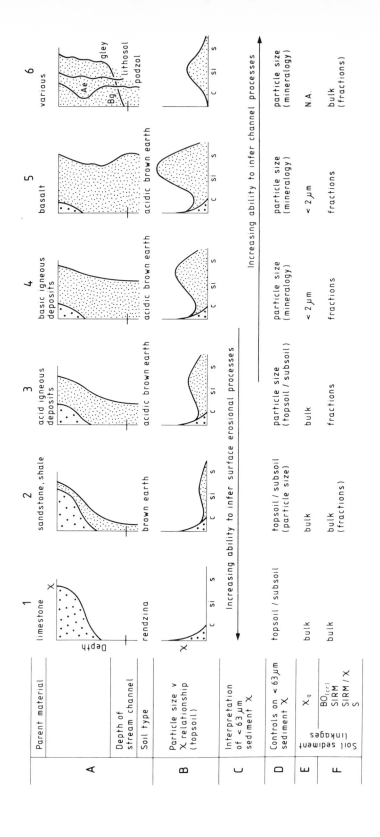

Figure 13.10 A model for the interpretation of χ in soil–sediment linkage studies. (A) Generalised χ profiles for relatively fertile and deep soils developed in a range of parent materials (1–5), and typical profiles for podzols, lithosols and gleys (6). Also shown is the likely proportion of bulk χ attributable to secondary minerals (large dots), and the depth to which fluvial processes may operate. (B) Typical χ/particle size relationships for the fraction ranges, clay (c), silt (si) and sand (s), showing the fraction in which secondary minerals (large dots) will dominate, where present. (C) A guide to the type of sediment transport process that can be inferred from bulk sample (fine) χ measurements of stream or lake sediment derived from soil types 1–6. (D) A summary of the controlling factors on <63 μm χ values for sediment derived from soil types 1–6. Topsoil/subsoil, particle size and mineralogy relate to sections in the text; minor or possible alternative controls are shown in brackets. (E) The basis for extending soil–sediment linkages by quadrature susceptibility (χ$_q$) measurements: bulk or <2 μm fraction based comparisons, except category 6 where the measurements are not applicable. (F) As for (E) in relation to (Bo)cr, SIRM, SIRM/χ, S and other demagnetisation parameters; bulk or fraction based comparisons.

Where sediment sources are diverse, the model cannot be used directly, and selection of 'mean' or 'dominant' soil type categories may depend upon field observation and evaluation of likely sediment sources. In such cases, magnetic mineralogical comparisons (see text) will maximise the chance of constructing linkages. Example of use: the soils in an upland lake catchment are developed in acid igneous glacial deposits of variable thickness and texture, and range from acid brown earths to peaty gleys; bulk enhancement occurs in some valley floor soils under permanent pasture (cf. category 3). In this case, χ of bulk fine stream sediment and lake sediment will largely be controlled by particle size (D3, D6), and interpretation of 'channel processes' should proceed on the basis of empirically derived χ/particle size relationships for different soil types. With the large extent of gleys, it may be possible to 'fingerprint' potential sediment sources with other magnetic measurements expressed on a bulk basis (see E/F6 and E/F3), though fraction comparisons would be preferable. Tracing of enhanced topsoil by χ$_q$ is unlikely, except perhaps following cultivation of the valley soils, and then only on <2 μm lake sediments or <2 μm stream sediments sampled in the lower reaches.

of soil χ profiles on different substrates, showing the likely proportion of secondary minerals present and χ/particle size relationships. It summarises the different controls on sediment χ in terms of the categories described in the text, and suggests additional magnetic measurements in terms of their expression on either bulk or fraction bases. Three general points are worth noting. First, the model is derived from empirical data obtained in temperate environments and might well need modification (especially in terms of the vertical extent of secondary magnetic mineral formation) in Mediterranean and tropical environments. Second, most catchments comprise more than one soil category type, thus making the model schematic, and restricting its use to that of a guide rather than a key to interpretation. Third, χ measurements may not be possible on material which is low in ferrimagnetic minerals or on very small samples (e.g. suspended sediment on filter paper), in which case SIRM measurements may be the only means of characterising magnetic mineral concentrations. Where SIRM is used alone, care should be taken to evaluate magnetic mineral grain size, in addition to particle size, as a control on bulk sample values (cf. Fig. 13.1).

Other applications

Finally, mention should be made of further applications. First, χ and SIRM have been used to correlate synchronous levels between lake sediment cores in order to build up a three-dimensional picture of sediment thicknesses from which historical records of sediment influx can be calculated (Bloemendal *et al.* 1979, Dearing *et al.* 1981, Oldfield *et al.* in press, Dearing 1982, 1983, Dearing *et al.* 1982, Foster *et al.* 1985). Where particle size exerts a control on bulk magnetic properties, care should be taken to ensure that all core material has a similar particle size distribution. Otherwise, an influx event which leads to the deposition of coarse sand in a marginal lake zone, and silt and clay in a central lake zone will give rise to a synchronous sediment layer characterised by both trough and peak values in the measured magnetic parameter. Mismatching of downcore variations is then a serious possibility in the absence of independent tests for synchroneity, such as microfossil analyses or absolute dating. Secondly, the use of naturally enhanced topsoil as a sediment tracer has recently been extended to bedload tracing by means of artificial enhancement caused by laboratory heat treatment. In a study in the upper reaches of the Rivers Severn and Wye in mid-Wales, enhanced bedload (defined by increases in χ and SIRM) was produced by growing magnetite and haematite in naturally paramagnetic mudstones and slates by rapid heating in air to 900°C, followed by quenching (Oldfield *et al.* 1981). Tracing the movements of different particle sizes enhanced in this way is possible through either single sample measurements of trapped sediment or magnetic search loop and ferrite probe scans of shoaled material (Arkell *et al.* 1983).

Acknowledgements

B. A. M. wishes to acknowledge financial support from the Natural Environment Research Council, through the tenure of a studentship. The authors thank Dr K. Edwards, Dr R. Bradshaw, Dr G. Turner and Mr G. Yates for allowing the use of their unpublished data; Dr R. Thompson for discussions and permission to reproduce Figures 13.7 and 13.9; and Mrs S. Appleton for cartographic assistance.

References

Aristovskaya, T. V. 1974. Role of microorganisms in iron mobilization and stabilisation in soils. *Geoderma* **12**, 145−50.

Arkell, B., G. Leeks, M. Newson and F. Oldfield 1983. Trapping and tracing: some recent observations of supply and transport of coarse sediment from upland Wales. *Spec. Publns. Int. Assoc. Sediment* **6**, 107−19.

Banerjee, S. K., J. King and J. Marvin 1981. A rapid method for magnetic granulometry with applications to environmental studies. *Geophys. Res. Lett.* **8**, 333−6.

Björck, S., J. A. Dearing and A. Jonsson 1982. Magnetic susceptibility of late Weichselian deposits in southeastern Sweden. *Boreas* **11**, 99−111.

Bloemendal, J. 1982. *The quantification of rates of total sediment influx to Llyn Goddionduon, Gwynedd.* Unpublished Ph.D. Thesis, University of Liverpool.

Bloemendal, J., F. Oldfield and R. Thompson 1979. Magnetic measurements used to assess sediment influx at Llyn Goddionduon. *Nature* **280**, 50−3.

Dearing, J. A. 1979. *The application of magnetic measurements to studies of particulate flux in lake-watershed ecosystems.* Unpublished Ph.D. Thesis. University of Liverpool.

Dearing, J. A. 1982. Core correlation and total sediment influx. In *Palaeohydrological changes in the temperate zone in the last 15 000 years. IGCP Project 158B. Lake and mire environments*, B. E. Berglund (ed.), Vol. 3, 1−23. Lund.

Dearing, J. A. 1983. Changing patterns of sediment accumulation in a small Lake in Scania, southern Sweden. *Hydrobiologia* **103**, 59−64.

Dearing, J. A. and R. Flower 1982. The magnetic susceptibility of sedimenting material trapped in Lough Neagh, and its erosional significance. *Limnol. Oceanogr.* **27**, 969−75.

Dearing, J. A., J. K. Elner and C. M. Happey-Wood 1981. Recent sediment flux and erosional processes in a Welsh upland lake-catchment based on magnetic susceptibility measurements. *Quat. Res.* **16**, 356−72.

Dearing, J. A., I. D. L. Foster and A. D. Simpson 1982. Timescales of denudation: the lake drainage basin approach. In *Recent developments in the explanation and prediction of erosion and sediment yield*, Proc. Exeter Symp. 351−60. IAHS Publn No.137.

Dickson, J. H., D. A. Stewart, R. Thompson, G. Turner, M. S. Baxter, N. D. Drndarsky and J. Rose 1978. Palynology, palaeomagnetism and radiometric dating of Flandrian marine and freshwater sediments of Loch Lomond. *Nature* **274**, 548−53.

Edwards, K. J. 1978. *Palaeoenvironmental and archaeological investigations in the Howe of Cromar, Grampian Region, Scotland.* Unpublished Ph.D. thesis, University of Aberdeen.

Flower, R., R. Nawas and J. A. Dearing 1984. Sediment supply and accumulation in a small Moroccan lake; a historical perspective. *Hydrobiologia* **112**, 81−92.

Foster, I. D. L., J. A. Dearing, A. D. Simpson and P. G. Appleby 1985. Estimates of contemporary and historical sediment yields in the Merevale catchment, Warwickshire, UK. *Earth Surf. Proc. Landforms* **10**, 45−68.

Harvey, A. M., F. Oldfield, A. F. Baron and G. W. Pearson 1981. Dating of Post-glacial landforms in the central Howgills. *Earth Surf. Proc. Landforms* **6**, 401−2.

King, J., S. K. Banerjee, J. Marvin and O. Ozdemir 1982. A comparison of different magnetic methods of determining the relative grain size of magnetite in natural materials: some results from lake sediments. *Earth Planet. Sci. Lett.* **59**, 404–19.

Le Borgne, E. 1955. Susceptibilité magnetique abnormal du sol superficiel. *Ann. Geophys.* **11**, 399–419.

Longworth, G., L. W. Becker, R. Thompson, F. Oldfield, J. A. Dearing and T. A. Rummery 1979. Mossbauer effect and magnetic studies of secondary iron oxides in soils. *J. Soil Sci.* **30**, 93–110.

Mullins, C. E. 1977. Magnetic susceptibility of the soil and its significance in soil science − a review. *J. Soil Sci.* **28**, 223–46.

Neumeister, H. and G. Peschel 1968. Magnetic susceptibility of soils and Pleistocene sediments in the neighbourhood of Leipzig. *Albrech Thaer Arch.* **12**, 1055–72.

Oldfield, F. 1983. The role of magnetic studies in palaeohydrology. In *Background to Palaeohydrology*, K. J. Gregory (ed.), 141–65. Chichester: Wiley.

Oldfield, F., P. G. Appleby and R. Thompson 1980. Palaeoecological studies of lakes in the highlands of Papua New Guinea I. The chronology of sedimentation. *J. Ecol.* **68**, 457–77.

Oldfield, F., P. G. Appleby and A. F. Worsley in press. Evidence of lake sediments for recent erosion rates in the highlands of Papua New Guinea. *Earth Surf. Proc. Landforms.*

Oldfield, F., T. A. Rummery, R. Thompson and D. E. Walling 1979. Identification of suspended sediment sources by means of magnetic measurements: some preliminary results. *Water Res. Res.* **15**, 211–18.

Oldfield, F., R. Thompson and D. P. E. Dickson 1981. Artificial magnetic enhancement of stream bedload: a hydrological application of superparamagnetism. *Phys. Earth Planet. Int.* **26**, 107–24.

Poutiers, J. 1975. *Sur les proprietes magnetiques de certain sediments continentals et marins; application.* These de Doctorat Université de Bordeaux.

Rummery, T. A., J. Bloemendal, J. A. Dearing, F. Oldfield and R. Thompson 1979. The persistence of fire-induced magnetic oxides in soils and lake sediments. *Ann. Geophys,* **35**, 103–7.

Schwertmann, U. and N. Kampf in press. Young iron oxides in Brasilian soil environments. *Geoderma.*

Schwertmann, U. and R. M. Taylor 1977. Iron oxides. In *Minerals in soil environments*, J. B. Dixon (ed.), 145–80. Madison: Soil Science Society of America.

Scoullos, M., F. Oldfield and R. Thompson 1979. Magnetic monitoring of marine particulate pollution in the Elefsis Gulf, Greece. *Mar. Pollution Bull.* **10**, 288–90.

Stober, J. C. and R. Thompson 1977. Palaeomagnetic secular variation studies in Finnish lake sediment and the carriers of remanence. *Earth Planet. Sci. Lett.* **371**, 139–49.

Stober, J. A. and R. Thompson 1979. An investigation into the source of magnetic minerals in some Finnish lake sediments, *Earth Planet. Sci. Lett.* **45**, 464–74.

Thompson, R. and D. J. Morton 1979. Magnetic susceptibility and particle-size distribution in recent sediments of the Loch Lomond drainage basin, Scotland. *J. Sed. Petrol* **49**, 801–12.

Thompson, R., R. W. Battarbee, P. E. O'Sullivan and F. Oldfield 1975. Magnetic susceptibility of lake sediments. *Limnol. Oceanogr.* **20**, 687–98.

Tite, M. S. and C. Mullins 1971. Enhancement of the magnetic susceptibility of soils on archaeological sites. *Archaeometry* **13**, 209–19.

Turner, G. M. 1979. *Geomagnetic investigations of some recent British sediments.* Unpublished Ph.D. thesis, University of Edinburgh.

Vadyunina, A. F. and V. F. Babanin 1972. Magnetic susceptibility of some soils in the USSR. *Sov. Soil Sci.* **4**, 588–99.

Vadyunina, A., V. F. Babanin and V. YA. Kovtun 1974, Magnetic susceptibility of the separates of some soils. *Soviet Soil Sci.* **6**, 106–10.

Walling, D. E., M. R. Peart, F. Oldfield and R. Thompson 1979. Suspended sediment sources identified by magnetic measurements. *Nature* **281**, 110–13.

Yates, G. 1983. *Magnetic mineral distributions in soils and their relationship to pedogenic processes.* Unpublished B. Sc. dissertation, University of Liverpool.

Part IV

SOILS AND DATING

14

Radiocarbon dating of surface and buried soils: principles, problems and prospects

J. A. Matthews

Introduction

The reliability of radiocarbon dates from soils is a topic of continuing controversy (Geyh *et al.* 1983, Matthews & Dresser 1983). Indeed, soil organic matter has for long been considered an inferior if not unsuitable material for ^{14}C assay. Nevertheless, the ubiquity of soils in the landscape, together with their importance as relict and buried palaeosols, suggest the value of the ^{14}C dating of soils as a geomorphological technique. The dating of soils has immense potential for the provision of an absolute timescale for landform development, for the rates of operation of earth-surface processes, and for the reconstruction of environmental change.

Many applications have been attempted, some of which will be referred to in this review, but full realisation of the potential requires a deeper understanding of the nature and causes of ^{14}C ages within surface and buried soils. There is a lack of detailed information about the variations to be expected and the factors controlling them. Consequently it is difficult to judge the accuracy of a particular date in its geomorphological context. By concentrating on these main areas of uncertainty, within which recent work has made some significant advances, this review should assist in the interpretation of existing ^{14}C dates from soils as well as suggest some further avenues for research into the technique and its applications.

Principles

Preferred materials for ^{14}C dating, such as charcoal, wood, peat and bone, are the remains of organisms that have suffered relatively little alteration since death in an environment conducive to their *in situ* preservation. This does not apply to soil organic matter, however, in which physically and chemically transformed plant and animal remains of many individuals that died at different times are poorly and differentially preserved.

Apparent mean residence time and soil age

Soil organic matter has been subjected to varying degrees of decomposition, humification and translocation, processes that do not necessarily reach an equilibrium in surface soil profiles; neither do they necessarily cease following burial. In a surface soil the net result of these processes is the production of soil organic matter of mixed age, continually rejuvenated by the addition of fresh plant remains from above. Thus any surface soil has a radiocarbon age which is a *relative age* (Gerasimov 1971) or an 'equivalent' age (Jenkinson 1969). Campbell *et al.* (1967) termed this age the *apparent mean residence time* (AMRT) of the soil organic matter (cf. Geyh *et al.* 1971, Scharpenseel & Schiffmann 1977a, Stout *et al.* 1981).

The *absolute age* of a surface soil, defined as the period of time since the beginning of soil formation at the site, will always be older than the AMRT of the uncontaminated soil organic matter within it. A ^{14}C date from a surface soil is therefore a *minimum* estimate of the absolute soil age and can be only a minimum age for geomorphological events contemporaneous with the initiation of soil development. In some contexts close minimum ages may be indicated but where organic matter turnover rates are high, and consequently AMRT is low, any estimate of absolute soil age may be extremely inaccurate.

Following burial of a soil, rejuvenation of the soil organic matter is interrupted. If soil burial terminates rejuvenation completely (true fossilisation), then a ^{14}C age determination from the buried soil reflects, additively, the AMRT of the soil prior to burial and the time elapsed since then (Scharpenseel & Schiffmann 1977b). Whilst retaining the quality of a minimum age for the initiation of soil development at the site, a ^{14}C date from a buried soil provides, therefore, a *maximum* estimate of the period of time for which the soil has been buried. The closeness of the estimate depends on the magnitude of the AMRT factor, because:

$$\text{Time elapsed since burial} = {}^{14}\text{C age} - \text{AMRT}$$

Interpretation will be complicated, however, if organic matter rejuvenation is able to continue after burial. Nevertheless, the prospects for this type of absolute age estimation, which may be directly applicable to contemporaneous geomorphological events, seem to be generally better than for the type of estimation associated with the initiation of soil formation.

Fractionation of soil organic matter

From the first, the pretreatment of soil samples for ^{14}C dating has attempted to separate soil organic fractions of different ages. In its simplest form this might involve the hand picking of roots and rootlets from the sample. Much more sophisticated chemical procedures may be employed, however (e.g. Goh & Molloy 1978, Scharpenseel 1979, Gilet-Blein et al. 1980, Kigoshi et al. 1980). Although these attempts have had considerable implications for the isolation of contaminants in both surface and buried soils, they have the additional and more fundamental aim of achieving closer minimum or maximum age estimates. The general principles outlined above apply also to any organic matter fraction within the soil. Thus the ^{14}C age of the *oldest* uncontaminated organic fraction in a surface (or buried) soil provides the closest approximation to the date at which soil formation was initiated at the site; the age of the *youngest* uncontaminated fraction in a buried soil provides the best estimate of the time elapsed since burial.

Organic materials found in soils may be classified into four major components: (a) living plant roots, soil animals and micro-organisms; (b) dead and partly decomposed biomass; (c) relatively simple organic compounds released during decomposition, such as carbohydrates, cellulose and lignin; and (d) humus (or humic) substances, which tend to be large acidic molecules with a complex chemical structure and a high molecular weight, formed during the process of humification by biochemical synthesis from the products of decomposition (Flaig et al. 1975, Haider et al. 1975). The humus substances form complexes between themselves and with clays and metal ions (Schnitzer & Khan 1972, Tate & Theng 1980). While the dead biomass usually decomposes rapidly and most of the simpler organic molecules have very short half lives of less than a few years, in many surface soils much of the humus complex persists for hundreds or thousands of years (Paul 1969, Sauerbeck & Johnen 1973, Goh 1980).

The classical fractionation procedure for humus substances, in which the first stage involves separation of an alkali-insoluble fraction (humin), has been widely used as a basis for the pretreatment of soil samples. This approach tends to produce fractions of increasing complexity in the order fulvic acid → hymatomelanic acid → brown humic acid → grey humic acid → humin (Kononova 1961, Goh 1980). Some applications of ^{14}C dating to soil organic fractions separated into these components found a progressive increase in age in the same sequence, which is generally explained in terms of the synthesis of each fraction from the preceding one during humification with the subsequent persistence of the more complex fractions owing to immobility and/or biochemical recalcitrance. Not all results produced the same age sequence, however (Scharpenseel 1971b, Rafter et al. 1972). Studies of organic matter labelled experimentally or by nuclear bomb ^{14}C enrichment (Nakhla & Delibrias 1967, Jenkinson 1971, Goh et al. 1976) cast further doubt on the ability of the classical fractionation procedure to produce age-differentiated fractions in a consistent age sequence.

Repeated acid hydrolysis, most notably using concentrated HCl, has also been found to produce age-differentiated fractions in some types of soil, the residual fraction after each step tending to become progressively older. This pretreatment has been preferred in many studies where the aim has been to isolate relatively old organic material that provides an approximation of the absolute age of a surface or buried soil (Scharpenseel *et al.* 1968, Scharpenseel 1977, Stout *et al.* 1981). The explanation is again that relatively complex, immobile, resistant and old humus substances require a more prolonged pretreatment for extraction. A combination of the classical and acid hydrolysis approaches, in which an acid hydrolysis precedes an alkaline extraction, has been used by Matthews (1980, 1981) and by Matthews and Dresser (1983) in an attempt to retain a degree of comparability with the traditional fractions while adopting the logic of employing less severe chemical treatments before more severe ones.

There remains, however, a lack of agreement on the best approach to soil organic matter fractionation which is symptomatic of our imperfect understanding of the nature of the organic substances involved. All the various attempts at fractionation have isolated complex organic mixtures rather than specific chemicals, and the contents of the fractions vary depending on such factors as the type of soil and the pH, concentration, temperature and duration of the treatment. Clearly, therefore, although progress has been made in separating age-differentiated soil organic fractions, these are arbitrary. There is as yet no fractionation scheme that either has been found empirically to produce consistent age-differentiation in all soil types or can be fully justified theoretically.

Problems

The complex nature of soil organic matter, imperfect knowledge of its components, and sub-optimal fractionation procedures, form the first and foremost set of problems which pervade all others in the ^{14}C dating of soils. Other problems may influence surface or buried soils or both.

Stage of soil development and state of soil equilibria
Initiation of soil development at a site results in the establishment and gradual differentiation of a soil profile, one aspect of which is the accumulation of soil organic matter. As a result of soil development, the ^{14}C age of the organic matter will increase through time. In theory a steady state equilibrium may be reached after which time the ^{14}C age of the organic matter will remain relatively constant provided external environmental factors do not change (Nikiforoff 1959, Lavkulich 1969). Many studies have obtained relatively young ^{14}C dates from soils that are known, on the basis of independent evidence, to have had a longer history of development (Guillet & Robin 1972, Beckmann & Hubble 1974). In these cases the ^{14}C dates have been interpreted as reflecting high rates of rejuvenation with rapid turnover of carbon within the soil, and equilibrium concepts may be applicable.

In many other cases, however, particularly where ^{14}C ages of many thousands of years have been obtained (Gerasimov & Chichagova 1971, Scharpenseel 1971a), the time that would be necessary for establishment of a steady state equilibrium is so long as to render the concept of little or no relevance. Furthermore, the suggestion that some organic matter components may be almost inert, such as biogenic opal (Wilding 1967), some microbial enzymes (Skujins 1976), some humus substances (Gerasimov 1974, Scharpenseel 1977) and charcoal (Goh & Molloy 1979), is a notion completely at variance with the concept of an equilibrium state in the sense described above.

Soils at an early stage of development (immature soils) are likely to yield relatively close estimates of both absolute soil age and the time elapsed since burial. Equilibrium soils with high rates of rejuvenation and hence a low AMRT are the least likely to yield ^{14}C ages that are close approximations to absolute soil age; equilibrium soils with a low rejuvenation rate and a high AMRT (or soils with inert fractions) are correspondingly unlikely to result in close minimum estimates of the time elapsed since burial.

Soil type and environment

AMRT of various organic matter fractions from surface soils has been found to vary greatly according to soil type (Herrara & Tamers 1971, Scharpenseel 1975, Bottner & Peyronel 1977). This was recognised early on, using the classical fractionation approach, and has since been confirmed using other pretreatments. Thus dates older than 6000 ^{14}C years BP have been obtained from chernozems (Ganzhara 1974, Rubilin & Kozyreva 1974) while podzols have generally yielded ages younger than 3000 ^{14}C years (Tamm & Holmen 1967, Hassko *et al.* 1974). It is also possible to explain some apparent discrepancies in the age sequence of fractions by reference to the kind of soil involved.

Some of these variations appear to be related to climate, drainage and vegetation, both between and within soil types. In hydromorphic soils, for example, residual fractions (irrespective of the precise pretreatment) would be likely to contain a high proportion of relatively undecomposed (young) plant remains as well as relatively old humus substances. Arctic-alpine podzols, where decomposition and other soil-forming processes are slow, may yield considerably older dates than podzols from lower altitudes (Ellis & Matthews 1984). Two podzols, located only 300 m apart, the one under climax forest and the other under secondary *Calluna* heathland, yielded ages of 180 ± 50 and 2100 ± 50 ^{14}C years respectively (Guillet & Robin 1972). This was attributed to vigorous rejuvenation of soil organic matter under the forest and the production of relatively resistant humus substances from *Calluna* litter under the heathland. Local and regional differences in the AMRT of particular fractions may vary, therefore, in response to environmental differences and should not be assumed constant even within the same soil type.

Age−depth gradients within soil profiles

Many surface soils have been shown to increase in age with depth. A large number of ^{14}C determinations on depth-controlled samples from five soil orders: spodosols (including podzols), Udolls (including chernozems), Udalfs, vertisols and plaggen soils were analysed by Scharpenseel (1972, 1975). Udolls ($n = 122$, $r^2 = 0.79$), Udalfs ($n = 86$, $r^2 = 0.55$) and vertisols ($n = 271$, $r^2 = 0.60$) showed a strong positive correlation between age and depth, while the relationships for spodosols ($n = 32$, $r^2 = 0.11$) and plaggen soils ($n = 34$, $r^2 = 0.04$) showed no significant correlation. Significant age–depth gradients were considered to form in those soils where vertical translocation and mixing processes were poorly developed. Strong translocation of carbon down the spodosol profiles, and deep mixing in the plaggen soils, were thought responsible for the absence of age–depth gradients in those soil orders.

The increase in age with depth found in chernozem profiles from the USSR was attributed by Ganzhara (1974) to a decrease in the rate of supply of fresh litter and a decrease in the level of microbial activity with depth in the soil. In a meadow chernozem profile, Rubilin and Kozyreva (1974) obtained youngest dates, not at the surface but at 10–20 cm depth, which they explained as the horizon of maximum concentration of living plant roots. It would appear that rejuvenation of the soil organic matter decreases with depth in most surface soils and that immobile and resistant fractions are most likely to persist at depth. Podzols appear to be an important exception, however, in that the organic matter in B (illuvial) horizons may not be older than similar material in A horizons (Tamm & Holmen 1967, Cruickshank & Cruickshank 1981, Ellis & Matthews 1984).

The existence of strong age–depth gradients in individual profiles is of great importance in the ^{14}C dating of soils. For surface or buried soils with steep age–depth gradients, close minimum estimates of absolute soil age are only possible if samples are available from the *deepest* levels of the soil profile; the oldest soil organic matter is most likely to be found there. Similarly, the date of burial of a palaeosol is most closely approximated by samples taken from as close as possible to the former soil *surface*, where the youngest soil organic matter is located.

Age–depth gradients within soil horizons

Most ^{14}C dates from soils have been based on bulk samples, because of the relatively low soil organic content. Almost invariably, therefore, the possibility of investigating any age–depth relationship within individual horizons has been excluded, an exception being the thick organic-rich surface horizons of chernozems (see above) and podzols.

By sampling at two depths, steep age–depth gradients of about 600 ^{14}C years cm^{-1} were revealed in the 5 cm thick FH horizon (the uppermost horizon) of a humo-ferric podzol buried beneath an end moraine near the Haugabreen glacier, Norway (Matthews 1980). Subsequently, more intensive sampling of 1.0 and 0.5 cm thick slices from the same horizon, where it was 14.5 cm thick, showed near-surface ages as young as 485 ± 60 ^{14}C years, basal ages up to 4020 ± 70 ^{14}C years, and an age reversal attributed to disruption or contamination in the top 1.0 cm (Fig. 14.1a) (Matthews & Dresser 1983). Eight 'humic acid' dates from depths of 1.0–14.5 cm revealed a strong *linear* age–depth gradient of 244 ^{14}C years cm^{-1}. This gradient becomes 290 calendar years cm^{-1} when using ages corrected according to the calibration of Clark (1975). Interpretation of the linear gradient as reflecting a relatively steady accumulation of organic matter since the initiation of the horizon in the mid-Holocene was supported by micro-

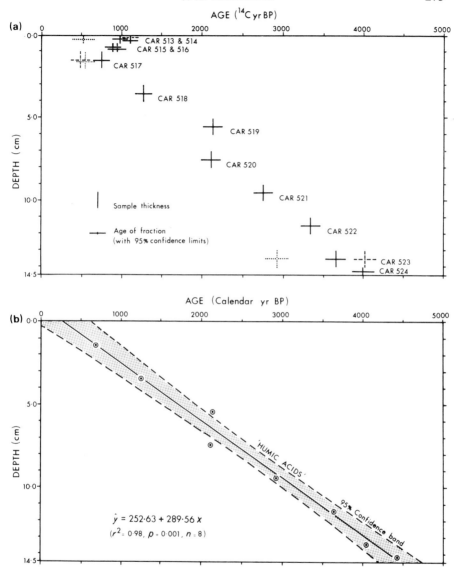

Figure 14.1   Intensive ^{14}C-dated surface organic (FH) horizon of a buried humo-ferric podzol at Haugabreen, Norway (Matthews & Dresser 1983). (a) ^{14}C dates from chemically fractionated samples: 'humic acids' (——); 'fulvic acids' (....); 'fine residue' (----). (b) The linear age−depth gradient based on 'humic acid' fractions at depths of 1.0−14.5 cm (age is the dependent variable, y).

morphological (Ellis & Matthews 1984) and palynological (Caseldine 1983) analyses. Extrapolation of the regression line, shown in Figure 14.1b, gave a predicted age at the presumed surface of the buried horizon of 253 ± 365 calendar years BP (with 95% confidence interval). This prediction is in agreement with independent evidence of burial of the soil when the glacier was at its 'Little Ice Age' limit in the mid-18th century.

These results, from the most detailed ^{14}C-dating study yet attempted on a single horizon, point clearly to the dangers of bulk sampling of whole horizons in well-developed soils, which may result in dating errors of the order of thousands of years. Little, if any, correction for AMRT may be necessary, however, where the maximum estimate of time elapsed since burial is based on a statistically significant age−depth relationship. The extent to which other kinds of soil or other horizons exhibit similar steep age−depth gradients remains to be seen. Preliminary dating at two depths within the uppermost 5 cm of the A horizon of a buried arctic-alpine Brown Soil has suggested gradients of about 700 ^{14}C years cm^{-1} (Matthews 1981), demonstrating that within-horizon age−depth gradients are not confined to the organic-rich surface horizons of podzols and chernozems.

Horizontal (spatial) variation in soil age
The reproducibility of ^{14}C dates from soils has rarely been tested by dating more than one sample from the same soil horizon or from similar depths in adjacent profiles. Five samples from a podzolic B horizon from Les Landes du Médoc, France, were found to vary systematically in age from 770 ± 80 to 2810 ± 70 ^{14}C years over a total horizontal distance of about 12 m (Righi & Guillet 1977). These age differences were interpretable, however, in terms of a short catenary sequence (cf. Martel & Paul 1974). Cruickshank and Cruickshank (1981) found quite good agreement between equivalent B horizons of podzols located at two sites about 30 m apart at Breen Wood, Northern Ireland. The upper parts of the horizons at the two sites yielded ^{14}C ages of 1065 ± 40 and 1080 ± 40 years, while the lower parts yielded ^{14}C ages of 1200 ± 45 and 1085 ± 40 years, respectively. From a single face from the same soil pit in an end moraine in front of Styggedalsbreen, Norway, two immediately adjacent samples from a buried Brown Soil yielded 'humic acid' dates that were statistically indistinguishable from each other, but 'fulvic acid' fractions from the same samples differed in age by over 800 ^{14}C years (Griffey & Matthews 1978).

These and other data suggest that wide variations in ^{14}C age over short distances cannot be ruled out, particularly as the formation and preservation of surface and buried soils is dependent on local environmental conditions. Nevertheless, horizontal variations in ^{14}C age would appear to be a less serious problem for the dating of soils than vertical age−depth gradients.

Bioturbation, physical and anthropogenic mixing
Because of their number, size, burrowing and casting behaviour, earthworms are the soil fauna having the most significant pedological effects in temperate soils (Stout & Lee 1980). The mixing that results from the presence of earthworms will tend to homogenise the age of the soil organic matter. Although some species are found at different depths within the soil profile, they are in general most numerous in surface horizons and may therefore destroy or prevent the development of near-surface age−depth gradients. Thus ^{14}C dates from the upper levels of the soil profile may be made older than usual (Stout & Goh

1980), which would reduce the accuracy of maximum estimates of date of burial. Larger burrowing animals, ants, termites in tropical soils, and many other members of the soil fauna play a similar role (Hole 1981). By carrying young material down the soil profile, deep bioturbation may also influence minimum estimates of absolute soil age.

Physical mixing processes include tree fall and uprooting in forest soils, frost heave, frost cracking and solifluction in periglacial soils, and swelling and desiccation cracking due to the wetting and drying of semi-arid soils. However, ^{14}C-dated profiles from gilgai mounds and depressions in Australia (Blackburn et al. 1979) indicated that swelling of the soil due to alternate wetting and drying was insufficient to destroy ^{14}C age−depth gradients. Vertisols often exhibit steep ^{14}C age−depth gradients below the maximum depth of surface desiccation cracking (Scharpenseel 1972). Perhaps the greatest problems for ^{14}C dating arise under periglacial conditions where severe mixing processes may extend throughout the whole soil profile. Zoltai et al. (1978) showed that the ages of buried materials, including soil organic matter, under 26 hummocks in Arctic and sub-Arctic Canada were not clearly related to depth or position under the hummocks (see also Ellis 1983). Brown (1965, 1969) discussed some of the problems of interpretation of surface and buried soils and peats associated with a variety of landforms in permafrost areas of Alaska.

Various agricultural practices, particularly tillage and use of the moldboard plough, have been reported as producing 'ageing' or even inversion of the uppermost soil layers (Allison 1973, Rubilin & Kozyreva 1974). Not all such practices produce the same effects, however, as has been shown in relation to some Canadian chernozems. There, cultivation reduced the ^{14}C age because of additions of young plant debris to the upper soil horizons (Paul 1969).

Minerogenic carbon in soil and subsoil

Most rock types and the sediments derived from them contain carbon that for the purposes of ^{14}C assay is infinitely old (Sutherland 1980). The parent materials and mineral components of soils are therefore potential sources of an ageing effect. The ^{14}C-dating errors introduced by given levels of infinitely old carbon are precisely known (Olsson 1974) so that where soil organic matter content is very low, such as at depth in most surface soil profiles and in many palaeosols, an ageing effect must be considered a possibility (cf. Olsson 1979). Mechanisms exist in some circumstances for the concentration of minerogenic carbon in soils (e.g. Bowler & Polach 1971, Hallet 1976, Sutherland 1980). However, if mild pretreatment procedures are used to the extent that mineral components of the soil are not altered chemically, then minerogenic carbon is unlikely to be a serious problem in non-calcareous soils. When calcareous material is involved this should of course be removed by acid extraction at an early stage of the pretreatment (see section on fractionation above).

Surface additions of allochthonous carbon

Allochthonous material may be carried on to a soil surface by a wide variety of sub-aerial processes, such as colluvial, alluvial or aeolian transport (Ruellan 1971). This has been found a serious problem in the ^{14}C dating of sequences of A0/A1 soil horizons in the Holocene deposits of the north Netherlands coast (Schoute *et al.* 1981). During marine regression phases soils develop on the exposed clays, becoming buried during subsequent marine transgressions. Some horizons yielded ^{14}C ages too old when judged according to independent lithostratigraphic correlations. That this was due to allochthonous carbon in overlying sediments was substantiated by ages of up to 5910 ± 90 ^{14}C years obtained from the organic component of coastal muds being deposited at the present time at various locations in the Netherlands. Schoute *et al.* (1981) suggested that alkaline extracts, which represent humus substances, give better indications of the true date of formation of the soil horizons than residual fractions in these cases.

Similar effects may be found in other environments. For example, a date of 1120 ± 90 ^{14}C years BP was obtained from organic matter deposited, presumably by wind, in a semi-permanent snowbank in Norway (Østrem 1964). Surrounding soils were also probably subject to additions of comparably old carbon. Surface peats from two sites in upland south Wales were found to be anomalously old owing to the addition of atmospheric pollution in the form of coal dust (Chambers *et al.* 1979). This produced a ^{14}C age of 3710 ± 90 years for a residual fraction in a near-surface peat sample, the 'humic acids' of which gave a much more realistic age of 70 ± 55 ^{14}C years BP.

Contamination by 'bomb' carbon

Atmospheric ^{14}C levels increased after the mid-1950s as a result of the detonation of thermonuclear devices since 1953. Maximum levels were reached in 1964, since when there has been a steady fall (Nydal & Lovseth 1970, Levin *et al.* 1980). In the present context the most important consequence is the world-wide contamination of surface soils by 'bomb' carbon. This has in turn reduced the usefulness of ^{14}C assay on surface soils as modern analogues for the interpretation of buried soils.

The scale of the 'bomb' effect has been demonstrated by ^{14}C measurements on soils sampled in the field before and after the rise in atmospheric 'bomb' carbon, and by tracing the pulse of the contamination into the soil ecosystem. Despite adequate aeration, a near-neutral pH and a mild climate, soils collected from the unmanured plot of Broadbalk at the Rothamsted Experimental Station, southern England, in the 1880s yielded a strong increase in age with depth from 1385 ± 140 at a depth of 0−23 cm to about 12 100 ^{14}C years at 206−229 cm (Jenkinson 1969, 1975). Such ages for the topsoil were widespread, having been verified by further samples from experimental farms in different parts of England and under different agricultural practices (Jenkinson 1969, 1972, 1975). By 1969 the ^{14}C age of the uppermost 23 cm of the Broadbalk plot was less than half of that of the prebomb samples (Jenkinson 1975, Jenkinson & Rayner 1977). 'Bomb' carbon had, moreover, entered soil organic matter fractions separated by the classical fractionation (fulvic acids, humic acids and humin) as well as those separated by 6N HCl acid hydrolysis (hydrolysate and residue) (Jenkinson 1975).

Extensive studies on ^{14}C enrichment of New Zealand soils by 'bomb' carbon (Goh *et al.* 1977, O'Brian & Stout 1978) has shown that the level of enrichment and the depth to which it occurs vary with such factors as soil type, climate, vegetation and soil faunal activity. The upper horizon of all surface soils exposed to the atmosphere is almost certainly contaminated to a significant degree, leading to underestimates of natural ^{14}C age and AMRT. Although it would appear that 'bomb' carbon is unlikely to have been a major influence on either the age of the oldest fractions at greatest depths in soil profiles, or palaeosols buried beyond the depth to which surface roots or translocated organic substances penetrate, some effect cannot be ruled out even in these cases.

Root penetration and concentration, percolation of organic acids
Root penetration and decomposition is a normal process in surface soils. Once soil burial has occurred, however, roots that penetrate through the overburden must be regarded as contaminants. Roots may concentrate in buried soil horizons (Gould *et al.* 1979) and decompose there over a long period of time. Scharpenseel (1971a) has provided two good examples of the possible scale of the rejuvenation produced by roots where burial is sufficiently shallow. In the first example, ^{14}C dates were obtained from the A horizon of a parabraunerde developed in '*Würm*' loess. Where this palaeosol was buried by about 3 m of pumice a ^{14}C age of about 10 600 years was obtained; the same horizon yielded younger ages of about 5000 and 7000 ^{14}C years where the overburden was 1−2 m thick. The second example relates to a Rhodoxeralf in a semi-arid climate near Tel Aviv, Israel. Where covered by a sand dune to a depth of about 2 m the buried soil profile yielded ages ranging from 10 130 to 14 790 ^{14}C years; the same soil buried by less than 1 m of sand gave ages from 8340 to 11 860 ^{14}C years; where the soil was exposed at the surface the range of ages was 7330 − 10 470 ^{14}C years.

Schneebeli (1976, p. 29) presented a diagram suggesting that the influence of organic acids penetrating from above to soils buried beneath moraines in the Alps extends to a depth of about twice the depth of the root zone. The depth of both zones will vary greatly with local conditions, however, so that more precise values cannot, as yet, be given.

Continued decomposition and removal after burial
Although the addition of plant remains at the surface and within the root zone may have ceased, decomposition of soil organic matter may continue after burial provided that moisture and aeration conditions are suitable (Paul 1969). With reference to a Canadian chernozem with an age of 2000 ^{14}C years, Paul calculated that decomposition and removal of the young 'active' fraction after burial would increase the AMRT by 200−400 years. Little is known, however, about the extent to which continued decomposition actually occurs in palaeosols or the rate of any ageing effect in reality.

Micro-organisms, particularly bacteria and actinomycetes, have been found in abundance in buried organic horizons by Cook (1969) who considered that they would transform the organic matter slowly. However, micro-organisms that are normally found heavily concentrated in the upper horizons of surface soils, such as fungi, nematodes and protozoa, will probably not survive in buried soils (Cook 1969, Gould *et al.* 1979). Some anaerobic forms, on the other hand, may increase at least temporarily after burial, and continued microbial metabolism may result in isotopic fractionation (Stout *et al.* 1981). The fact that microbial enzyme activity has been detected in soil and peat samples dated to 8715 ± 250 and 9550 ± 240 ^{14}C years, which were frozen in permafrost (Skujins & McLaren 1968) suggests that some decomposition can also occur after the death of the organisms. Lastly, it should be pointed out that physicochemical decomposition may take place in buried soils devoid of micro-organisms by the protolytic action of water (Laura 1975).

In summary, it appears that the influence of decomposition after burial will probably be small where anaerobic conditions prevail, particularly because of very slow decomposition rates and ineffective mechanisms for the differential removal of older or younger fractions.

Soil erosion and disturbance prior to or during burial

Removal of organic matter from soils within drainage basins is commonly reflected in lake sediments (Oldfield 1977), which may show ^{14}C age discrepancies of thousands of years (Vuorela 1980). Erosion of surface soils may therefore be producing modifications to their ^{14}C age, particularly where mismanagement of land use is widespread. During soil burial, erosion, folding or inversion of soil horizons may occur, as has been described from excavations beneath end moraines in Norway (Griffey & Worsley 1978, Matthews 1980) and in the Alps (Schneebeli 1976, Röthlisberger & Schneebeli 1979). Such effects are also likely where mass wasting processes are involved, such as in the dating of soils buried by talus (Innes 1983) or solifluction processes (Benedict 1966, Harris & Worsley 1974, Ellis 1979). The progressive increase in ^{14}C age of buried soil horizons with distance from the front of a solifluction lobe or terrace has been interpreted in terms of variable rates of solifluction in North America (Benedict 1966, Alexander & Price 1980) and the Alps (Gamper 1982). However, an alternative explanation could be the differential erosion of the upper surface of the buried soil layer.

Some burial events are obviously more destructive than others and some sections of buried soils will be less well preserved than others. Unless soil sampling is made in relation to undisturbed profiles there may be severe problems for the ^{14}C dating of buried soils. This is particularly important if a soil with a steep age–depth gradient is truncated, a possibility that can be investigated by micromorphological and other pedological analyses (Griffey & Ellis 1979).

Polygenetic, relict and exhumed soils

The characteristics of many surface and buried soils reflect more than one period of soil formation; that is, they are polygenetic (Butler 1959) or polymorphic

(Simonson 1978). Some or all of the features of a surface soil, including its organic component, need not therefore be the product of present environmental conditions. Some soils are relict (Ruellan 1971), i.e. formed when environmental conditions were different from the present, but their effects nevertheless persist in the present-day soil; others represent former buried soils that have been exhumed (Ruhe & Daniels 1958) and now lie at the surface. Two or more periods of episodic soil formation may be clearly visible in the form of distinct horizons or soil structures (Crampton 1974). Serious consequences for the ^{14}C dating of soils are, however, most likely where the causal environmental change was gradual and if such features are not obvious, when there may be a failure to recognise them.

Prospects

Although the problems of the ^{14}C dating of soils often reach formidable proportions, there are strong grounds for cautious optimism in relation to some applications. Most of the reasons for this conclusion have been mentioned above in relation to the specific problems of the technique. It is suggested here that, provided certain criteria are satisfied, close estimates of either absolute soil age or the time elapsed since burial are possible. Many surface and buried soils, and most applications, do not and have not satisfied all of the relevant criteria, however. For *surface* soils, the prospects for making close *minimum* estimates of *absolute soil age* are greatest where:

(a)   the soil is at an early stage of development or, as an inferior alternative, if the age of the soil organic matter has reached a steady state equilibrium, the rejuvenation rate is low and AMRT is high;
(b)   relatively old or inert organic fractions can be isolated and used for dating (e.g. old humus substances or charcoal);
(c)   sampling is confined to the deepest zones of soil profiles with steep age—depth gradients;
(d)   the soil is deep in relation to root penetration and other translocation and mixing processes (including natural and anthropogenic turbation) that carry young surface material (including 'bomb' carbon) down the soil profile;
(e)   regional and local environmental conditions were conducive to the preservation of organic matter throughout the period of formation of the soil;
(f)   the development of the soil profile has been free of natural or anthropogenic allochthonous additions of carbon;
(g)   minerogenic carbon from the soil matrix has not become inextricably mixed with soil organic matter;
(h)   the soil organic matter is representative of an intact monogenetic soil;

For *buried* soils, close *minimum* estimates of *absolute soil age* are most likely where the following criteria apply, in addition to criteria (a) to (h) applying prior to burial:

(i)   burial effectively sealed the soil organic matter from subsequent con-tamination (particularly root penetration), decomposition and removal of organic matter;
(j)   disturbance during burial was minimal;

Close *maximum* estimates of the *time elapsed since burial* are most likely where:

(a)   at the time of burial the soil was at an early stage of development or, if the age of the soil organic matter had reached a steady state equilibrium, the rejuvenation rate was high and AMRT was low;
(b)   relatively young organic fractions can be isolated and used for dating (e.g. young humus substances or undecomposed plant remains from the former soil surface);
(c)   sampling is confined to the uppermost layer of the former soil surface in a buried profile with a steep age−depth gradient;
(d)   before the soil was buried, mixing processes that carried older organic material towards the soil surface were minimal;
(e)   regional and local environmental conditions were conducive to the pre-servation of organic matter in the last phase of soil formation;
(f)   natural or anthropogenic allochthonous additions of carbon did not ac-company the development of the soil profile (especially in the later stages);
(g)   minerogenic carbon from the soil matrix has not become inextricably mixed with the soil organic matter;
(h)   the soil organic matter is representative of an intact monogenetic soil.
(i)   burial effectively sealed the soil organic matter from subsequent contami-nation, decomposition and removal of organic matter;
(j)   disturbance during burial (particularly any erosion of the uppermost part of the soil profile) was minimal;

The older the absolute age of either a surface or a buried soil, and the greater the period of time for which a soil has been buried, the less likely it is that the relevant criteria will be satisfied or that it will be possible to perceive whether or not some of them apply. Where imperfect knowledge limits the application of the criteria listed above, the uncertainty may be reduced if ^{14}C dates can be shown to be reproducible. This applies both in the sense of different fractions showing similar ages and in terms of the replication of dates using more than one sample.

In some cases the possibility of deriving an absolute soil age can be ruled out altogether. This is especially true of surface soils that have reached a steady state equilibrium with their environment and where the rejuvenation rate is high. The AMRT of such soils may be very low in relation to the age of the surfaces on which they are found. This has been demonstrated with respect to some podzols (Guillet & Robin 1972) and krasnozems (Beckmann & Hubble 1974), the latter having yielded quite similar ^{14}C ages from volcanic surfaces of widely differing ages in Australia. A similar pessimistic conclusion was reached by Gilet-Blein et al. (1980) and by Geyh et al. (1983), who found differences of over 10 000 years between ^{14}C ages from buried soils and 'known' ages derived from lithostratigraphic, pedostratigraphic, archaeological and palynological data.

Other applications from the study of Quaternary buried palaeosols have nevertheless yielded close minimum absolute age estimates, the most optimistic conclusions being possible where the soils were in an early stage of development prior to burial, such as soils developed during short intervals of relatively mild climate in the Lateglacial (Konecka-Betly 1977, Krajewski 1977, Heine 1983). Close estimates can also be contemplated where relatively resistant or inert fractions persist in the soil. Perhaps the most significant advances in the search for inert fractions in soil have been in the use of carbonised wood (Polach & Costin 1971) and charcoal (Goh & Molloy 1979). There are, unfortunately, some major assumptions in the use of these materials for the estimation of absolute soil ages (Stout et al. 1981): (a) that the wood preserved as charcoal derived from vegetation synchronous with the first stage of soil development; (b) that the sample was not transported to its stratigraphic position in the soil; (c) that soluble contaminants of differing age have not been absorbed, and (d) that the relationship of the sample to tree age can be established.

Soils in an early stage of development probably also hold the greatest potential for obtaining accurate estimates of the time elapsed since burial of a soil and, moreover, such estimates appear to be generally better founded than estimates of absolute soil age. Soils in an early stage of development found in lateral moraines built up by intermittent superimposition or accretion have made a major contribution to the establishment of an intricate glacier variation chronology for the Alps (Röthlisberger & Schneebeli 1979, Röthlisberger et al. 1980). Furthermore, detailed ^{14}C dating of thin soil slices in the surface organic horizon of a podzolic soil buried about 230 years ago (Matthews & Dresser 1983, Matthews 1984, see Fig. 14.1) suggests considerable possibilities for estimating the time elapsed since burial from age−depth gradients in thicker soils that have developed over a much longer period of time than those dated in the Alps. Although Figure 14.1 shows that steep age−depth gradients in the buried surface horizon could result in errors of thousands of years, it also demonstrates the level of accuracy that may be feasible given sufficiently well-preserved soils with careful field sampling and sophisticated pretreatment procedures.

In order to increase the prospects of the technique, much fundamental research is required into the problems that have been outlined in this review. There will always be problems, associated in particular with the nature of soil inorganic matter, its inherent variability, its persistence and its susceptibility to contamination. Nevertheless, in geomorphology there is a wide spectrum of possible applications ranging from historical reconstructions with an emphasis on description and form, to highly analytical work requiring precise knowledge of mechanisms and processes. Much of this work would benefit immensely from a more accurate chronological control and an extended timescale, to which soil dating can contribute. Furthermore, as many of the potential uses of the ^{14}C dating of soils in geomorphology involve soils in the early stages of formation and also soils that were buried relatively recently, the subject is well placed to take advantage of present knowledge and future advances.

Acknowledgements

I thank Dr P. Quentin Dresser and Dr John L. Innes for their comments on the manuscript.

References

Alexander, C. S. and L. W. Price 1980. Radiocarbon dating of the rate of movement of two solifluction lobes in the Ruby Range, Yukon Territory. *Quat. Res.* **13**, 365–79.

Allison, F. E. 1973. *Soil organic matter and its role in crop production.* Amsterdam: Elsevier.

Beckmann, G. G. and G. D. Hubble 1974. The significance of radiocarbon measurements of humus from krasnozems (ferralsols) in subtropical Australia. In *Transactions of the 10th International Congress on Soil Science*, Vol. VI, 362–71.

Benedict, J. B. 1966. Radiocarbon dates from a stone-banked terrace in the Colorado Rocky Mountains, USA. *Geogr. Ann.* **48(A)**, 24–31.

Blackburn, G., J. R. Sleeman and H. W. Scharpenseel 1979. Radiocarbon measurements and soil micromorphology as guides to the formation of gilgai at Kaniva, Victoria. *Austral. J. Soil Res.* **19**, 1–15.

Bottner, P. and A. Peyronel 1977. Dynamique de la matière organique dans deux sols, méditerranéens étudiée à partir de techniques de datation par le radiocarbone. *Rev. Écol. Biol. Sol.* **14**, 385–93.

Bowler, J. M. and H. A. Polach 1971. Radiocarbon analyses of soil carbonates: an evaluation from palaeosols in southeastern Australia. In *Paleopedology: origin, nature and dating of paleosols*, D. H. Yaalon (ed.), 97–108. Jerusalem: International Society of Soil Scientists and Israel Universities Press.

Brown, J. 1965. Radiocarbon dating, Barrow, Alaska. *Arctic* **18**, 37–48.

Brown, J. 1969. Buried soils associated with permafrost. In *Pedology and Quaternary research*, S. Pawluk (ed.), 115–27. Edmonton: University of Alberta Press.

Butler, B. E. 1959. *Periodic phenomena in landscapes as a basis for soil studies.* Soil Publn No. 14. Melbourne: Commonwealth Scientific and Industrial Research Organisation, Australia.

Campbell, C. A., E. A. Paul, D. A. Rennie and K. J. McCallum 1967. Factors affecting the accuracy of the carbon-dating method in soil humus studies, *Soil Sci.* **104**, 81–5.

Caseldine, C. J. 1983. Pollen analysis and rates of pollen incorporation into a radiocarbon-dated palaeopodzolic soil at Haugabreen, southern Norway. *Boreas* **12**, 233–46.

Chambers, F. M., P. Q. Dresser and A. G. Smith 1979. Radiocarbon dating evidence on the impact of atmospheric pollution on upland peats. *Nature* **282**, 829–31.

Clark, R. M. 1975. A calibration curve for radiocarbon dates. *Antiquity* **49**, 251–66.

Cook, F. D. 1969. Significance of micro-organisms in paleosols, parent materials and ground water. In *Pedology and Quaternary research*, S. Pawluk (ed.), 53–61. Edmonton: University of Alberta Press.

Crampton, C. B. 1974. A bisequal periglacial section in the Miramichi Valley, eastern Canada. *Can. J. Soil Sci.* **54**, 111–13.

Cruickshank, J. G. and M. M. Cruickshank 1981. The development of humus–iron podsol profiles, linked by radiocarbon dating and pollen analysis to vegetation history. *Oikos* **36**, 238–53.

Ellis, S. 1979. Radiocarbon dating evidence for the initiation of solifluction *ca.* 5500 years B.P. at Okstindan, north Norway. *Geogr. Ann.* **61(A)**, 29–33.

Ellis, S. 1983. Stratigraphy and ¹⁴C dating of two earth hummocks, Jotunheimen, south central Norway. *Geogr. Ann.* **65(A)**, 279–87.

Ellis, S. and J. A. Matthews 1984. Pedogenic implications of a ¹⁴C-dated paleopodzolic soil at Haugabreen, southern Norway. *Arctic Alpine Res.* **16**, 77–91.

Flaig, W., H. Beutelspacher and E. Rietz 1975. Chemical composition and physical properties of humic substances. In *Soil components*, Vol. 1, J. E. Gieseking (ed.), 1–211. Berlin: Springer.

Gamper, M. 1982. Postglaziale solifluktionsphasen am Albulapass (Oestliche Schweizer Alpen). *Phys. Geog.* **1**, 171–86.

Ganzhara, N. F. 1974. Humus formation in chernozem soils. *Soviet Soil Sci.* **6**, 403–7.

Gerasimov, I. P. 1971. Nature and originality of paleosols. In *Paleopedology: origin, nature and dating of paleosols*, D. H. Yaalon (ed.), 15–27. Jerusalem: International Society of Soil Scientists and Israel Universities Press.

Gerasimov, I. P. 1974. The age of recent soils. *Geoderma* **12**, 17–25.

Gerasimov, I. P. and O. A. Chichagova 1971. Some problems in the radiocarbon dating of soil. *Soviet Soil Sci.* **3**, 519–27.

Geyh, M. A., J. H. Benzler and G. Roeschman 1971. Problems of dating Pleistocene and Holocene soils by radiometric methods. In *Paleopedology: origin, nature and dating of paleosols*, D. H. Yaalon (ed.), 63–75. Jerusalem: International Society of Soil Scientists and Israel Universities Press.

Geyh, M. A., G. Roeschmann, T. A. Wijmstra and A. A. Middeldorp 1983. The unreliability of ¹⁴C dates obtained from buried sandy podzols. *Radiocarbon* **25**, 409–16.

Gilet-Blein, N., G. Marien and J. Evin 1980. Unreliability of ¹⁴C dates from organic matter of soils. *Radiocarbon* **22**, 919–29.

Goh, K. M. 1980. Dynamics and stability of organic matter. In *Soils with variable charge*, B. K. G. Theng (ed.), 373–93. Lower Hutt: New Zealand Society of Soil Science.

Goh, K. M. and B. P. J. Molloy 1978. Radiocarbon dating of paleosols using soil organic matter components. *J. Soil Sci.* **29**, 567–73.

Goh, K. M. and B. P. J. Molloy 1979. Contaminants in charcoals used for radiocarbon dating. *N. Z. J. Sci.* **22**, 39–47.

Goh, K. M., T. A. Rafter, J. D. Stout and T. W. Walker 1976. The accumulation of soil organic matter and its carbon isotope content in a chronosequence of soils developed on aeolian sand in New Zealand. *J. Soil Sci.* **27**, 89–100.

Goh, K. M., J. D. Stout and T. A. Rafter 1977. Radiocarbon enrichment of soil organic matter fractions in New Zealand soils. *Soil Sci.* **123**, 385–91.

Gould, W. D., R. V. Anderson, J. F. McClellan, D. C. Coleman and J. L. Gurnsey 1979. Characterization of a paleosol: its biological properties and effect on overlying soil horizons. *Soil Sci* **128**, 201–10.

Griffey, N. J. and S. Ellis 1979. Three *in situ* paleosols buried beneath Neoglacial moraine ridges, Okstindan and Jotunheimen, Norway. *Arctic Alpine Res.* **11**, 203–14.

Griffey, N. J. and J. A. Matthews 1978. Major Neoglacial expansion episodes in southern Norway: evidences from moraine ridge stratigraphy with ¹⁴C dates on buried palaeosols and moss layers. *Geogr. Ann.* **60(A)**, 73–90.

Griffey, N. J. and P. Worsley 1978. The pattern of Neoglacial glacier variations in the Okstindan region of northern Norway during the last three millenia. *Boreas* **7**, 1–17.

Guillet, B. and A. M. Robin 1972. Interprétation de datations par le ¹⁴C d'horizons Bh de deux podzols humo-ferrugineux, l'un formé sous calluna, l'autre sous chênaie-hêtraie. *C. R. Hebdomadare Sci. Acad. Sci. Paris* **274**, 2859–62.

Haider, K., J. P. Martin and Z. Filip 1975. Humus biochemistry. In *Soil biochemistry*, Vol. 4, E. A. Paul and A. D. McLaren (eds.), 195–244. New York: Marcel Dekker.

Hallet, R. 1976. Deposits formed by sub-glacial precipitation of $CaCO_3$. *Geol Soc. Am. Bull.* **87**, 1003–15.

Harris, C. and P. Worsley 1974. Evidence for Neoglacial solifluction at Okstindan, north Norway. *Arctic* **27**, 128–44.

Hassko, B., B. Guillet, R. Jaegy and R. Coppens 1974. Nancy natural radiocarbon measurements III. *Radiocarbon* **16**, 118–30.

Heine, K. 1983. Ein Aussergewöhnlicher Gletschervorstorstoss in Mexico vor 12 000 Jahren. *Catena* **10**, 1–25.

Herrera, R. and M. A. Tamers 1971. Radiocarbon dating of tropical soil associations in Venezuela. In *Paleopedology: origin, nature and dating of paleosols*, D. H. Yaalon (ed.), 109–15. Jerusalem: International Society of Soil Scientists and Israel Universities Press.

Hole, F. D. 1981. Effects of animals on soil. *Geoderma* **25**, 75–112.

Innes, J. L. 1983. Stratigraphic evidence of episodic talus accumulation on the Isle of Skye, Scotland. *Earth Surf. Proc. Landforms* **8**, 399–403.

Jenkinson, D. S. 1969. Radiocarbon dating of soil organic matter. In *Rothamsted Experimental Station, Report for 1968*, Part 1, 73. Harpenden: Rothamsted Experimental Station.

Jenkinson, D. S. 1971. Studies on the decomposition of C^{14} labelled organic matter in soil. *Soil Sci.* **111**, 64–70.

Jenkinson, D. S. 1972. Radiocarbon dating of soil organic matter. In *Rothamsted Experimental Station, Report for 1971*, Part 1, 84–5. Harpenden: Rothamsted Experimental Station.

Jenkinson, D. S. 1975. The turnover of organic matter in agricultural soils. In *Welsh Soils Discussion Group Report 16*, 91–105.

Jenkinson, D. S. and J. H. Rayner 1977. The turnover of soil organic matter in some of the Rothamsted classical experiments. *Soil Sci.* **123**, 298–305.

Kigoshi, K., N. Suzuki and M. Shiraki 1980. Soil dating by fractional extraction of humic acid. *Radiocarbon* **22**, 853–7.

Konecka-Betly, K. 1977. Soils of the dune areas of central Poland in Late Glacial and Holocene. *Folia Quaternaria* (Kraków) **49**, 47–62.

Kononova, M. M. 1961. *Soil organic matter: its nature, its role in soil formation and in soil fertility.* Oxford: Pergamon.

Krajewski, K. 1977. Late-Pleistocene and Holocene dune-forming processes in the Warsaw–Berlin pradolina. *Acta Geogr. Lodziensis* **39**, 82–7 (English summary).

Laura, R. D. 1975. The role of protolytic action of water in the chemical decomposition of organic matter in soil. *Pedologie* **25**, 157–70.

Lavkulich, L. M. 1969. Soil dynamics in the interpretation of paleosols. In *Pedology and Quaternary research*, S. Pawluk (ed.), 25–37. Edmonton: University of Alberta Press.

Levin, I., K. O. Münich and W. Weiss 1980. The effect of anthropogenic CO_2 and ^{14}C in the atmosphere. *Radiocarbon* **22**, 379–91.

Martel, Y. A. and E. A. Paul 1974. The use of radiocarbon dating of organic matter in the study of soil genesis. *Soil Sci. Soc. Am. Proc.* **38**, 501–6.

Matthews, J. A. 1980. Some problems and implications of ^{14}C dates from a podzol buried beneath an end moraine at Haugabreen, southern Norway, *Geogr. Ann.* **62(A)**, 185–208.

Matthews, J. A. 1981. Natural ^{14}C age/depth gradient in a buried soil. *Naturwissenschaften* **68**, 472–4.

Matthews, J. A. 1984. Limitations of ^{14}C dates from buried soils in reconstructing glacier variations and Holocene climate. In *Climatic changes on a yearly to millennial basis: geological, historical and instrumental records*, N. A. Mörner and W. Karlén (eds), 281–290. Dordrecht: Reidel.

Matthews, J. A. and P. Q. Dresser 1983. Intensive ^{14}C dating of a buried palaeosol horizon. *Geol. Föreningens Stockholm Förhandlingar* **105**, 59–63.

Nakhla, S. M. and G. Delibrias 1967. Utilisation du carbonne-14 d'origine thermonucleaire pour l'étude de la dynamique du carbone dans le sol. In *Radioactive dating and methods of low-level counting*, International Atomic Energy Agency (ed), 169−76. Vienna: IAEA.

Nikiforoff, C. C. 1959. Reappraisal of the soil. *Science* **129**, 186−96.

Nydal, R. and K. Lovseth 1970. Prospective decrease in atmospheric radiocarbon. *J. Geophys. Res.* **75**, 2271−8.

O'Brian, B. J. and J. D. Stout 1978. Movement and turnover of soil organic matter as indicated by carbon isotope measurements. *Soil Biol. Biochem.* **10**, 309−17.

Oldfield, F. 1977. Lakes and their drainage basins as units of sediment-based ecological study. *Prog. Phys. Geog.* **1**, 460−504.

Olsson, I. U. 1974. Some problems in connection with the evaluation of ¹⁴C dates. *Geol. Föreningens Stockholm Förhandlingar* **96**, 311−20.

Olsson, I. U. 1979. A warning against radiocarbon dating of samples containing little carbon. *Boreas* **8**, 203−7.

Østrem, G. 1964. Problems of dating ice-cored moraines. *Geog. Ann.* **47(A)**, 1−38.

Paul, E. A. 1969. Characterization and turnover rate of soil humic constituents. In *Pedology and Quaternary research*, S. Pawluk (ed.), 63−76. Edmonton: University of Alberta Press.

Polach, H. A. and A. B. Costin 1971. Validity of soil organic matter radiocarbon dating: buried soils in Snowy Mountains, southeastern Australia as example. In *Paleopedology: origin, nature and dating of paleosols*, D. H. Yaalon (ed.), 89−96. Jerusalem: International Society of Soil Scientists and Israel Universities Press.

Rafter, T. A., H. S. Jansen, L. Lockerbie and M. M. Trotter 1972. New Zealand radiocarbon reference standards. In *Proceedings of the 8th International radiocarbon dating conference*, Lower Hutt, New Zealand, Vol. 2, 625−75.

Righi, D. and B. Guillet 1977. Datations par le carbone-14 naturel de la matiere organique d'horizons spodiques de podzols des Landes du Médoc (France). In *Soil organic matter studies*, Vol. 2, International Atomic Energy Agency (ed.), 187−92. Vienna: IAEA.

Röthlisberger, F. and W. Schneebeli 1979. Genesis of lateral moraine complexes, demonstrated by fossil soils and trunks; indicators of postglacial climatic fluctuations. In *Moraines and varves: origin, genesis, classification*, Ch. Schlüchter (ed.), 387−419. Rotterdam: Balkema.

Röthlisberger, F., P. Haas, H. Holzhauser, W. Keller, W. Bircher and F. Renner 1980. Holocene climatic fluctuations − radiocarbon dating of fossil soils (fAh) and woods from moraines and glaciers in the Alps. *Geog. Helvetica* **35**, 21−52.

Rubilin, Y. V. and M. G. Kozyreva 1974. The age of Russian chernozems. *Soviet Soil Sci.* **6**, 383−92.

Ruellan, A. 1971. The history of soils: some problems of definition and interpretation. In *Paleopedology: origin, nature and dating of paleosols*, D. H. Yaalon (ed.), 3−13. Jerusalem: International Society of Soil Scientists and Israel Universities Press.

Ruhe, R. V. and R. B. Daniels 1958. Soils, paleosols, and soil horizon nomenclature. *Soil Sci. Am. Proc.* **22**, 66−9.

Sauerbeck, D. and B. Johnen 1973. Radiometrische untersuchungen zur humusbilanz. *Landwirtschaftliche Forschung Sonderheft* **30**, 137−45.

Scharpenseel, H. W. 1971a. Radiocarbon dating of soils: problems, troubles, hopes. In *Paleopedology: origin, nature and dating of paleosols*, D. H. Yaalon (ed.), 77−88. Jerusalem: International Society of Soil Scientists and Israel Universities Press.

Scharpenseel, H. W. 1971b. Radiocarbon dating of soils. *Soviet Soil Sci.* **3**, 76−83.

Scharpenseel, H. W. 1972. Natural radiocarbon measurement of soil and organic matter fractions and on soil profiles of different pedogenesis. In *Proceedings of the 8th international radiocarbon dating conference, Lower Hutt, New Zealand*, Vol. 2, 382−93.

Scharpenseel, H. W. 1975. Natural radiocarbon measurements on humic substances in the light oɪ carbon cycle estimates. In *Humic substances: their structure and function in the biosphere*, D. Povoledo and H. L. Goltermann (eds), 281−92. Wageningen: Centre for Agricultural Publishing and Documentation.

Scharpenseel, H. W. 1977. The search for biologically inert and lithogenic carbon in recent soil organic matter. In *Soil organic matter studies*, Vol. 2, International Atomic Energy Agency (ed.), 193–201. Vienna: IAEA.

Scharpenseel, H. W. 1979. Soil fraction dating. In *Proceedings of the 9th international radiocarbon conference* (Los Angeles), R. Berger and H. E. Suess (eds), 277–83.

Scharpenseel, H. W. and H. Schiffmann 1977a. Radiocarbon dating of soils, a review. *Zeitschr. Pflanzenernährung Düngung Bodenkunde* **140**, 159–74.

Scharpenseel, H. W. and H. Schiffmann 1977b. Soil radiocarbon analysis and soil dating. *Geophys. Surv.* **3**, 143–56.

Scharpenseel, H. W., C. Ronzani and F. Pietig 1968. Comparative age determination on different humic-matter fractions. In *Isotopes and radiation in soil organic-matter studies*, International Atomic Energy Agency (ed), 67–73. Vienna: IAEA.

Schneebeli, W. 1976. Untersuchungen von Gletscherschwankungen im Val de Bagnes. *Die Alpen (Zeitschr. Schweizer Alpen-Club)* **52**, 5–57.

Schnitzer, M. and S. U. Khan 1972. *Humic substances in the environment.* New York: Marcel Dekker.

Schoute, J. F. Th., J. W. Griede, W. G. Mook and W. Roeleveld 1981. Radiocarbon dating of vegetation horizons, illustrated by an example from the Holocene coastal plain in the northern Netherlands. *Geol. Mijn.* **60**, 453–9.

Simonson, R. W. 1978. A multiple process model of soil genesis. In *Quaternary soils*, W. C. Mahaney (ed.), 1–25. Norwich: Geo Abstracts.

Skujins, J. J. 1976. Extracellular enzymes in soil. In *Critical reviews in microbiology*, Vol. 4, I. Laskin and H. Lechevalier (eds), 383–421. Cleveland, CRC Press.

Skujins, J. J. and A. S. McLaren 1968. Persistence of enzymatic activity in stored and geologically preserved soils. *Enzymologia* **34**, 213–25.

Stout, J. D. and K. M. Goh 1980. The use of radiocarbon to measure the effects of earthworms on soil development. *Radiocarbon* **22**, 892–6.

Stout, J. D. and K. E. Lee 1980. Ecology of soil micro- and macro-organisms. In *Soils with variable charge*, B. K. G. Theng (ed.), 353–72. Lower Hutt: New Zealand Society of Soil Scientists.

Stout, J. D., K. M. Goh and T. A. Rafter 1981. Chemistry and turnover of naturally occurring resistant organic compounds in soil. In *Soil biochemistry*, Vol. 5, E. A. Paul and J. N. Ladd (eds), 1–73. New York: Marcel Dekker.

Sutherland, D. G. 1980. Problems of radiocarbon dating deposits from newly deglaciated terrain: examples from the Scottish Lateglacial. In *Studies in the Lateglacial of north-west Europe*, J. J. Lowe, J. M. Gray and J. E. Robinson (eds), 139–49. Oxford: Pergamon.

Tamm, C. O. and H. Holmen 1967. Some remarks on soil organic matter turnover in Swedish podzol profiles. *Meddelelser Norske Skogforsøksvesen* **85**, 67–88.

Tate, K. R. and B. K. G. Theng 1980. Organic matter and its interactions with inorganic soil constituents. In *Soils with variable charge*, B. K. G. Theng (ed.), 225–49. Lower Hutt: New Zealand Society of Soil Scientists.

Vuorela, I. 1980. Old organic material as a source of dating errors in sediments from Vanajavesi and its magnification in the pollen stratigraphy. *Ann. Bot. Fennici* **17**, 244–57.

Wilding, L. P. 1967. Radiocarbon dating of biogenic opal. *Science* **156**, 66–7.

Zoltai, S. C., C. Tarnocai and W. W. Pettapiece 1978. Age of cryoturbated organic materials in earth hummocks from the Canadian Arctic. In *Proceedings of the 3rd international conference on permafrost,* Edmonton, Canada, Vol. 1, 325–31.

15

Soil chronosequences on Neoglacial moraine ridges, Jostedalsbreen and Jotunheimen, southern Norway: a quantitative pedogenic approach

A. Mellor

Introduction

The qualitative concept of the soil *chronosequence* had its foundation in the 19th-century Russian school of pedology. This theme was formalised by Jenny (1941, 1946), who identified a soil *chronofunction* as the quantitative solution of the relationship:

$$S = f(T):Cl,O,R,P... \qquad (15.1)$$

where the soil (S) and the properties that define it are functions of time (T), with the variables of climate (Cl), organisms (O), relief (R) and parent material (P) held constant. Many of the conceptual developments and problems relating to such temporal studies of soils have been extensively reviewed (e.g. Vreeken 1975, Bockheim 1980).

Many soil chronosequence studies on Neoglacial moraine ridge sequences have been conducted in Alaska (e.g. Crocker & Major 1955, Ugolini 1968) and neighbouring areas of the Canadian Rockies (e.g. Jacobson & Birks 1980). Some investigations have also been undertaken in the European Alps (e.g. Fitze 1981), the Elbrus Mountains (Gennadiyev 1978) and Iceland (Boulton & Dent 1974, Romans *et al.* 1980). However, little published material is available from Scandinavia; Stork (1963) has briefly described soils on a recessional moraine sequence in northern Sweden using a limited number of soil properties, while Alexander (1970) conducted a preliminary pedological investigation in northern Norway. In southern Norway, Faegri (1933) gave a generalised account of soil

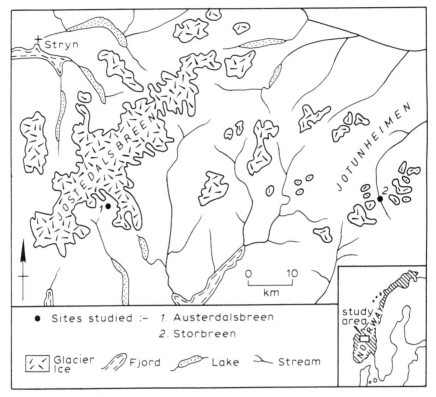

Figure 15.1 Location map.

profiles on moraine ridges close to Jostedalsbreen (Fig. 15.1), and Merrington (1972) briefly examined soil development in relation to plant colonisation within the 12 year pioneer zone at Storbreen (Fig. 15.1).

Many published soil chronosequence studies are qualitative, and little detailed work appears to have been carried out on the quantitative aspects of soil chrono-functions, probably for two main reasons. First, dating control is often poor (Bockheim 1980) and, second, inadequate site selection and sampling design cause lack of environmental control. Bockheim (1980), however, has constructed chronofunctions using selected linear and non-linear models applied to several published data sets. The results suggested that the logarithmic function

$$Y = a + b \log T \qquad\qquad (15.2)$$

yielded the best relationship when 15 soil properties (Y) were regressed on time (T). Sondheim *et al.* 1981 have, in contrast, suggested a logistic relationship for the variation through time of scores from a principal components analysis of soils on a Canadian prograded beach chronosequence. Although the qualitative information gained from soil chronosequences has been applied to the relative dating of moraine ridges deposited by different Neoglacial glacier advances in North America (e.g. Mahaney *et al.* 1981), very few studies consider the application of quantitative chronofunctions in such a geomorphological context.

This study aims to contribute to current research on soil chronofunctions by using examples from Scandinavia. Within the area of interest in southern Norway there is a widespread occurrence of compact, well-preserved and accurately dated Neoglacial moraine ridge sequences (e.g. Matthews 1975), two of which form the subject of this investigation. Chronofunctions have been defined for a variety of soil morphological, physical and chemical properties, using linear and non-linear models. These results are supplemented with surface textural, clay mineralogical and micromorphological information in order to identify the nature and rate of organic, weathering and translocatory soil-forming processes. This quantitative approach, based on a more detailed soil analytical framework than has previously been available, allows comparison of rates of soil development between the two sites, and inference concerning the effect of environmental factors on pedogenic processes. The valuable contribution of Neoglacial geomorphology to an understanding of arctic-alpine pedogenic rates must be stressed, and is reflected in the accurate dating control available in the Neoglacial zone.

The study area

The study area is approximately 100–200 km from the west coast of southern Norway (Fig. 15.1). Its glacial geomorphology includes the Jostedalsbreen ice cap, with its associated outlet valley glaciers, to the west, and the Jotunheimen mountain range to the east, characterised by a more local distribution of niche, cirque and valley glaciers. The region is thought to have been vacated by the Weichselian Fenno-Scandian continental ice sheet during the late Preboreal Chronozone, approximately 9000 years BP (Andersen 1980).

Several glaciers in the area possess well-defined moraine ridge sequences within a few kilometres of their snouts. Historical, lichenometric and radiocarbon dating evidence (e.g. Matthews 1977, 1982) strongly suggests that the outermost ridge of the majority of these moraine sequences was formed by a Neoglacial glacier advance that culminated in the mid-18th century during the 'Little Ice Age'. Successive stillstands and slight readvances superimposed on the general retreat since then have often left a series of recessional moraine ridges. This study focuses on the moraine sequences within the glacier forelands of Austerdalsbreen and Storbreen (Fig. 15.1).

The moraine ridges studied in these two areas are from 3 to 10 m in height and are generally asymmetrical in form, the proximal slopes tending to be less steep than the distal slopes, with angles of 15–20° and 30–35°, respectively. Both moraine sequences have been mapped in detail and dated lichenometrically

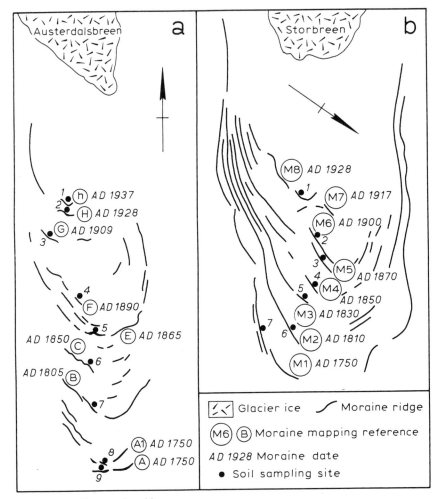

Figure 15.2 The moraine ridge sequences.

and historically (King 1959, Matthews 1975, R. W. Bickerton personal communication), and are illustrated in Figures 15.2a and b.

The moraine ridges of both sequences are composed of poorly sorted bouldery and stony till with a matrix dominated by sand and coarse silt. Lithological composition differs between the two sites, however, because of a difference in bedrock geology of the source areas. At Austerdalsbreen the till has been derived from the acidic 'basal gneiss' of granitic to granodioritic composition characteristic of Jostedalsbreen (Holtedahl 1960), while at Storbreen the parent material is derived from a more basic pyroxene-granulite gneiss which dominates the Jotunheimen mountains (Battey & McRitchie 1973).

Although no climatic data are available for the glacier forelands studied, Green and Harding (1980) have constructed mean altitudinal temperature gradients for southern Norway which allow mean annual temperature to be predicted, within reasonable limits of accuracy, for any glacier foreland of known altitude in the region. The Austerdalsbreen and Storbreen forelands are at mean altitudes of 350 and 1225 m respectively, giving them predicted mean annual temperatures of 5 and −1 °C. Precipitation values of approximately 2250 and 1500 mm per annum are estimated (Østrem & Ziegler 1969).

The glacier forelands of Austerdalsbreen and Storbreen lie respectively in the regional sub-alpine birch woodland and low alpine altitudinal vegetation zones of Dahl (1975). However, the complex nature of rapid vegetation succession within such youthful glacier foreland environments precludes simple zonal classification, although the succession will probably progress towards these regional types, which are locally observed beyond the limits of Neoglacial activity.

Materials

Soil sampling was conducted on a horizon basis following a procedure modified from that of Hodgson (1974). The visually best-developed profile on each ridge was sampled, to avoid unrepresentative cases of limited soil development owing to excessive exposure as on moraine crests, or impeded drainage and late-lying snow-patch influences as on the lower slopes. Visual criteria included solum depth and horizon thickness, differentiation and colour. After a detailed preliminary examination of soil profile variation on individual ridges, it became evident that the best-developed profiles tended to exist under an optimum combination of high slope stability, vegetation cover and proportion of fines (<2 mm) in the parent material. These criteria were therefore used in the field to determine sampling site positions. Generally the less steep proximal slope provided the optimum site conditions required. Thus consistency of sampling location occurred, which ensured that variations in soil-forming factors other than time were minimised in each glacier foreland (Figs 15.2a & b).

Although there exists a classification system for Norwegian mountain soils (Ellis 1979), it is difficult to apply to such youthful soils, developed over a maximum time period of only 230 years (Figs 15.2a & b). However, comparison with the older soils beyond the Neoglacial zones is possible. At Storbreen, these are analogous to the Brown Soils described by Ellis (1979), with an organic-rich mineral horizon directly overlying an illuvial horizon, the latter being enriched by translocated sesquioxides and organic matter. At Austerdalsbreen, the soils have been classified as Humo-Ferric Podzols using the Canadian system of classification, which has been considered the most appropriate for Norwegian mountain soils (Ellis 1979). A typical profile would comprise an almost exclusively organic horizon overlying a bleached eluvial mineral horizon depleted of sesquioxides, this in turn being underlain by an illuvial horizon enriched by translocated sesquioxides and organic matter. It is probable that the youthful Neoglacial soils represent stages of development towards these, their older counterparts.

Methods

To examine organic, weathering and translocatory pedogenic processes, detailed physical and chemical laboratory analyses were conducted on samples from each horizon, following procedures modified from Bascomb (1974). Soil properties considered were particle size, reaction (pH), organic carbon (C), cation exchange capacity (CEC), exchangeable bases and pyrophosphate- and dithionite-extractable iron and aluminium. Clay mineralogy was analysed on a Philips X-ray diffractometer using Fe-filtered Co Kα radiation at a scanning speed of 1° 28 min⁻¹. Treatments included air-drying, low temperature (50°C) glycolation and heating to 550 °C. Dominant particle size fractions were subjected to surface texture analysis using a Cambridge Stereoscan 600 scanning electron microscope. Fifty grains were counted at set intervals on traverses across the mounting stubs (Bull 1981), and the presence or absence of weathering features was noted. Araldite-impregnated samples collected in the vertical plane (FitzPatrick 1970) were subjected to quantitative micromorphological examination using the descriptive scheme of FitzPatrick (1977, 1980) previously considered successful in its application to Norwegian arctic-alpine soils by Ellis (1983). Counts of 500 per horizon were made, giving an acceptable absolute error of approximately $1-4\%$ at the 95% confidence interval (Van der Plas & Tobi 1965).

Chronofunctions were defined by regressing soil properties (Y) against time (T). In some cases, soil properties were expressed as ratios between values in two horizons in the same profile. This was particularly necessary where absolute values were very small (e.g. percentages of clay), and was desirable as a method of controlling for the effect of small-scale variability in the parent material; the ratios provide indices of *relative* variation between profiles. Four statistically linear models were employed:

$$Y = a + bT \tag{15.3}$$

$$Y = a + b \log T \tag{15.4}$$

$$\log Y = a + bT \tag{15.5}$$

$$\log Y = a + b \log T \tag{15.6}$$

These represent, respectively, linear, single logarithmic, exponential and power function relationships. Two additional non-linear functions were tested:

$$Y = a - (b/T) \tag{15.7}$$

$$Y = a/(1 + \exp(c-bT)) \tag{15.8}$$

These are asymptotic and logistic relationships. In each equation, the b parameter represents the rate of change of the soil property, this being more rapid with larger values of b, and in a direction indicated by the sign of b. In Equations 15.3 to 15.6 the a parameter specifies an intercept value of the soil property, while in Equations 15.7 and 15.8 it represents a limiting value. Polynomial models were rejected because of the difficulty of explaining them in theoretical pedogenic terms, and because of their limited success in fitting the trends in empirical tests. Having identified chronofunctions for each site, comparisons of the pedogenic rates thereby defined allowed assessment of the influence of environmental factors on the processes of soil formation.

Results and discussion

Organic processes
Figures 15.3a and b illustrate diagrammatically the temporal trends exhibited by some soil morphological attributes. At both Austerdalsbreen and Storbreen the organic-rich surface horizons display a significant increase with age. On the youngest moraines a surface crust only a few millimetres thick occurs, with a sparse cover of vascular plants. On the older moraines, however, the contri-

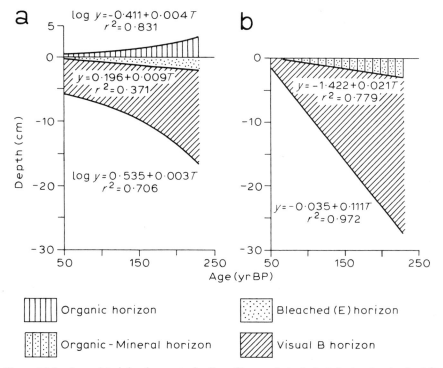

Figure 15.3 Age-related development of soil profile morphological attributes showing best-fit chronofunctions: (a) at Austerdalsbreen; (b) at Storbreen.

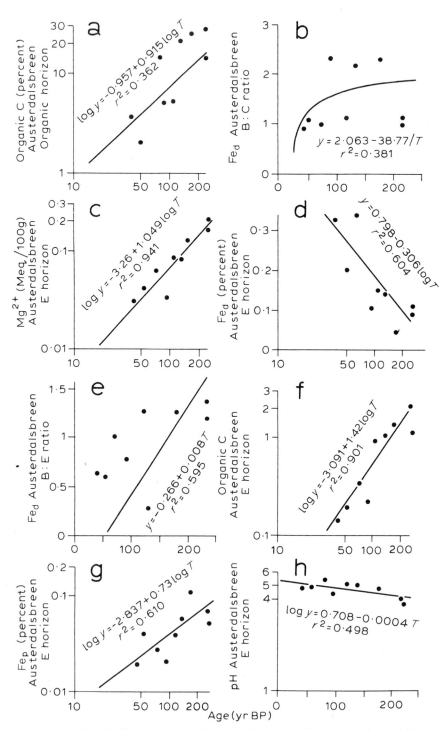

Figure 15.4 Selected soil property chronofunctions showing best-fit equations: Austerdalsbreen.

bution of vascular plants increases and an increased litter supply and more penetrative root network aid in the development of a surface organic-rich horizon. This has been observed in other chronosequence studies (e.g. Ugolini 1968), but appears never to have been expressed before as a chronofunction. The trends exhibited by the morphological data (Figs 15.3a & b) are supported at Austerdalsbreen by a significant increase with age of organic C within the surface organic-rich horizon (Fig. 15.4a). At Storbreen this relationship proved non-significant, perhaps because considerable local variability occurs in the supply of organic material to the soil surface in this more sparsely vegetated and severe climatic environment. Also, the Brown Soils characteristic of this area possess more mineral-rich organic horizons than the podzolic soils at Austerdalsbreen.

Birkeland (1974) argued that organic matter probably reaches a steady state condition more rapidly than any other soil property, possibly within only 200 years. Jacobson and Birks (1980) showed that surface organic matter percentages on recent Canadian end moraines increased rapidly over the initial 100–150 years of soil development, then increased more gradually for the next 100 years. The data for Austerdalsbreen and Storbreen suggest that after 230 years of soil development in this particular environment, the organic component has not reached a steady state. Organic-rich horizons of soils beyond the Neoglacial limits commonly have organic C values exceeding 30%. Thus extrapolation of the organic C model (Fig. 15.4a) suggests a marked decline in the rate of increase some time after 230 years from the average value of approximately 15% achieved by then.

Micromorphological examination of the organic-rich surface horizon at Austerdalsbreen shows a predominance of fresh to slightly decomposed organic material with a spongy structure (Fig. 15.5a) similar to that recognised in northern Norway by Ellis (1983). At Storbreen, however, the organic material is of a more fragmentary nature although still only slightly decomposed, and shows a higher degree of incorporation with mineral material. This results in a composite structure including granular and crumb elements (Fig. 15.5b) that have also been previously recognised in arctic-alpine soils (e.g. Pawluk & Brewer 1975, Ellis 1983). Faecal pellets produce the granular structural form, while the crumb structure consists of loosely bound aggregates of both organic and mineral components. In the majority of cases, however, the percentage point count for faecal pellets is relatively low (generally less than 6%), suggesting a low degree of organic matter decomposition.

The micromorphological data show very few age-related trends, and although the percentage of organic matter and faecal pellets appeared initially to show a relationship with age in the surface organic-rich horizons, this was statistically insignificant ($p > 0.05$). This may reflect the problems of sampling the thin organic-rich horizons, which result in unrepresentative data and a high degree of scatter. Generally, organic matter declines markedly with depth in the soil profile, with values of up to 30% at the surface declining to less than 1% below a depth of 20 cm. This relationship coincides with a marked increase in the percentage of

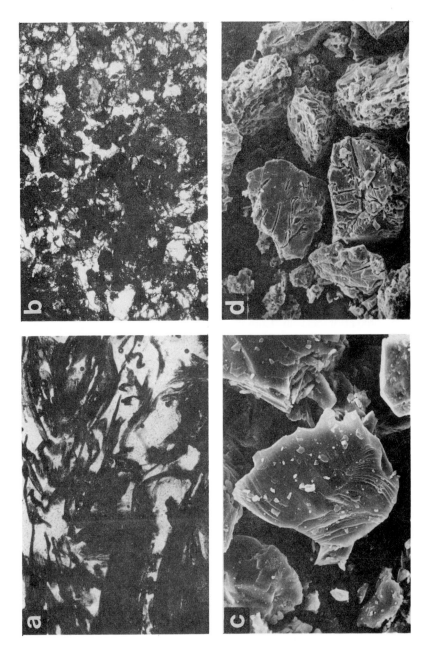

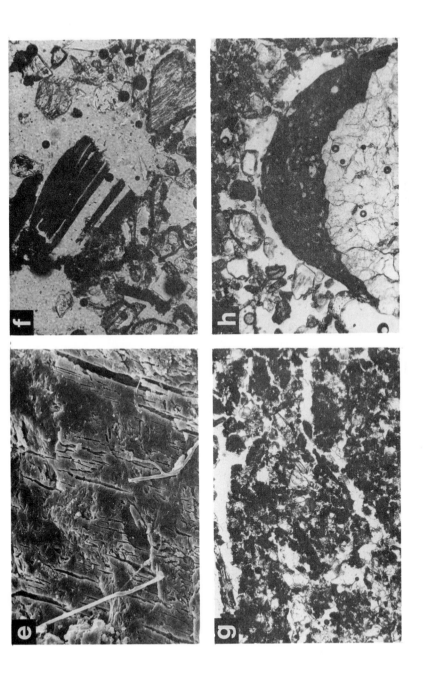

Figure 15.5 Petrological and scanning electron microscope photomicrographs: (a) organic-rich horizon showing spongy microstructure (frame length = 4 mm); (b) mineral-rich organic horizon showing granular and crumb structural elements (frame length = 4 mm); (c) glacially comminuted quartz grains; (d) chemically etched feldspar grains; (e) crystallographically controlled chemical etching of feldspar grains; (f) mica grain showing expansion along cleavage planes (frame length = 2 mm); (g) bleached horizon showing granular structural elements (frame length = 4 mm); (h) illuvial coating of fine sand and silt on upper surface of detrital grain (frame length = 2 mm). All petrological microscope photomicrographs are viewed in plain polarised light.

detrital grains and matrix, which is probably a reciprocal effect. At Austerdals-breen an increase in the percentage of pores in the organic-rich horizon probably corresponds with the spongy structure noted in Figure 15.5a. This feature is not so clearly apparent in the soils at Storbreen, reflecting the less porous, granular structure characteristic of organic material of a more highly incorporated nature (Fig. 15.5b). These observations indicate the potential of quantitative micro-morphological investigations in supporting soil morphological and organic C information.

Weathering processes

At both Austerdalsbreen and Storbreen the morphological data (Figs 15.3a & b) show a significant temporal increase in the thickness of what appears visually to be the B horizon. This trend may be attributed to the increased intensity and depth of operation of weathering processes with increasing age. This conclusion is supported by a significant increase in the ratio of dithionite-extractable iron (Fe_d) in the B horizon to that in the C horizon at Austerdalsbreen (Fig. 15.4b): this was not apparent at Storbreen. Both Ugolini (1968) and Jacobson and Birks (1980) provide morphological evidence for a rapid increase in the B horizon thickness over the first 100 years of soil development on moraine sequences in Alaska and Canada respectively, with a steady state apparently being attained beyond that time period. This evidence is not supported in these Norwegian examples, which show a consistent increase in B horizon thickness over the first 230 years of soil development. The Fe_d data do, however, show a significant tendency towards steady state within 230 years (Fig. 15.4b). Significant increases with age of CEC and exchangeable potassium and magnesium (e.g. Fig. 15.4c) are seen in the bleached E horizon at Austerdalsbreen, but corresponding in-creases in calcium and sodium are not observed. This may be attributed to base release by pedogenic weathering processes from the acidic gneiss parent material, which is comparatively rich in potassium and magnesium but poor in calcium and sodium (Holtedahl 1960).

The soil parent materials at both glacier forelands are tills dominated by sand and coarse silt, often in quantities exceeding 80% of the total fine fraction (< 2 mm). This is considered to relate to the fact that the rocks from which the till is formed were most readily comminuted into this particle size category (Haldorsen 1978). There is no evidence to suggest an increase of clay with depth, and time-related trends were also lacking in the particle size data. These obser-vations, along with the very low clay percentages in all the soils examined (often $< 1\%$), indicate that weathering into clay-sized material is of an extremely low order, as suggested in other studies of arctic-alpine soils (Ellis 1980a).

A study of weathering within and between sand and silt fractions in Norwegian soils using heavy mineral ratios (Ellis 1980a) suggested very limited weathering, although perhaps masked by the inherent mineralogical variability of the till parent materials. Mineralogical data have been used in the treatment of soil chronosequences (e.g. Ruhe 1956, Locke 1979), but normally the time scale is much greater than that available at Austerdalsbreen and Storbreen. It was felt, therefore, that such techniques would lack the sensitivity to pick out such weathering trends as might be apparent in only 230 years of soil development. As an alternative approach, the dominant particle size fraction was subjected to a high magnification surface texture analysis using scanning electron micro-scopy. This was considered a more sensitive method of identifying incipient weathering features. Preliminary observations of sand and silt grains show the fresh, sharp and angular features (Fig. 15.5c) commonly associated with glacial comminution of bedrock (e.g. Krinsley & Doornkamp 1973). Evidence for chemical weathering is only present in soils developed beyond the limits of Neoglacial glacier activity, and the chemical composition results from an EDAX−link facility suggest that even here it is only the more weathering-susceptible feldspars, pyroxenes and amphiboles that are subject to alteration. This takes the form of a type of chemical etching (Fig. 15.5d) which appears to be crystallographically controlled (Fig. 15.5e), a feature previously described and attributed to selective dissolution by organic acids (Dearman & Baynes 1979). Micromorphological evidence for weathering is limited to the presence of a very small number (less than 0.2%) of biotite grains showing expansion along clea-vage planes (Fig. 15.5f), attributable to hydration and noted in arctic-alpine soils by Ellis (1983).

Results of X-ray diffractometry show distinct peaks for air-dried samples at approximately 7, 10, 12 and 14 Å at Austerdalsbreen and Storbreen, with an additional peak at approximately 8.4 Å in soils from the latter site. Glycolation shows no peak displacement, suggesting an absence of expanding clay minerals (Brindley & Brown 1980). Heating to 550 °C has no effect on the 8.4 Å peak seen at Storbreen, but in all cases it results in disappearance of the 7 Å peak and collapse of the 12 and 14 Å peaks to approximately 10 Å, thus reinforcing the 10 Å peak. The 7, 8.4, 10 and 14 Å peaks are considered, therefore, to rep-resent kaolinite, amphibole, mica and vermiculite respectively, and as they are present at all depths in all soil profiles studied, it is suggested that they are largely inherited from the parent material. The 12 Å peak is thought to represent hydro-biotite (Kapoor 1972), although the 24 Å peak commonly associated with this regular biotite−vermiculite interstratification is rarely seen in the examples studied. Evidence for weathering of clay minerals is apparent only in the altera-tion of the mica phase to this hydrobiotite phase. This feature has been recognised in lowland Norwegian soils (Kapoor 1972), but at Austerdalsbreen and Storbreen it is only evident in the soils developed on the oldest (AD 1750) Neoglacial moraines, where a gradual increase in hydrobiotite and a reciprocal decrease in mica occur towards the surface of the soil profiles. Evidence for weathering overall, however, indicates that the process is of limited importance on this timescale in these environments, with few significant age- or depth-related trends being apparent.

Translocatory processes
Morphological data show an age-related increase in the thickness of the E horizon at Austerdalsbreen (Fig. 15.3a), suggesting progressive depletion of certain

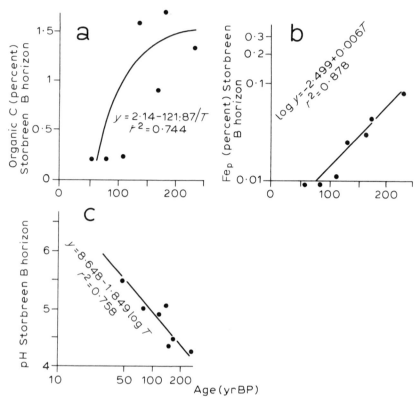

Figure 15.6 Selected soil property chronofunctions showing best-fit equations: Storbreen.

constituents by translocatory processes. Apart from depletion of material from this horizon, there is evidence to suggest that certain constituents may be translocated into it from above. Organic C and pyrophosphate-extractable iron and aluminium (Fe_p and Al_p values (e.g. Figs 15.4f & g) all show significant increases with age within this horizon, while reaction shows a corresponding decrease (Fig. 15.4h). Micromorphological evidence suggests that the organic C characteristics may be caused by translocation of faecal pellets into the bleached layer from the lower part of the organic-rich horizon to form granular structural elements (Fig. 15.5g). The increase in Fe_p and Al_p in the E horizon (e.g. Fig. 15.4g) suggests the formation of organo-metallic complexes whose mobility may be responsible for transfer of these constituents through the E horizon. This process probably also occurs in the B horizon, although here no significant trend was apparent. These observations may be due to the fact that these soils are youthful and therefore the operation of certain organically related processes tends to be concentrated near the surface. The decrease in reaction (Fig. 15.4h) suggests that the translocatory input of these types of organic material from above is probably responsible in part for the increase in acidity with age observed in the E horizon. At Storbreen significant increases in organic C and Fe_p (Figs 15.6a &

b), together with a corresponding decrease in reaction (Fig. 15.6c) in the B horizon, may be explained in similar terms as the result of translocatory input from the overlying organic-rich horizon.

Another important point to consider is the presence of significant depletion of Fe_d from the bleached horizon at Austerdalsbreen (Fig. 15.4d). It has been suggested that iron may be translocated in an inorganic form (e.g. Moore 1976, Childs *et al.* 1983) into the B or even C horizons of podzolised soils, after initially being released during the early stages of till formation (Ellis 1980b). The enrichment of Fe_d in the B horizon is illustrated in Figure 15.4e where the ratio of Fe_d in the B horizon to that in the bleached layer shows a significant increase with age. Since Fe_d contents are noticeably higher than those of Fe_p it is therefore considered that at Austerdalsbreen variations in Fe_d are primarily responsible for subsurface horizon differentiation in terms of colour. Absence of a bleaching effect in the soils at Storbreen may be attributed to the presence of a more basic, iron-rich parent material than that at Austerdalsbreen, and could also be explained by the reduced translocation potential resulting from lower rainfall and a longer duration of seasonally frozen ground. Early in the discussion of weathering it was suggested that the increase in the Fe_d B : C horizon ratio (Fig. 15.4b) may be due to the effect of *in situ* weathering. It appears, however, in the light of evidence presented in the latter part of this discussion that pedogenic weathering processes are of a very low order in this environment, and from the above observations it seems more likely that the relative increases in Fe_d observed in the B horizon at Austerdalsbreen are due to the operation of translocatory rather than weathering processes.

Micromorphological observations reveal the presence of illuvial coatings of fine sand and silt on the upper surfaces of both mineral grains and lenticular-shaped peds (Fig. 15.5h). This feature has been recognised in the subsurface horizons of many cold environment soils, for example in Norway (Ellis 1983) and Canada (Mermut & St Arnaud 1981), where it has been interpreted as the result of translocation following the melting of ice in the soil. A 10−20% increase with depth of illuvial coatings occurs from the top of the mineral soil to the upper part of the parent material in both study areas, even in the youngest soils. This suggests that these coatings form during the early post-depositional stages of the till. Supportive evidence has been produced from micromorphological investigation of young soils in Iceland (Boulton & Dent 1974, Romans *et al.* 1980).

Comparison of pedogenic rates

A comparison of pedogenic rates between the maritime and relatively continental climatic environments of Austerdalsbreen and Storbreen, similar to that made by Jacobson and Birks (1980) in North America, was not possible using the best-fit chronofunctions. This was because of the varying mathematical forms taken by the chronofunctions. In some cases, however, comparisons could be made between chronofunction models which were mathematically equivalent and only slightly less efficient in describing the data than the best-fit functions. This approach was not successful in the context of data relating to organic and weathering processes, but was more suitable for assessment of translocatory processes. For example, in comparing pH decline and Al_p gain in the B horizons with increasing age, using a simple linear model, lower rates are observed at Austerdalsbreen than at Storbreen. pH declines at approximately 0.003 units per year at the former site and 0.006 units per year at the latter, while Al_p increases by 0.002% per year and 0.006% per year.

This comparative approach can also be applied to consideration of profile morphological characteristics at the two sites (Fig 15.3a & b). After 50 years of pedogenesis, the rate of increase in B horizon thickness at Storbreen exceeds that at Austerdalsbreen, whereas the organic-rich surface horizons, despite being of different types, appear to develop at similar rates. Within the first 50 years of soil development, however, both the organic-rich and B horizons have attained greater thicknesses at Austerdalsbreen than at Storbreen, together with a greater degree of horizon differentiation. These features may be explained in part by the effects of environmental factors. More severe climatic conditions and consequent low rates of vegetation colonisation at Storbreen may contribute to lower initial rates of soil development than is the case in the more climatically favourable environment at Austerdalsbreen. Similar conclusions were drawn by Jacobson and Birks (1980) in a comparison of soil development on two moraine ridge sequences in Alaska and Yukon Territory. Changes in soil properties appeared to be delayed by 50 − 100 years at the inland continental site compared to the coastal site, this being attributed to environmental differences between the two locations. On moraines older than approximately 50 years at Storbreen, however, an increased vegetation cover and the likelihood of a more stable ground surface allow a rapid penetration of soil-forming processes into a parent material which is iron-rich and more weatherable in a mineralogical sense than the acidic gneiss at Austerdalsbreen.

Evidence presented in this study suggests that, despite a relatively rapid profile morphological expression, pedogenic processes on the whole operate slowly, with rates being more rapid at Austerdalsbreen than Storbreen during the first 100 years of soil development, but with the position reversing after this time. Micromorphological evidence indicates that translocation of fine sand and silt is one of the most rapidly occurring processes, being well established in even the youngest soils of these glacier forelands.

Conclusions

The data presented in this study suggest that in the low-altitude, well-vegetated and maritime climatic environment at Austerdalsbreen, podzolisation is the dominant pedogenic process. In the high-altitude, sparsely vegetated and more continental climatic regime at Storbreen, Brown Soils similar to those described by Ellis (1979) are prevalent. Despite the differences in profile morphology, the processes of soil development are similar at the two sites, with two exceptions: first, organic matter accumulation is limited at Storbreen, probably because of the sparse vegetation cover and low rainfall; and second, a bleached horizon was absent in the soils at Storbreen for the same reasons, and because of the longer duration of seasonally frozen ground and the more iron-rich basic parent material. Pedogenic rates in the first 100 years of soil development appear to be faster at Austerdalsbreen, although subsequently this is reversed. Generally, the rate of development is slow, and translocation of fine sand and silt to form coatings on the upper surfaces of larger mineral grains is one of the most rapid processes occurring in the soils of these glacier forelands.

The study of soil chronofunctions has proved useful in identifying the nature and rate of change of organic, weathering and translocatory pedogenic processes at the two sites examined, and 40% of all chronofunctions fitted were statistically significant at the 95% confidence level. The quantitative application of micromorphological information to the examination of soil chronosequences has proved particularly useful; this approach has not been extensively exploited previously. The successful application of this quantitative approach is largely dependent on the nature of the geomorphic environment investigated, with the compact and well-preserved moraine ridge sequences providing accurate dating control in this particular case.

It is suggested that simple mathematical models should be employed to construct chronofunctions which can be explained in theoretical pedogenic terms. The results from this study indicate that linear, semi-logarithmic, exponential, power and asymptotic models are more efficient than the logistic model in describing time-related trends observed in the data. However, this reflects the high degree of scatter, the relatively small number of dated points, and, most importantly, the shortness of the time span available in the chronosequences of both study areas.

The chronofunction approach used here could be applied to the prediction of ages of undated moraine ridges in other areas with similar environments to those experienced in this region of Scandinavia. In this context, soil chronofunctions could supplement historical and lichenometric dating. Clearly, however, a confident and accurate chronological picture rests on the simultaneous use of a range of dating methods. A particular danger in applying chronofunction models lies in the unwarranted extrapolation of fitted curves beyond the range of the available dates, especially when the soil property considered does not attain a steady state within the time period of study, a condition which appears to be the case in the majority of soil indices examined here. Indeed, Bockheim (1980) has applied statistically linear chronofunctions to a range of published data and disputes the steady state concept even for chronosequences exceeding 10^6 years in age. However, it is suggested that a logistic model may prove more applicable over such long time spans, within which an approach to steady state may occur (Birkeland 1974, Sondheim et al. 1981). The application of soil chronofunctions may prove particularly useful in studies where the land-surfaces lie beyond the range of other, more conventional dating techniques (e.g. Evans & Cameron 1979).

Geomorphological and pedological aspects of the study of soils on Neoglacial moraine ridges are closely related. Geomorphologically, the glacier foreland provides both the dating control needed for successful establishment of a chronosequence and the range of site conditions within which local variability of soil properties can be defined in relation to slope stability on individual ridges. Pedologically, the chronofunction approach provides an additional method for dating moraine ridges, and the detailed interpretation of the relative roles of weathering and translocatory processes on such recently exposed surfaces provides an explanation of the low levels of proglacial stream solute load and chemical denudation in arctic-alpine areas underlain by resistant rocks.

Acknowledgements

This research was carried out while the author was in receipt of a NERC Research Studentship at the University of Hull. Thanks are due to Dr S. Ellis and Dr K. S. Richards for helpful criticism of the manuscript, and to Dr J. A. Matthews, leader of the Jotunheimen Research Expedition alongside which fieldwork was conducted. I am also grateful to Dr A. G. Fraser and Mr A. Sinclair for help and advice relating to X-ray diffraction and scanning electron microscope techniques respectively. Illustrative material was prepared and assistance in the laboratory was provided by members of the technical staff of the Departments of Geography and Geology at the University of Hull.

References

Alexander, M. J. 1970. A study of some soils in the Austre Okstindsbredal area. In *Okstindan research project, preliminary report 1968*, P. Worsley (ed.), 25−31. Reading: Reading University.

Andersen, B. G. 1980. Deglaciation of Norway after 10 000 BP. *Boreas* **9**, 211−16.

Bascomb, C. L. 1974. Physical and chemical analyses of <2 mm samples. In *Soil Survey laboratory methods*, B. W. Avery and C. L. Bascomb (eds), 14−41. Soil Survey Tech. Monogr. 6. Harpenden: Rothamsted Experimental Station.

Battey, M. H. and W. D. McRitchie 1973. A geological traverse across the pyroxene granulites of Jotunheimen in the Norwegian Caledonides. *Norsk. Geol. Tidsskr.* **53**, 237−65.

Birkeland, P. W. 1974. *Pedology, weathering and geomorphological research.* London: Oxford University Press.

Bockheim, J. G. 1980. Solution and use of chronofunctions in studying soil development. *Geoderma* **24**, 71−85.

Boulton, G. S. and D. L. Dent 1974. The nature and rates of post-depositional changes in recently deposited till from south-east Iceland. *Geogr. Ann.* **56A**, 121−34.

Brindley, G. W. and G. Brown (eds) 1980. *Crystal structures of clay minerals and their X−ray identification.* Mineralogical Soc. Monogr. 5.

Bull, P. A. 1981. Environmental reconstruction by electron microscopy. *Progress Phys. Geog.* **5**, 368−97.

Childs, C. W., R. L. Parfitt and R. Lee 1983. Movement of aluminium as an inorganic complex in some podzolised soils, New Zealand. *Geoderma* **29**, 139−55.

Crocker, R. L. and J. Major 1955. Soil development in relation to vegetation and surface age at Glacier Bay, Alaska. *J. Ecol.* **43**, 427−48.

Dahl, E. 1975. Flora and plant sociology in Fennoscandian tundra areas. In *Fennoscandian tundra ecosystems. Part 1.* F. E. Wielgolaski (ed.), 62–7. Berlin: Springer-Verlag.

Dearman, W. R. and F. J. Baynes 1979. Etch-pit weathering of feldspars. *Proc. Ussher Soc.* **4**, 390–401.

Ellis, S. 1979. The identification of some Norwegian mountain soil types. *Norsk. Geogr. Tidsskr.* **33**, 205–11.

Ellis, S. 1980a. An investigation of weathering in some arctic-alpine soils on the north-east flank of Okskotten, north Norway. *J. Soil Sci.* **31**, 371–85.

Ellis, S. 1980b. Physical and chemical characteristics of a podzolic soil formed in Neoglacial till, Okstindan, northern Norway. *Arctic Alpine Res.* **12**, 65–72.

Ellis, S. 1983. Micromorphological aspects of arctic-alpine pedogenesis in the Okstindan mountains, Norway. *Catena* **10**, 133–48.

Evans, L. J. and B. A. Cameron 1979. A chronosequence of soils developed from granitic morainal material, Baffin Island, N.W.T. *Canadian J. Soil Sci.* **59**, 203–10.

Faegri, K. 1933. Über die längen variationen einiger gletscher des Jostedalsbre und die dadurch bedingten pflanzensukzessionen. *Bergens Mus. Årb.* **7**, 1–255.

Fitze, P. 1981. Zur bodentwicklung auf morainen in den Alpen. *Bull. Bodenkundliche Gesellschaft Scweiz.* **5**, 29–34.

FitzPatrick, E. A. 1970. A technique for the preparation of large thin sections of soils and unconsolidated materials. In *Micromorphological techniques and applications*, A. Osmond and P. Bullock (eds), 3–13. Harpenden: Rothamsted Experimental Station.

FitzPatrick, E. A. 1977. *The preparation and description of thin sections of soils.* Aberdeen: University of Aberdeen.

FitzPatrick, E. A. 1980. *Soils: their formation, classification and distribution.* New York: Longman.

Gennadiyev, A. N. 1978. A study of soil formation by the chronosequence method as exemplified by the soils of the Elbrus region. *Soviet Soil Sci.* **10**, 707–16.

Green, F. H. W. and R. J. Harding 1980. The altitudinal gradients of air temperature in southern Norway. *Geogr. Ann.* **62A**, 29–36.

Haldorsen, S. 1978. Glacial comminution of mineral grains. *Norsk Geol. Tidsskr.* **58**, 241–3.

Hodgson, J. M. (ed.) 1974. *Soil Survey field handbook.* Soil Survey Tech. Monogr. 5. Harpenden: Rothamsted Experimental Station.

Holtedahl, O. (ed.) 1960. *The geology of Norway.* Oslo: Norske geologiske undersøkelse, nr. 208.

Jacobson, G. L. and H. J. B. Birks 1980. Soil development on recent end moraines of the Klutlan glacier, Yukon Territory, Canada. *Quat. Res.* **14**, 87–100.

Jenny, H. 1941. *Factors of soil formation, a system of quantitative pedology.* London: McGraw-Hill.

Jenny, H. 1946. Arrangement of soil series and soil type according to functions of soil forming factors. *Soil Sci.* **61**, 375–91.

Kapoor, B. S. 1972. Weathering of micaceous clays in some Norwegian podzols. *Clay Min.* **9**, 383–94.

King, C. A. M. 1959. Geomorphology in Austerdalen, Norway. *Geogr. J.* **125**, 357–69.

Krinsley, D. H. and J. C. Doornkamp 1973. *Atlas of quartz sand surface textures.* Cambridge: Cambridge University Press.

Locke, W. W. 1979. Etching of horneblende grains in arctic soils: an indicator of relative age and palaeoclimate. *Quat. Res.* **11**, 197–212.

Mahaney, W. C., B. D. Fahey and D. T. Lloyd 1981. Late Quaternary glacial deposits, soils and chronology, Hell Roaring valley, Mount Adams, Cascade Range, Washington. *Arctic Alpine Res.* **13**, 339–56.

Matthews, J. A. 1975. Experiments on the reproducibility and reliability of lichenometric dates, Storbreen gletscherforveld, Jotunheimen, Norway. *Norsk Geogr. Tidsskr.* **29**, 97–109.

Matthews, J. A. 1977. A lichenometric test of the 1750 end-moraine hypothesis: Storbreen gletscherforveld, southern Norway. *Norsk Geogr. Tidsskr.* **31**, 129–36.

Matthews, J. A. 1982. Soil dating and glacier variations: a reply to Wibjörn Karlén. *Georgr. Ann.* **64A**, 15–20.

Mermut, A. R. and R. J. St Arnaud 1981. Microband fabric in seasonally frozen soils. *Soil Sci. Soc. Am. J.* **45**, 578−86.

Merrington, O. 1972. Pioneer stage soil development on deglaciated terrain in front of Storbreen. *Horizon* **21**, 78−9.

Moore, T. R. 1976. Sesquioxide-cemented soil horizons in northern Quebec: their distribution, properties and genesis. *Canadian J. Soil Sci.* **56**, 333−44.

Østrem, G. and T. Ziegler 1969. *Atlas over breer i sør Norge.* Norges vassdrags og elektrisitetsvesen hydrologisk avedeling Meddelse, 20.

Pawluk, S. and R. Brewer 1975. Micromorphological and analytical characteristics of some soils from Devon and King Christian Islands, North West Territories. *Canadian J. Soil Sci.* **55**, 349−61.

Romans, J. C. C., L. Robertson and D. L. Dent 1980. The micromorphology of young soils from south-east Iceland. *Geogr. Ann.* **62A**, 93−103.

Ruhe, R. V. 1956. Geomorphic surfaces and the nature of soils. *Soil Sci.* **82**, 441−5.

Sondheim, M. W., G. A. Singleton and L. M. Lavkulich 1981. Numerical analysis of a chrono-sequence, including the development of a chronofunction. *Soil Sci. Soc. Am. J.* **45**, 558−63.

Stork, A. 1963. Plant immigration in front of retreating glaciers, with examples from the Kebnekajse area, northern Sweden. *Geogr. Ann.* **45A**, 1−22.

Ugolini, F. C. 1968. Soil development and alder invasion in a recently deglaciated area of Glacier Bay, Alaska. In *The biology of alder*, J. M. Trappe, J. F. Franklin, R. I. Tarrant and G. M. Hansen (eds), 115−40. Portland: US Forestry Service.

Van der Plas, L. and A. C. Tobi 1965. A chart for judging the reliability of point counting results. *Am. J. Sci.* **263**, 87−90.

Vreeken, W. J. 1975. Principal kinds of chronosequence and their significance in soil history. *J. Soil Sci.* **26**, 378−94.

16

A late Pleistocene–Holocene soil chronosequence in the Ventura basin, southern California, USA

T. K. Rockwell, D. L. Johnson, E. A. Keller and G. R. Dembroff

Introduction

The concept of relative soil profile development of the individual soils in a chronosequence (i.e. a suite of soils on the landscape that differ in age and stage of development) is becoming a widely used tool for temporal control of geological deposits when radiometric or other age determinations are not available. Soils develop continuously through time providing that the soil remains at the surface and that no regressive pedogenesis (erosion or profile rejuvenation) occurs, making them useful in assessing relatively old deposits (10^5 years) as well as younger ones ($10^1 - 10^4$ years). There are, however, certain pedogenic and geological variables that need to be clarified or assessed in order to arrive at reasonable age assignments, and it is therefore necessary to age-calibrate a soil chronosequence regionally before it can be applied to sites where absolute dates are lacking. With such constraints, estimates of soil ages using their relative development are generally good to about ± 25% of the true age of the deposit (Birkeland 1974).

The Ventura basin, an east–west trending structural and depositional feature in the Transverse Ranges province of southern California (Figs 16.1 & 16.2), is ideally suited to the study and development of a soil chronosequence because:

(a) fast rates of vertical tectonism have isolated many more geomorphic surfaces (fluvial terrace and alluvial fan surfaces), than would normally be expected in most other areas for similar time periods (Figs 16.3 and 16.4);

(b) the alluvial parent material for most of the soils is compositionally nearly the same; and

(c) radiometric (^{14}C), dendrochronological and chemical (amino-acid race-

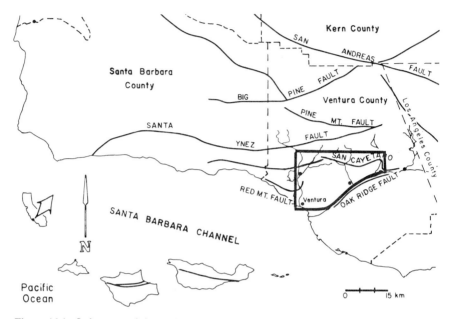

Figure 16.1 Index map of the study area.

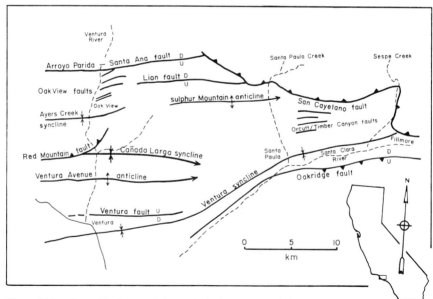

Figure 16.2 Generalised map of the geological structure within the study area.

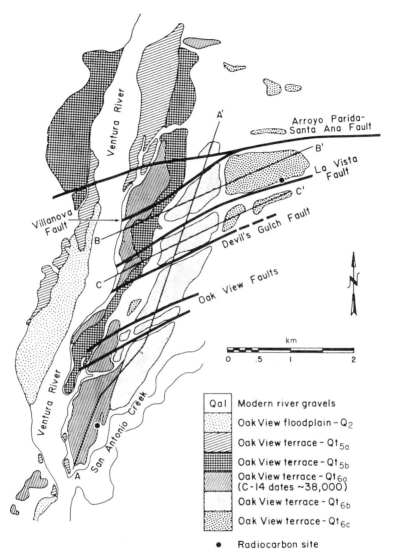

Figure 16.3 Quaternary depositional units and faults in the Oak View area.

misation) age determinations coupled with age estimates based on offset of alluvial deposits by active faults with known slip rates provide good temporal control for the soils to about 200 000 years BP.

In many tectonic studies (e.g. Dibblee 1968, Davis & Burchfield 1973, Crowell 1975), bedrock stratigraphy has been used to determine the sense and amount of fault displacement over geologically significant periods of time. In the Ventura basin, the bedrock is fairly well known and most of the significant geological structures are recognised, but the Late Quaternary history of de-

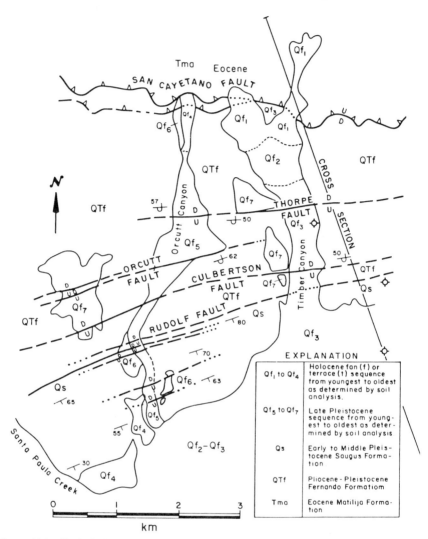

Figure 16.4   Geological map of Orcutt and Timber Canyons, north flank of Santa Clara trough.

formation had not been determined, owing to, in part, the lack of age control for the Late Pleistocene alluvial deposits. In order to develop a chronology to establish recency and rates of deformation, soils developed on stable geomorphic surfaces such as fluvial terraces and alluvial fans were studied and analysed. Once their relative or absolute ages are established, soils may be used as time stratigraphic units to determine amount and timing of faulting or folding. Problems associated with soil stratigraphic correlation are similar to those of bedrock correlation, such as the lateral variations in soil and parent material characteristics (facies of a catena) which are comparable to facies changes in sedimentary units. This sometimes makes correlation difficult, but generally the soils in the Ventura basin display minimal lateral variations, show consistent trends in pedogenic development with increasing age and, in addition, some of the soils have been dated. Thus a useful soil chronosequence exists.

In this chapter, the general aspects of the Ventura basin soil chronosequence are presented, and its usefulness is considered. However, because soils in the Ventura basin appear to be developing at a faster rate than other soils of equivalent age in California, a new model of pedogenesis must be outlined briefly in order to explain the reasons for interregional variation in rates of soil formation as well as local variations within the Ventura study area itself.

Soil evolution models: significant variables and the Ventura basin soils

Soils evolve as a result of a complex interaction of many natural processes which tend to both augment and retard (or reverse) horizonation of the soil profile. Factors that promote horizonation include accumulation of organic matter in the A1 horizon; depletion (eluviation) of organic matter and clay from the A1 and A2 or E (eluvial) horizons; illuviation of clay in the B horizon; and accumulation of carbonates in the B or C horizons. Processes which retard or reverse horizonation (regressive pedogenesis) are those that cause profiles to erode or rejuvenate themselves (Hole 1976, Johnson & Watson-Stegner in press).

While a number of models of soil formation have been formulated (see reviews in Johnson & Watson-Stegner in press, Smeck *et al.* 1983), four are particularly relevant to this study and demand consideration. According to perhaps the most widely known soil model, soil formation (S) is a function (f) of five soil environmental factors: climate (Cl), organic matter or biota (O), topography or relief (R), parent material (P) and time (T), plus other less important factors that were assumed to be independent variables (Jenny 1941). Relations of these factors to soil formation may be expressed as

$$S = f(Cl, O, R, P, T) \qquad (16.1)$$

Wilde (1946) viewed soil formation as a dynamically evolving process consisting of interdependent soil environmental factors that interact through time, and he expressed this concept thus:

$$S = \int (G, E, B) \, dt \qquad (16.2)$$

where G is parent material (i.e. geological substrate), E represents environmental factors and B represents biological factors.

According to another model, which also emphasises pedogenic dynamics and change (Simonson 1959), soil evolution may be conceived as a function of an interacting set of processes involving removals (R), additions (A), transformations (T_1) and translocations (T_2) of materials to, within and from the profile:

$$S = f(A, R, T_1, T_2) \qquad (16.3)$$

Runge (1973) modified the 'five factors' model where water and organic matter were emphasised as the organising and retarding vectors of pedogenesis in a model expressed as:

$$S = f(W, O, T) \qquad (16.4)$$

where W is soil energy related to leaching potential of water, O is organic matter production and nutrient cycling and T is time. According to this model, concentrated flow of water through a porous substrate, with minimal organic matter production, results in relatively strong profile development.

More recently, Johnson and Rockwell (1982) have advanced a conceptual model similar to that of Wilde, in which strength and rate of soil formation can be expressed most simply as:

$$S = f(P, D)dt \qquad (16.5)$$

where S is the strength of soil formation of an entire solum or a single attribute of the solum, and P and D are two interactive sets of various passive and dynamic factors of soil formation (the term 'factors' here includes those phenomena traditionally called factors and those called processes by pedologists; see Fig. 16.5). The derivative 'dt' represents change in both passive and dynamic factor sets with time, such as increased Quaternary precipitation. In this dynamic rate model, which we think best explains the strong development of the soils in the Ventura basin, the derivative of S with respect to time (dS/dt) is the rate of soil development.

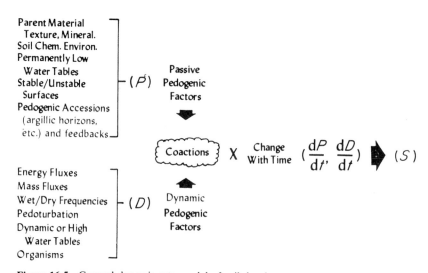

Figure 16.5 General dynamic rate model of soil development.

Barring significant regressive pedogenesis, soil profile development should increase through time according to Equation 16.5, and as idealised in Figure 16.6a. However, the rate of development defined by dS/dt varies, as idealised in the simple (but probably unrealistic) case shown in Figure 16.6b. It is likely that both passive and dynamic factors of soil development will change and interact through time, perhaps with an occasional regressive pulse, to produce several maxima in the rate of soil formation as suggested in Figure 16.7. The maxima in Figure 16.7b should hypothetically reflect Pleistocene and Holocene climatic, biological and physical changes (i.e. in the passive and dynamic factors) that affect rates of soil development. For example, an increase in precipitation increases the energy/mass flux and probably the wet/dry frequency (D factors), which would increase the rate of soil development. On the other hand, as illuvial clay content increases in the B2t horizon the effects of increasing D factors may be reduced owing to the profile becoming gradually plugged with clay, a change in part of the P factor set which characterises the changed soil condition (i.e. an evolved pedogenic accession and associated feedbacks). Thus, over time, there may be several to many maxima and minima in the rate of profile development related to Pleistocene and Holocene changes in passive and, especially, dynamic factors of soil formation, plus a few regressive pulses. For a more comprehensive and detailed discussion of the dynamic rate model and its ramifications, see Rockwell (1983).

Soil properties as indicators of soil profile development

Specific soil properties observed in the field and others assessed through analytical laboratory work indicate the degree or strength of soil profile development. Total soil depth, B horizon distinctness and thickness, soil colour (both mixed matrix and clay film colours), estimated texture, structure, consistence, boundary conditions and abundance, thickness and location of clay films (argillans) are properties observable in the field. All these were used to delineate individual members of the soil chronosequence. The soils were mapped and correlated with representative localities; in addition to using field descriptions containing the above data, these soils were analysed in the laboratory for particle size distribution (sand, silt and clay fractions), soil pH and acidity, cation exchange capacity, base saturation, calcium carbonate content, organic matter content and concentration of soluble salts, exchangeable bases and, to a limited extent, free iron oxides.

The Ventura basin soil chronosequence

The soils developed on alluvium in the Ventura basin are principally Entisols, Inceptisols, Mollisols and Alfisols (Soil Survey Staff 1975). Soils in younger alluvium evolve rapidly from Xerofluvents to Fluventic Haploxerolls to Typic Haploxerolls as dark, generally thick A horizons and cambic B horizons develop

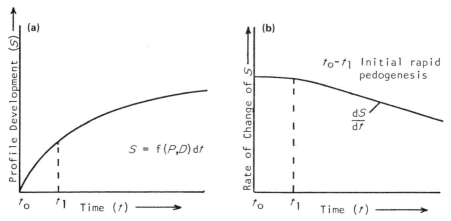

Figure 16.6 (a) The idealised increase in soil profile development with time, suggesting that it slows but does not stop in late development. (b) The simplest case of differentiating S with respect to time to yield the rate of change of soil development S through time.

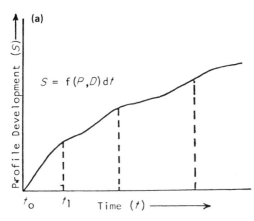

Figure 16.7 (a) A more realistic model, modified from Figure 16.6a, for the increase of soil profile development with time, accounting for changes in P and D factors over time (for example, higher precipitation during the Late Pleistocene). (b) The differentiation of soil development with respect to time for the hypothetical example shown in (a). The model assumes that regressive pedogenesis has been minimal, and that soil development slows but does not reach a steady state.

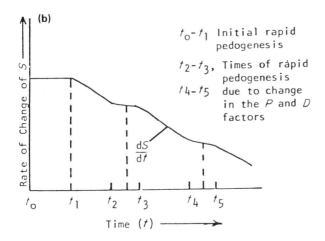

in a few hundred to a few thousand years. The coarse-grained mid- to Early Holocene and Late Pleistocene soils with weak argillic horizons are generally Mollic Haploxeralfs and Typic Argixerolls; fine-grained members of similar age have strongly developed argillic horizons. These in turn continue to develop with time into Typic Haploxeralfs, Typic Palexerolls, Mollic Palexeralfs, and finally Typic Palexeralfs. A few of the very oldest soils resemble Ultisols, although base saturation data are not available to confirm this. This progression from no soil to strongly developed soil is summarised in Table 16.1.

Table 16.1 Measures and indices of relative age of soils from the Ventura basin chronosequence. Soil nomenclature follows Soil Survey Staff (1975).

Geomorphic surface		Classification	Hue	Chroma	Colour index[1]	Clay[2] XB/XA	Clay film index[3]	Estimated age in years BP
Qt$_1$	Car body	Xerofluvent	AC profile, no B horizon				0	10–20
Qt$_2$	Sespe	Fluventic Haploxeroll	AC profile, no B horizon				0	85–200[4]
Qt$_3$	Orcutt 0	Pachic Xerumbrept	10YR	3	4	0.6	0	500–5000[5]
Qt$_4$	Orcutt 1	Typic Argixeroll	10YR	4	5	1.3	3.0	8000–12 000[6]
Qt$_{5a}$	Honor Farm	Typic Argixeroll	10YR	4	5	ND	4.0	15 000–20 000[7]
Qt$_{5a}$	Shell 2	Typic Argixeroll	10YR	3.5	4.5	ND	4.5	15 000–20 000[7]
Qt$_{5b}$	Orcutt 2	Mollic Haploxeralf	10YR	4	5	1.1	6.0	25 000–30 000[6]
Qt$_{5b}$	Bankamericard	Typic Argixeroll	10YR	4	5	ND	7.0	30 000[7]
Qt$_{6a}$	Oak View[10]	Mollic Palexeralf	7.5YR	5	7	1.4	7.25	38 000[8]
Qt$_{6b}$	Apricot	Mollic Palexeralf	7.5YR	6	8	1.5	5.5	54 000±10 000[9]
Qt$_{6c}$	La Vista[10]	Typic Palexeralf	7.5YR	7	9	1.6	7.0	92 000±13 000[9]
Qt$_{6c}$	Orcutt 3	Typic Palexeralf	5YR	4	6	1.6	7.5	80 000–100 000[6]
Qt$_7$	Timber Canyon 4	Typic Palexeralf	5YR	6	9	ND	8.0	160 000–200 000[6,11]

[1]   Colour index is computed by adding chroma number to hue (of moist mixed sample), where 10YR = 1, 7.5YR = 2, 5YR = 3. Indices from different profiles on same geomorphic surface are averaged. To determine colour, a large air-dried bulk sample was passed through a 2 mm sieve, then fractionated in a mechanical splitter, moistened, hand homogenised to a putty consistency and rolled to a sphere; this was then pulled into halves, and colour noted from one freshly broken surface.

[2]   Ratio of the mean percentage of clay in the B horizon to that in the A horizon (computed from particle size graphs).

[3]   This index is based on clay film information contained in the profile descriptions and is computed by adding the percent frequency of clay film occurrence to their thickness, as follows: percentage frequency – very few = 1, few = 2, common = 3, many = 4, continuous = 5; thickness – thin = 1, moderately thick = 2, thick = 3. For example, in the B2t horizon of La Vista 2 there are 'many to continuous (4.5) moderately thick and thick (2.5) clay films'. The index would be 7.0.

[4]   This age is based on the inclusion of an abraded brick fragment in the C horizon of the Qt$_2$ soil at Sespe Creek. A photograph taken in 1898 shows that the terrace was already present.

[5]   This age estimate is collectively based on tree ring ages of a number of mature oaks growing on the Orcutt 0 surface, the degree of soil profile development, and a ^{14}C date on charcoal collected from a presumed buried soil in the lower part of the Timber Canyon profile.

[6]   Age estimate based in part upon relative amount of displacement on flexural-slip faults between older surfaces in Orcutt and Timber Canyons, and one ^{14}C date. Also based upon soil correlation to well-dated soils along the Ventura River.

[7]   Age based on ^{14}C dates from correlative terraces along the lower Ventura River.

[8]   Based on two ^{14}C dates on charcoal collected at the base of the Oak View Terrace below Oliva 1.

[9]   Age based upon relative amount of displacement on the Arroyo Parida fault.

[10]   The measures were taken from the buried soil portion of the profile; only the buried soil portion of the profiles of Oliva 1 and La Vista 3 are correlated to the Qt$_6$ geomorphic surfaces.

[11]   Older and more developed soils grouped with Qt$_7$ have been sampled and described. Thus, a 160 000 years age estimate is a minimum for Qt$_7$ soils, but appears correct for Timber Canyon 4.

Table 16.2 Average soil properties of the Orcutt and Timber Canyon soils. Properties are averaged from at least three pedons for each surface.

Average B horizon property	Q_3	Q_4	Q_{5b}	Q_{6c}	Q_{7a}
moist mixed colour[1]	10YR 4/3	10YR 4/4	7.5–10YR 4/4	5–7.5YR 4/4–4/6	5–7.5YR 5/6
moist clay film colour[1] (reddest)	–	5YR 4/3	5YR 3/3	5YR 4/6–2.5YR 3/6	2.5YR 4/8
Thickness and abundance of clay films[2]	–	1–2 n	2–3 mk	3–4 mk	4k
texture[3]	SL	L–SL	SL–SCL	SCL–C	CL
consistence[4]	so–sh	sh	h	h–eh	vh–eh
structure (for dry pedons)[5]	1,msbk	1–2,m–csbk	2–3,msbk	3,msbk–3,cpr	3,cpr
B horizon thickness (cm)	35–115	150–200	260–380	›340	570–2000
B2t thickness (cm)	–	45–110	120–260	140–340	300–380
contact between A and B horizons	clear	clear	clear	abrupt	abrupt
Presence of A2 or E horizon?	no	no	yes	yes	yes

[1] Munsell colour notation.
[2] n = thin, mk = moderately thick, k = thick; 1 = few, 2 = common, 3 = many, 4 = continuous.
[3] SL = sandy loam, L = loam, SCL = sandy clay loam, CL = clay loam, C = clay.
[4] so = soft, sh = slightly hard, h = hard, vh = very hard, eh = extremely hard.
[5] 1 = weak, 2 = moderate, 3 = strong; msbk = medium sub-angular blocky, m–csbk = medium to coarse sub-angular blocky, cpr = coarse prismatic.

The main soil members of the chronosequence, some of which are in Orcutt and Timber Canyons near Santa Paula (see Table 16.2 for examples) and some of which are along the Ventura River, are described in the following sections. The chronosequence comprises seven primary members, many of which are subdivided into two or three submembers based on geomorphic position and soil properties (Table 16.1). Because the Ventura basin is actively undergoing north–south crustal shortening with resultant faulting and folding, and because each area may respond differently to this deformation, individual correlative members of the chronosequence may to some degree differ in age from those in the dated localities. Nevertheless, the soil profile characteristics for any one deposit are generally distinctive enough to assign that deposit and its soil to at least a major member, if not a submember, of the soil chronosequence.

The soils are designated by numbers 1 – 7, with increasing numerical value corresponding to increasing age. Additionally, the letters 'f' and 't' denote fan and terrace deposits, respectively.

The historically deposited alluvium (presumably similar to the original parent material of all the soils in the immediate locality) is moderately effervescent in 10% HCl and has a mixed matrix colour varying between dark greyish-brown and olive brown (2.5Y 4/4, 2.5Y 4/3 and 10YR 4/2 moist; 2.5Y 6/3, 2.5Y 6/2 and 10YR 6/3 dry).

Q_1 − late historic floodplain and channel alluvium

The Q_1 soils are Entisols with little soil profile development. The alluvial parent material of these soils was deposited in the last few decades during major flood events, such as the twin '100-year' floods of 1969. The profile typically is moderately calcareous throughout, although in some of the older Q_1 soils, the A horizon may be partly leached of calcium carbonate. Darkened surfaces are apparent in some Q_1 soils indicating accumulation of organic matter in the A horizon; for example, in the 'car body' soil along the Ventura River, 1% organic matter has accumulated in the top few centimetres. Also apparent in Q_1 soils is a slight 'tightness' (under physical hand probing) of the soil texture, presumably as a result of organic accumulation.

Q_2 − early historic or late prehistoric low terrace deposits

The soils associated with surfaces of Q_2 age are generally Entic to Fluventic Haploxerolls, with recognisable mollic epipedons overlying C horizons. Carbonates are leached to the C horizon, at a depth of 80 cm on the Qt_2 terrace along Sespe Creek a few kilometres east of Santa Paula, and a weak Cca may be present (at Sespe Creek, the Cca occurs between 140 and 205+ cm and contains thin, discontinuous, white, highly effervescent carbonate coatings on many of the stones).

Q_3 − late Holocene alluvium

Soils on these deposits are Typic or Pachic Haploxerolls (depending on base saturation) that display well-expressed mollic epipedons and cambic B horizons. They are leached of carbonates well into the C horizon, at least to 220 cm at the Qt_3 exposure in Orcutt Canyon, and may have some evidence of incipient illuviation in the B horizon (a few incipient clay-iron coatings, which do not qualify as clay films). Particle size analysis for the Orcutt 0 soil profile in Orcutt Canyon (Qt_3) shows an increase of clay in the B horizon consistent with incipient illuviation even though clay films are not, as yet, apparent.

Weak sub-angular blocky structure may be apparent in the B horizon of some Q_3 coarse-grained soils. The soil colour in the B horizon has 10YR hues and relatively low (<3) moist chromas. The B horizon at well-drained sites is between 30 and 120 cm thick, depending on primary stratification and local focusing of water energy, but is generally between 50 and 100 cm.

Q_4 − latest Pleistocene to Early Holocene Argixerolls and Haploxeralfs

Deposits supporting Q_4 soils are fairly widespread throughout the Ventura basin, although present to a lesser degree along the Ventura River. These soils are typically Argixerolls or Haploxeralfs. The Q_4 soils described below are developed on coarse-grained alluvium, although some fine-grained members do exist which exhibit some aspects of stronger profile development.

These soils are leached of $CaCO_3$ well into the C horizon (>350 cm in Orcutt Canyon) and have weakly developed argillic B horizons with few to common thin, dark, reddish-brown (5YR hues) clay films on ped interfaces, within pores, and on clast matrix interfaces. The overall mixed colour has a hue of 10YR and a chroma equal to or less than 4. Structure in the B2t horizon is commonly massive or weak to moderate, medium and coarse sub-angular blocky, whereas it is 'loose and single grain' in the C horizon; structure often did not appear in deep hand-excavated exposures until the soils were allowed to dry out. The thickness of the B horizon is generally of the order of 150 − 200 cm; a B1 horizon is commonly present and boundaries are generally clear to gradual.

Q_5 − Late Pleistocene Haploxeralfs and Argixerolls

The Q_5 deposits, generally widespread in coarse clast alluvium throughout the Ventura basin, are those that support Typic Haploxeralfs, Argixerolls and some Palexeralfs and Palexerolls (the older Q_5 soils) with moderately well-developed argillic horizons. These soils are characterised by B horizons from 200 to 350 cm thick that have B2t horizons with moderate to strong sub-angular, blocky structure, common to many moderately thick clay films, sandy loam to sandy clay texture, and average mixed soil colour of 10YR to 7.5YR 4/4 m and commonly, incipient to well-developed albic (A2) horizons.

Younger correlatives intergrade with Q_4 soils, whereas older ones intergrade with Q_6 suggesting a fairly wide range of profile development and age. Two distinct submembers, Q_{5a} and Q_{5b} (the younger and older, respectively), are recognised. These range in age from 15 000−20 000 to 20 000−30 000 years BP, dates which are discussed below. In some places there are, however, more than one Q_{5a} or Q_{5b} terrace along the Ventura River because of the very fast rates of uplift. In these cases, differentiation is based on geomorphic and elevational position rather than strength of soil profile development.

Q_6 − Late Pleistocene Palexeralfs and Palexerolls

The Q_6 soils are well-developed Palexerolls and Palexeralfs which have been divided into three submembers (a, b and c from youngest to oldest) based on geomorphic and elevational position along the Ventura River, and on subtle soil property differences within the B horizon. These soils are locally extensive but are less common than the Q_5 and younger soils, presumably owing to the relatively higher uplift and erosion that have limited their preservation.

These soils are characterised by a generally thick B horizon (>400 cm), with a mixed soil colour of 7.5YR 4/4 m to 5/6 m, although some redder correlatives exist owing to inclusion in the parent material of disproportionate amounts of reddish Sespe Formation clasts (Oligocene red beds) or other highly iron-bearing clasts. Clay films are usually thick and can be continuous in the B2t depending on the age of a submember and its geomorphic position relative to runon, runoff and the relative energy level of its location. Clay film colours vary from 7.5YR and 5YR 4/4 m to 5/8 m, with a few 2.5YR hues in the Q_{6c} submember. Textures vary from site to site, from a gravelly sandy clay loam to gravelly clay. The structures of the Q_5 soils were often observed to be massive in a moist state, but upon drying several days or weeks later, were observed to be generally 'strong sub-angular blocky to prismatic'. Apparently, coarse clast parent material interferes with structural expression in a moist state and requires drying to be visible.

The Q_{6c} submembers intergrade to Q_{7a} soils and exhibit brighter and redder colours than the Q_{6b} and Q_{6a} soils. Also, distinct reduction zones where clay has totally plugged the matrix are apparent in all Q_{6c} and some Q_{6b} soil profiles. The texture cannot be used confidently to discriminate between submembers unless slope, water energy and parent material are considered.

Q_7 – early Late Pleistocene to mid-Pleistocene Palexeralfs

The Q_7 soils, Palexeralfs developed on the oldest fluvial and fan surfaces, are subdivided as before (a, b, c) based upon geomorphic and elevational position and soil profile characteristics. The development of these soils appears to have slowed relative to the younger soils. Thus, small changes in profile characteristics may span a considerable length of time.

The representative profile of Q_{7a} is in Timber Canyon on old fan alluvium. This soil is very deep with the B horizon extending to about 750 cm; clay films are present very deeply into the C horizon, at least to 20 m along preferred vertical soil-water channelways. Clay films generally have 2.5YR hues with chromas as high as 8 and are generally continuous and thick where not destroyed by shrinking and swelling, in which case pressure faces are common. The mixed moist B2t horizon colour is from 5YR to 7.5YR 5/6. The Q_{7a} soils are only slightly more developed than the Q_{6c} soils in that they have redder clay films and redder and brighter overall colours; otherwise, most characteristics are broadly similar.

Older Q_7 members are well expressed in Sisar Canyon in the Upper Ojai Valley on the hanging wall of the San Cayeteno fault and along the Ventura River on the hanging wall of the Red Mountain fault. In Sisar Canyon in the Upper Ojai Valley, the older Q_7 soils have dominantly 2.5YR clay film hues (with a few 10R hues) and bright (6 – 8) chromas, and the mixed moist B2t soil colour is 2.5YR to 5YR 4/6 m. The texture of the B2t horizon is clay to sandy clay, the structure is very strong, very coarse prismatic, the A – B boundary is abrupt and the consistence is extremely hard when dry.

The Sisar Canyon Q_7 soils are interesting because the different surfaces on the hanging wall of the San Cayeteno fault (Fig. 16.8) must span a considerable length of time, perhaps several hundred thousand years, based on relative spacing of the fan and terrace levels, yet the soils are remarkably similar in profile development. Thus, soil profile development appears to be slowing with time, at least with respect to profile changes that are apparent to the pedologist's eye.

One trait of all the Q_7 soils studied is the presence of precipitated silica (presumed pedogenic) in the C horizon. The older Q_7 soils also contain silica concentrations or coatings within some pores in the B2t horizon. This, along with the strong red colour and apparent extreme weathering of non-siliceous material in the profile, suggest that these soils verge on Ultisols, although no base saturation data on them are available.

The major properties of the Orcutt/Timber Canyon members of the soil chronosequence are summarised in Table 16.2 as examples of the entire chrono-

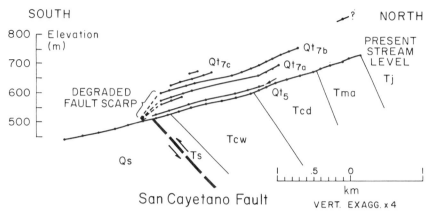

Figure 16.8   Alluvial fan and terrace profiles on the hanging wall of San Cayateno fault at Sisar Canyon. Qs = Saugus Formation of the upper Ojai Valley; Ts = Sespe Formation; Tcw = Coldwater Formation; Tma = Matilija Formation; Tj = Juncal Formation.

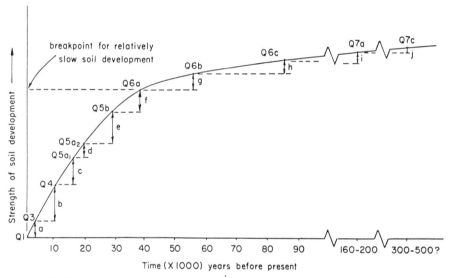

Figure 16.9   Theoretical relationship of strength of soil development with time. The development vectors, *a* through *j*, indicate the relative strength of soil development between soils in the Ventura River area. Soils younger than Qt$_{6a}$ are easily differentiated but older soils are more similar in their characteristics (after Keller *et al.* 1981).

sequence. Based on these properties, the soils in the Ventura basin appear to develop rapidly at first but then slow in relative change in profile development through time (Fig. 16.9). The younger soils (Late Pleistocene and Holocene) are thus fairly distinctive and easily differentiated, whereas the older soils, although dissimilar to the younger soils, appear to develop more slowly, although this could be more apparent than real. Older Q$_7$ soils span a longer period of time relative to the younger soils.

Age control and calibration of the chronosequence

Age control for the chronosequence comes from several ^{14}C dates (Table 16.3) on charcoal from burn zones stratified within the alluvial deposits upon which the soils have developed, from correlation to a dated marine terrace, from dendrochronology and, for the youngest soils, from inclusion of historical debris stratified within the parent deposits (Wood & Johnson 1978). Much of the age control comes from sites along the Ventura River, but several additional dates are from Orcutt and Timber Canyons.

From the Q_1 deposits have been collected or noted such historical articles as iron fence posts, concrete blocks, a 1952 Dodge car body, stock animal bones, and milled wood. In some cases, eye-witness accounts of historic floods have documented the emplacement of some of the Q_1 deposits.

An abraded brick fragment, with one corner intact except for slight rounding due to fluvial transport, was found stratified with the C horizon of a Q_2 deposit along Sespe Creek. An early photograph of the area taken in 1898 shows that the terrace was then present; thus it is at least 86 years old. The Spanish settled this region approximately 200 years ago (the missions of San Buenaventura and Santa Barbara were both established in 1782) so the Q_2 terrace along Sespe Creek is between 86 and 200 years old. Other correlative Q_2 members vary in age depending on local fluvial response to microclimatic or historical perturbations responsible for entrenchment and terrace isolation, or flood accretion of sufficient alluvium to bury an older soil. As there is no distinct older limit to this soil member, many sites may be prehistoric, perhaps several hundreds of years old.

Table 16.3 Radiocarbon (^{14}C) dates for soils in the Ventura basin chronosequence.

Site	Site number[1]	Geomorphic surface	Sample number[3]	Age (radio-carbon years)
Timber Canyon 1 (buried soil)	R1	Qf$_3$ (two soils)	UW−571	9 900±1300
La Vista 3	R2	Qt$_4$	ISGS−768	9 960±200
St Augustine[2]	R3	Qt$_4$	UW−704	10 410±90
Getty 1	R4	Qt$_{5a1}$	UW−721	15 880±210
Shell 3	R5	Qt$_{5a2}$	UW−741	20 040±590
Shell 1	R6	Qt$_{5b}$	UW−710	29 700±1250
Oak View	R6	Qt$_{6a}$	UW−570	39 360±2610
Oak View	R8	Qt$_{6a}$	ISGS−799	36 600±1100
Beach Fault Trench[2]	R9	Qf$_{6a}$	ISGS−718	45 400±1200

[1] These are site reference numbers.
[2] These sites and soils are out of the study area but provide additional control on rates of soil development in coastal southern California.
[3] All UW numbers refer to those samples run in the laboratory at the University of Washington. ISGS samples were run in the laboratory of D. D. Coleman at the Illinois State Geological Survey.

The Q_3 soils probably range in age from about 500 to 5000 years old, based on dendrochronological data from the Q_3 terrace in Orcutt Canyon (the largest oak trees growing on the surface date between 500 and 550 years BP), and on a ^{14}C date on a buried soil below the Q_3 fan surface in Timber Canyon. This buried soil is dated at about 10 000 years BP (see Table 16.3) and represents two Q_3 soils of approximately equal development.

The coarse-grained Q_4 deposits are undated. However, the age of the Q_3 and the youngest of the Q_5 deposits, which are fairly well dated, together with the degree of soil profile development of the Q_3, Q_4 and Q_5 soils, allows for chronosequence bracketing and estimating a reasonable age for the Q_4 deposits and soils at about 8000 − 12 000 years BP. One fine-grained Q_4 soil is dated by ^{14}C at 9960±200 radiocarbon years BP (Table 16.3), corroborating the above estimate.

The Q_5 deposits are dated by three independent (^{14}C) dates on different Q_5 terraces along the Ventura River (Table 16.3) where high rates of uplift have preserved many Late Pleistocene surfaces. The lowest Q_{5a} terrace is dated at 15 880±210 radiocarbon years BP (Table 16.3) and provides a minimum age for Q_5 alluvium. The highest Q_{5b} terrace is dated at 29 700±1250 radiocarbon years BP, providing an upper age limit. A third date of 20 040±590 radiocarbon years BP (Table 16.3) on an intermediate terrace (Q_{5a2}) provides additional chronological control. Based on these dates, the Q_{5a} deposits date generally from 15 000 to 20 000 years BP and the Q_{5b} deposits date from 20 000 to 30 000 years BP. Undifferentiated Q_5 deposits are then assigned an age range of 15 000−30 000 years BP.

Two independent ^{14}C dates, by different laboratories on charcoal from the base of the Q_{6a} terrace along the Ventura River near Oak View, date the youngest Q_6 deposits at about 38 000 years BP (see Table 16.3). Control on the ages of the older Q_6 terraces comes from assuming a constant slip rate on the Arroyo Parida fault and measuring the fault scarp heights (Table 16.4). The basis of the validity of this assumption is threefold. First, the slip rates based on the dated Q_{5b} and Q_{6a} surfaces are about 0.37 ± 0.02 mm yr^{-1} indicating an approximately constant slip rate for at least the last 38 000 radiocarbon years BP. Secondly, the fault is not a bedding plane fault as are others in the area, so the slip should not decrease significantly along the fault over the short distance considered. Finally, based on the first two arguments, even if the slip rate were not constant, it probably has not varied markedly; thus, the age assignments for Q_{6b} (54 000±10 000 years BP) and Q_{6c} (92 000±13 000 years BP) are believed to be approximately correct. This is supported by the correlation of the Q_{6c} terrace at the mouth of the Ventura River with a similarly deformed marine terrace dated at about 80 000 − 105 000 years BP by the amino-acid racemisation method on marine shells (Dembroff 1983, Wehmiller et al. 1978).

Similar to the older Q_6 soils, age control for Q_{7a} surfaces comes from assuming slip rates on faults in Orcutt and Timber Canyons and measuring relative scarp heights of faults which cut multiple surfaces (Table 16.5). Based on the assigned Q_{5b} alluvial fan in Orcutt Canyon, Q_{7a} is between 160 000 and 200 000 years BP. Such an age is consistent with the age of the bedrock (1 million years) which is

Table 16.4 Slip rates for faults in the Oak View area.

Fault	29 700 years ±1 250[1]		38 000 years ±1 500[2]		54 000 years ±10 000[3]		92 000 years ±13 000[3]	
	Dv[4]	Rv[5]	Dv	Rv	Dv	Rv	Dv	Rv
Arroyo Parida–Santa Ana[6]	11 ±0.3	0.37 ±0.02	14 ±0.3	0.37 ±0.02	20 ±3	ND[9]	34 ±3	ND[9]
Villanova[7]	9 ±0.3	0.30 ±0.02	11 ±0.3	0.29 ±0.02	ND[8]	ND[8]	ND[8]	ND[8]
La Vista[7]	11 ±0.3	0.37 ±0.02	15 ±0.3	0.39 ±0.02	41 ±3	0.76 +0.24 −0.17	98 ±3	1.07 +0.20 −0.17
Devil's Gulch[7]	ND[8]	ND[8]	18 ±0.3	0.47 ±0.02	37 ±3	0.69 +0.24 −0.16	ND[8]	ND[8]
Oak View[7]	ND[8]	ND[8]	ND[8]	ND[8]	19 ±3	0.35 +0.15 −0.10	ND[8]	ND[8]

[1] Based on ^{14}C date.
[2] Based on approximate average of two ^{14}C dates and their errors.
[3] Estimated: based on rate of displacement of 0.37 ± 0.02 mm yr^{-1} for the Arroyo Parida–Santa Ana fault during the last 38 000 years.
[4] Dv = vertical displacement (metres) from topographic maps ($Qt_{5b} \pm0.3$ m, $Qt_{6a,b,c} \pm3$ m).
[5] Rv = vertical slip rate (mm yr^{-1}).
[6] Fault that cuts section.
[7] Flexural slip fault.
[8] ND = not determined: field evidence insufficient to estimate vertical displacement.
[9] ND = not determined: slip rate is assumed to be 0.37 mm yr^{-1}, but not determined independently of the slip rate to estimate the ages of the 6b and 6c surfaces.

Table 16.5 Fault displacement and tilting of alluvial fans in Orcutt and Timber Canyons. Determinations made from USGS 7.5 minute topographic maps with 40 ft (12 m) contour interval and by hand level and tape measurements. Measurements are believed to be good in all cases to ± 3 m for the vertical displacements and ±0.2° for the slope determination.

Geomorphic surface	Displacement on faults (m)			Tilting		Estimated age[1] (years BP)
	Thorpe	Culbertson	Rudolph	Present slope	Degrees tilted	
Qf3	4.5	2	6	6°	0°	4 000–5000
Qf5	14	4.5	24–27	8.2°	2.2°	25 000–30 000
Qf6			61	11.1°	5.1°	80 000–100 000
Qf7	98	37		17°	11°	160 000–200 000

[1] Age estimates based on a ^{14}C date from a buried soil under Qf3, and on correlation to well-dated soils along the Ventura River.

overturned and underlies the Q_{7a} Timber Canyon deposits with about a 90° unconformity.

The age assignments for the various members of the chronosequence are summarised in Table 16.1. Based on these age assignments, undated deformed terraces and alluvial fans may be age-evaluated in order to assess rates of tectonic deformation, as long as the soil has not regressed and the deposits are preserved intact.

Summary and application

A soil chronosequence comprising seven primary members ranging in age from 200 000 years BP to the present has been studied and dated for the Ventura basin region. Good temporal control on the chronosequence allows, in most cases, undated deposits with intact soils to be assigned tentative age ranges. This is particularly important in assessing rates of active deformation when active faulting or folding deforms deposits that contain no radiometrically datable material. In all cases, caution must be taken to ensure that the soil is in a stable geomorphic position and that soil regression has not occurred. Such places as at the top or bottom of a fault scarp are inadvisable locations at which to assess the soil development, owing to the likelihood of erosion or burial of the soil. In addition, slope conditions may channel water towards or away from such sites. Similarly, as sites for soil analysis, the apexes of active folds should be avoided because folding may cause soil regression and a significant change in the dynamic factors of soil formation. Soil chronosequences are powerful tools which can be used in nearly all parts of the world to assess ages of alluvial deposits, once a chronology has been locally or regionally evaluated and calibrated.

References

Birkeland, P. 1974. *Pedology, weathering and geomorphological research*. New York: Oxford University Press.

Crowell, J. C. 1975. *The San Andreas fault in southern California*. Calif. Div. of Mines and Geol., Special Report No. 118, 7–27.

Davis, G. A. and B. C. Burchfield 1973. Garlock fault: an intra-continental transform structure, southern California. *Geol. Soc. Am. Bull.* **84**, 1407–22.

Dembroff, G. R. 1983. *Tectonic geomorphology and soil chronology of the Ventura Avenue anticline, Ventura, California*. Unpublished M.Sc. thesis, University of California, Santa Barbara.

Dibblee, T. W. 1968. Displacements on the San Andreas fault system in the San Gabriel, San Bernadino and San Jacinto Mountains, southern California. In *Proceedings of the conference on geologic problems of the San Andreas Fault system*, Vol. XI, 260–78.

Hole, F. D. 1976. *Soils of Wisconsin*. Madison: University of Wisconsin Press.

Jenny, H. 1941. *Factors of soil formation*. New York: McGraw-Hill.

Johnson, D. L. and T. K. Rockwell 1982. Soil geomorphology: theory, concepts and principles with examples and applications on alluvial and marine terraces in coastal California. *Geol Soc. Am. Abs. Programs* **14**(4), 176.

Johnson, D. L. and D. Watson-Stegner in press. The integrated effects of the processes of horizonation and haploidization in soil genesis: a model. *Geoderma.*

Keller, E. A., D. L. Johnson, T. K. Rockwell, M. N. Clark and G. R. Dembroff 1981. *Quaternary stratigraphy, soil geomorphology and tectonic geomorphology of the Ojai−Santa Paula area, California.* Friends of the Pleistocene Fieldtrip Guidebook, Pacific Cell.

Rockwell, T. K. 1983. *Soil chronology, geology, and neotectonics of the north-central Ventura basin, California.* Unpublished Ph.D. thesis, University of California, Santa Barbara.

Runge, E. C. A. 1973. Soil development sequences and energy models. *Soil Sci.* **115**, 183−93.

Simonson, R. W. 1959. Outline of a generalized theory of soil genesis. *Proc. Soil Sci. Soc. Am.* **23**, 152−6.

Smeck, N. E., E. C. A. Runge and E. E. Mackintosh 1983. Dynamics and genetic modelling of soil systems. In *Pedogenesis and soil taxonomy. I Concepts and introductions*, L. P. Wilding, N. E. Smeck and G. F. Hall (eds), 51−81. Amsterdam: Elsevier.

Soil Survey Staff 1975. *Soil Survey manual.* USDA Handbook 18. Washington: US Dept of Agriculture.

Wehmiller, J. F., K. R. Lajoie, A. M. Sarna-Wojcicki, R. F. Yerkes, G. L. Kennedy, T. A. Stephens and R. F. Kohl 1978. Amino-acid racemization dating of Quaternary mollusks, Pacific coast, United States. In *Short papers of the 14th international conference on geochronology, cosmochronology, isotope geology*, R. E. Zartman (ed.), 445−8. US Geol Survey, Open-file Report 78−701.

Wilde, S. A. 1946. *Forest soils and forest growth.* Waltham: Chronica Botanica.

Wood, W. R. and D. L. Johnson 1978. A survey of disturbance processes in archaeological site formation. *Adv. Archaeol. Method Theory* **1**, 315−81.

17

Pedogenic and geotechnical aspects of Late Flandrian slope instability in Ulvådalen, west-central Norway

S. Ellis and K. S. Richards

Introduction

Geomorphological studies of alpine slopes have recently emphasised the role of extreme events – rockfalls, landslides, mudflows and debris slides, flows and avalanches (e.g. Rapp & Strömquist 1976, Selby 1976, Larsson 1982). While glacially-oversteepened slopes in areas of orographically-intensified rainfall and spring snow-melt are always liable to instability, this focus is symptomatic of current geomorphological thinking. A neocatastrophist philosophy views landform evolution as an episodic sequence of erosional and depositional events (Schumm 1975, Brunsden & Thornes 1979), and denudation is not considered simply as the net sediment yield, but rather as the morphological expression of individual events whose impact reflects the relationship between recovery and return periods (Wolman & Gerson 1978). Catastrophic events may dominate morphological development of alpine slopes; indeed, in arid mountains even constructional landforms such as debris fans are caused by rare, extreme debris flows (Beaty 1974). However, a period of stability is often a necessary prerequisite for an extreme event, and little is known about the relationships between preparatory periods of weathering and pedogenesis, the return periods of meteorological events capable of triggering slope failure, the return periods of given magnitudes of slope event on a specific slope profile, and the subsequent recovery periods.

Soil studies can contribute to the resolution of these issues in several ways. The recognition of buried palaeosols is itself indicative of episodic landform development, and Curtis (1975) even suggests that soil horizonation may frequently reflect burial of prior soils rather than vertical pedogenesis. Butler (1967), describing several Australian examples of soil periodicity, notes that erosion and deposition are markedly intermittent processes separated by 'periods of uneventfulness', during which soil profiles are formed at the ground surface.

Successive buried palaeosols thus reflect stable periods punctuated by depositional episodes. Chronosequences based on analytical soil properties and radiocarbon dates permit assessment of the relative temporal importance of stable
and unstable phases, and where deposits are formed by successive extreme slope
events, it is possible to estimate their return period at a site (Rapp & Nyberg
1981). Precise dating is also necessary to justify association of slope failure with
secular climatic variations. Deep-seated landslides in Britain may have been
common in the Late Weichselian, but have also been dated in the Flandrian
(Simmons & Cundill 1974, Tallis & Johnson 1980). Historical evidence certainly
records intensified landslide, rockfall and avalanche activity during the 'Little
Ice Age', c. AD 1650−1760, in Norway (Grove 1972), but any individual catastrophic slope process may be the result of an extreme, random meteorological
event occurring at any time, rather than a systematic climatic fluctuation
(Innes 1983).

The pedological processes differentiating the soil profile generate discontinuities at which vertical permeability changes abruptly, including the ground
surface, soil horizon boundaries, bedrock−soil interface and jointed rock−
intact rock interface. Their development changes the potential for different
slope processes, with lateral flow at these discontinuities variously promoting
runoff, mudflows, landslides and rockfalls (Starkel 1976). For example, podzol
development may proceed until the illuvial horizon impedes vertical permeability and encourages surface saturation and shallow sliding. Thus, the eventual
occurrence of slope erosion involves interaction of a long-term pedological
process and a short-term meteorological event, and an effective meteorological
trigger at a late stage in pedological evolution may be ineffective at an earlier
stage.

Ulvådalen: an example of Late Flandrian slope instability

Soil investigations have been used to approach these issues in a study of catastrophic slope erosion in Ulvådalen, western Norway. Ulvådalen is a western
tributary of the major Romsdal valley. Its valley floor lies at approximately 800 m,
and the northern flanks where the field sites are located (62°15′N, 7°58′E) rise to
Kabbetind at 1338 m (Fig. 17.1). The gneissic bedrock is mantled extensively
by an acid, sandy and gravelly till exposed by the Weichselian inland ice sheet,
which vacated the area around 9000 years BP (Andersen 1980). Nearby meteorological stations record a mean annual temperature and precipitation of approximately 1 °C (Lesjaverk, 630 m) and 760 mm (Verma, 263 m), respectively.
The vegetation in Ulvådalen is of the sub-alpine type (Dahl 1975), comprising a
dense cover of birch trees (*Betula pubescens* subsp. *tortuosa*) with a ground cover
dominated by *Vaccinium myrtillus, V. uliginosum, V. vitis-idaea, Calluna vul-
garis, Empetrum nigrum* and *Deschampsia flexuosa*. Birch trees extend upslope
to about 1000 m, above which occur the alpine vegetation zones of tundra herbs
and shrubs.

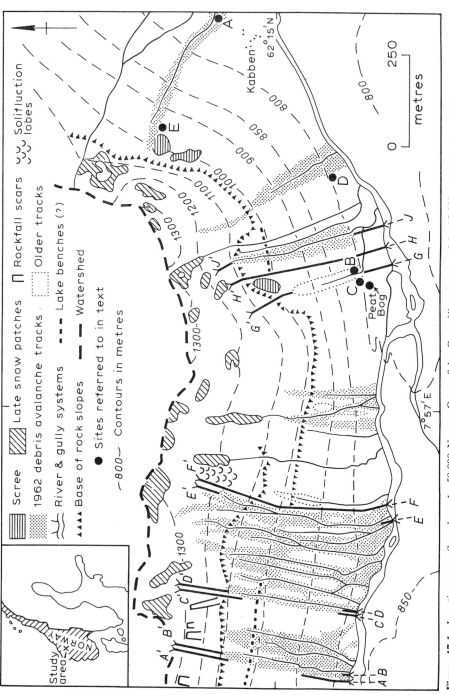

Figure 17.1 Location map (based on 1 : 50 000 Norges Geografiske Oppmåling maps and 1 : 15 000 aerial photographs).

On 26 June 1960 a heavy thunderstorm triggered a series of 28 expanding sheet slides on the south-facing valley side between Kabben and Ulvådalsvatn (Rapp 1963), on slopes of 30−35° beneath a rock face at 1100 m. About 50% of the surface was stripped by these events, and the mean denudation of 200 mm in the area affected contrasts with average annual normal denudation of less than 1 mm (Starkel 1976). A limiting rainfall intensity−duration curve for alpine debris slides (Caine 1980) suggests that a minimum 24-hour intensity of 4.3 mm h^{-1} is needed to trigger such events. The implied 100 mm of rain required in 24 hours is well in excess of the 12−19 mm recorded at nearby gauges, suggesting either extreme localisation of the thunderstorm cell, or sensitivity of these slopes to lower rain intensity because of the large catchment area of impermeable rock at the head of the debris-covered slopes, supplying additional moisture through rapid runoff response.

Although the 1960 event was unique in erosional terms, and the scars are still morphologically dominant features 23 years later (Fig. 17.2), Rapp (1963) noted evidence of overgrown tracks of earlier slope failures located between those caused in 1960. Ulvådalen therefore provided a location in which it was possible to investigate (a) the dating of earlier phases of Flandrian slope erosion, (b) the temporal aspects of pedogenesis on slope deposits of differing age, and (c) the soil geotechnical aspects of slope instability. A search revealed three sites associated with distinctive slope erosional features and slope deposits on which renewed pedogenesis occurred following instability. In addition, a stable site was identified on which uninterrupted Flandrian stability and podzolisation could be assumed.

The first site was an older slide track which had a complete ground cover of vegetation − unlike the 1960 tracks, which have been maintained as bare ground, except on their margins, by surface runoff and gully erosion. The vegetation was, however, less dense and less tall than that of adjacent stable areas, suggesting that the slide was insufficiently old for complete re-establishment of the former vegetation cover. The oldest birch trees appeared to be first generation, and the mean age of the five oldest birches cored in 1979 was 50 years. A similar coring procedure on a 1960 slide deposit indicated that here the mean of the five oldest birches was 15 years; colonisation therefore began in favourable locations about 5 years after the slide event. The older, vegetated slide is therefore dated dendrochronologically as about 55 years in age. This slide differed from the 1960 features in that it began well below the upslope bedrock outcrop, within the birch woodland, and extended down slope a lesser distance, failing to reach the River Ulvåa. Its lower margins were marked by a pronounced rampart at the edge of a depositional lobe; the event thus appears to have been a less intense translational slide which was not converted into a rapid flow or avalanche by high moisture content after failure.

The second site was overgrown by dense birch woodland, and could not be dated dendrochronologically as the trees were unlikely to be first generation. This feature appeared to be the downslope limit of a major rockfall from the upper slope. Large, angular blocks up to 1−2 m in diameter had come to rest adjacent to a small topogenous peat bog whose excavation revealed a layer of inorganic silt at a depth of 34—38 cm. Continued excavation to bedrock at 135 cm revealed no further similar layers. The inorganic silt was therefore interpreted as fine, inwashed sediment deposited prior to vegetation recolonisation over a short time period soon after the rockfall disturbed the slope. Accordingly, samples for ^{14}C dating were taken from 32—34 cm and 38—40 cm depths immediately above and below the inwash layer. The upper sample comprised well-humified peat containing small wood fragments, while the lower sample was of poorly-humified peat lacking such fragments. The upper sample was separated into component fractions using the procedure of Kihl (1975), while the lower sample remained unfractionated. All samples were digested in 2 M HCl for 24 hours at 80 °C prior to ^{14}C analysis. Results in Table 17.1 include conversion to calendar years according to Clark's (1975) system. With the exception of the fine organic fraction, there is no statistically significant difference between the dates, suggesting a short period of silt inwash into the bog following slope instability at about 500 calendar years BP. The 'modern' age of the fine (<125 μm) fraction may reflect younger fine organic downwash, decreasing the sample age, or contamination by fine rootlets of plants growing at the surface; these could not readily be removed from the peat during pretreatment.

The third site provided evidence of less catastrophic slope instability, with no surface manifestation. At about 950 m in an area of dense birch tree cover, on a slope of 18° above a mid-profile convexity adjacent to the easternmost 1960 slide, excavation revealed an *in situ* buried soil covered by a sheet of sandy, gravelly debris 40—70 cm thick. Two samples of the 5 cm thick palaeosol surface organic-rich horizon were collected for ^{14}C analysis, one from each side of the excavated trench. Prior to dating, modern rootlets were removed and samples concentrated into the clay-humus fraction (Kihl 1975). One sample was digested in 2 M HCl for 24 hours at 80 °C while the other remained untreated. The dates obtained (Table 17.1) give an age range of 1530—2140 calendar years BP. However, this

Table 17.1 Ulvådalen radiocarbon dates.

Site	Depth (cm)	Material dated	Lab. no.	^{14}C age (years BP)	Calibrated age (years BP)
peat bog	32—34	peat	SRR—1694	485±45	524 (590—440)
	32—34	fine organic (<125 μm)	SRR—1695	modern	modern
	32—34	wood fragments	SRR—1696	400±80	480 (570—310)
	38—40	peat	SRR—1697	380±60	470 (540—310)
palaeosol	40—45	clay-humus	SRR—1692	1705±45	1674 (1760—1530)
	40—45	clay-humus (HCl digested)	SRR—1693	1975±60	1920 (2140—1750)

cannot be assumed to represent the age of the slope wash responsible for burial, because the residence time of the palaeosol organic matter must be subtracted. This is discussed further after a consideration of the temporal aspects of pedogenesis on the other dated surfaces.

Establishing a chronosequence

Various dated surfaces have been used to establish soil 'chronosequences' (Jenny 1946), including fluvial deposits, till, mining debris, volcanic material, aeolian sand, raised shoreline sediments and slope deposits (cf. Bockheim 1980). Temporal studies of pedogenesis in Scandinavia have, however, concentrated on moraine ridges, generally above the altitudinal tree limit (e.g. Ellis 1980a,

Figure 17.2 The south-facing valley side of Ulvådalen. Note the debris slide/ avalanche scars, the mid-profile convexity, the runnels and scars on the rock free face and the lake shoreline or lateral drainage channel feature.

Table 17.2 Soil profile field and laboratory data.

Site	Horizon	Depth (cm)	Colour (Munsell)	% sand (2000–60 μm)	% silt (60–2 μm)	% clay (<2 μm)	pH (H₂O)	% Organic C	% Fe$_p$	% Fe$_d$	% Al$_p$	% Al$_d$
A	LFH	0–2	10YR 4/1	—	—	—	3.6	2.68	—	—	—	—
	Cu	2+	2.5Y 6/4	71.2	23.2	5.6	3.9	1.41	—	—	—	—
B	LFH	0–5	5YR 2.5/2	—	—	—	4.3	3.54	—	—	—	—
	Ea	5–8	10YR 7/2	71.6	26.8	1.6	4.4	1.00	0.05	0.07	0.04	0.06
	Bw	8–21	10YR 6/6	74.9	23.5	1.6	4.5	0.72	0.12	0.30	0.10	0.12
	Cu	21+	10YR 6/4	68.4	30.8	0.8	4.7	0.38	0.03	0.27	0.10	0.12
C	LFH	0–4	5YR 3/2	—	—	—	4.1	27.39	—	—	—	—
	Ea	4–16	10YR 7/1	80.7	19.1	0.2	4.5	1.17	0.06	0.10	0.02	0.03
	Bw	16–32	10YR 5/4	85.3	14.2	0.5	4.9	2.50	0.14	0.37	0.11	0.15
	Cu	32+	10YR 6/4	80.2	19.6	0.2	4.6	0.57	0.08	0.29	0.11	0.15
D	LFH	0–8	5YR 3/1	—	—	—	4.4	30.37	—	—	—	—
	Ea	8–16	10YR 7/1	70.3	28.6	1.1	3.9	0.68	0.02	0.07	0.04	0.07
	Bs	16–45	7.5YR 4/4	64.9	30.4	4.7	3.7	4.44	0.53	0.88	0.64	1.42
	Cu	45+	2.5Y 6/4	60.0	38.8	1.2	4.4	1.46	0.10	0.37	0.48	0.65
E[1]	LFH	0–5	10R 4/1	—	—	—	—	12.56	—	—	—	—
	Ea	5–23	10YR 6/1	71.5	28.3	0.2	—	1.75	0.11	0.27	0.11	0.10
	Bs	23–36	7.5YR 4/4	74.2	25.2	0.6	—	3.08	0.89	0.74	0.54	0.46
	Cu	36–40	10YR 5/2	81.6	18.2	0.2	—	1.00	0.08	0.25	0.19	0.23

[1] Site E is the overlying podzol at the palaeosol site.

Ellis & Matthews 1984, Mellor this volume). Ulvådalen therefore provides a distinctive case study of pedogenesis on a different type of surface, below the tree limit in a less hostile climatic environment. The sites chosen for pedological investigation were all freely drained, on slopes of around 2° with a southerly aspect and at approximately 800 m. Site A was situated on the easternmost 1960 slide deposit, site B on the slide material dated by tree-ring counts as 55 years old, site C on material disturbed and deposited by the rockfall about 500 years ago, and site D on till on a stable slope (Fig. 17.1). This latter site represents the stage of pedogenesis reached since deglaciation about 9000 years BP. Data from a fifth site, E, relate to the surface soil developed over the palaeosol; this site is higher (950 m) and steeper (18°) than the others (Fig. 17.1), and its dating is less secure.

With the exception of site A, all soils were podzolic, possessing a surface organic (LFH) horizon (Avery 1980) overlying a bleached eluvial (Ea) horizon beneath which occurred an illuvial (Bs or Bw) horizon and a pedologically little-altered Cu horizon of parent material. Site A showed only a surface organic horizon directly overlying the Cu horizon. Bulk samples were analysed for particle size distribution, reaction (pH), organic carbon, iron and aluminium content (Bascomb 1974), and thin sections were prepared after Araldite impregnation (FitzPatrick 1970). The profile field and laboratory data are summarised in Table 17.2. Graphical and regression techniques have commonly been used to examine the rate of operation of pedogenic processes reflected in such data from soil chronosequences (Bockheim 1980), although in the present investigation the limited number of dated sites often precludes the latter.

The first soil property examined is the surface horizon organic carbon (C) content. This increases markedly in the first 500 years of pedogenesis (Fig. 17.3a) as rapid vegetation colonisation occurs on disturbed surfaces, with associated litter production and organic matter accumulation and decomposition occurring at the soil surface. Highly decomposed organic material is evidenced by meso-faunal faecal pellets seen in thin section, but this is generally exceeded by the presence of fresh or only partly decomposed material (Fig. 17.4a), indicating net organic matter accumulation. After 500 years the rate of increase of surface organic C slows markedly, suggesting that accumulation and decomposition are more in balance. These trends accord with results for other chronosequence investigations in cold environments (e.g. Crocker & Dickson 1957, Mahaney 1974, Jacobson & Birks 1980). Organic C may therefore be a 'rapidly adjusting' soil property, attaining a steady state within 10^3 years (Yaalon 1971), and an exponential asymptotic function almost exactly fits the four data points plotted in Figure 17.3a. Organic C accumulation is also apparent in the B horizons of the three podzols at sites B, C and D (Table 17.2), and an increase with time is again apparent (Fig. 17.3b), although quantities are much lower, and the approach to steady state less apparent.

Other properties show less evident temporal trends. Soil reaction values (Table 17.2), for example, reflect the low potential for translocation of bases in the highly acid parent material, and clay percentages are consistently low at all

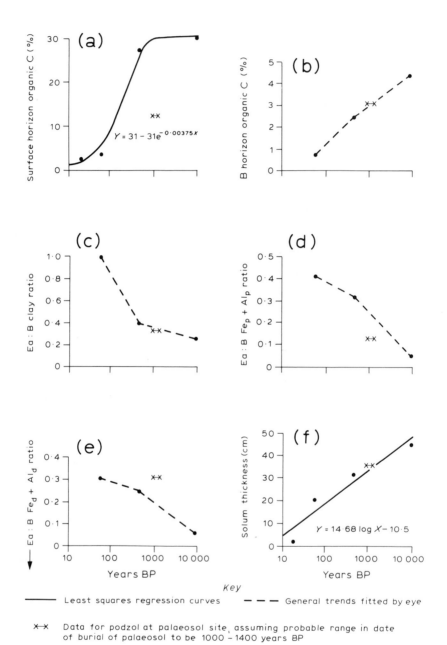

Figure 17.3 Chronosequence data for the four dated soil profiles (for some soil properties only three data points are available, because of the lack of development of the youngest soil).

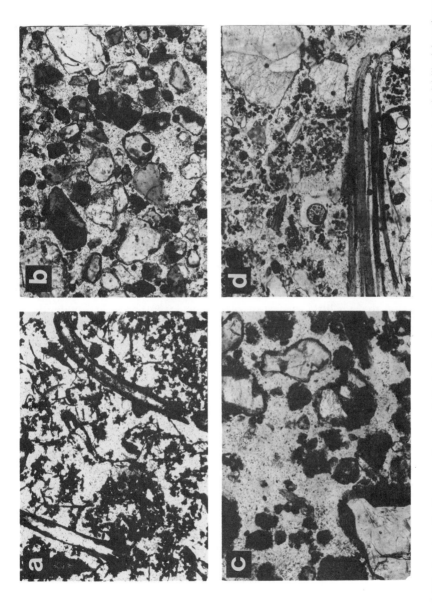

Figure 17.4 (a) Fresh and partly-decomposed organic matter in the surface organic horizon of site B (frame length = 4 mm); (b) illuvial coatings surrounding mineral grains in the Bw horizon of site C (frame length = 2 mm); (c) mineral grain illuvial coatings in the Bs horizon of site D (frame length = 2 mm); (d) a biotite grain showing expansion along its cleavage planes in the Bs horizon of site E; note also the granular and crumb microstructural elements (frame length = 2 mm). All photomicrographs are viewed in partially crossed polarised light.

depths in all profiles because of the clay-deficient nature of the till, and the low degree of weathering into this size fraction during pedogenesis. This is supported by the predominantly fresh, unweathered appearance of mineral grains in all horizons viewed in thin section. Clay translocation may, however, be apparent, although the process is difficult to quantify because of inherent clay content variation in the parent material (Evans 1978). As an index of translocation, a ratio was used of the quantity of clay in the eluvial (Ea) horizon to that in the illuvial (B) horizon of each podzolic profile (Fig. 17.3c). This index allows for variable parent material content by measuring relative concentration in the illuvial horizon, and its evident temporal decrease suggests a progressive translocation during pedogenesis, but at a decreasing rate over time. Supportive micromorphological evidence for progressive translocation is seen in B horizons, where development of illuvial coatings surrounding mineral grains increases with increasing soil age (Figs 17.4b & c).

Since sesquioxides are mobile during podzol development, these should also exhibit temporal trends. Pyrophosphate-extractable iron (Fe_p) and aluminium (Al_p) are generally considered to represent sesquioxides in an organically bound form (Childs *et al.* 1983), and a similar translocation index to that used for clay also suggests progressive translocation, but decreasing in rate over time (Fig. 17.3d). A similar trend is apparent using this index for dithionite-extractable sesquioxides (Fe_d + Al_d) (Fig. 17.3e). This could indicate that B horizons are experiencing progressively greater *in situ* weathering than Ea horizons, although in thin section there is little suggestion of chemical weathering at any depth; almost all mineral grain surfaces are smooth and lack discolouration. Weathering is evidenced only by the very occasional hydration of biotite grains, producing expansion along cleavage planes (Fig. 17.4d). It seems more probable, therefore, that Fe_d and Al_d represent sesquioxides released during formation of the till and subsequently translocated in an inorganic form. Inorganic translocation of sesquioxides has been suggested for other podzols in Norway and North America (Moore 1976, Ellis 1980a), and has recently come to be considered as a significant component of podzolisation (Anderson *et al.* 1982, Childs *et al.* 1983). Illuvial coatings in B horizons, seen in thin section, therefore probably contain both organically and inorganically translocated sesquioxides in addition to organic matter and clay. Also apparent in these horizons are granular and crumb microstructures (Fig. 17.4d) thought to result from illuviation of organic matter and sesquioxides, and identified in other Norwegian podzols (Ellis 1983, Ellis & Matthews 1984). It is the progressive illuviation of sesquioxides and organic matter which results in the darkening (decrease in colour value) of B horizons with increasing soil age (Table 17.2). The rate of clay translocation appears to decrease markedly after about 500 years, perhaps approaching a steady state. This could reflect the initially high porosity of loosely accumulated slope debris which subsequently compacts. However, the sesquioxides appear to represent 'slowly adjusting' soil properties which, although experiencing a decreasing translocation rate over time, show no evidence of reaching a steady state within 10^3 years (Yaalon 1971).

Solum thickness is a function of both surface accumulation of organic material and the translocation of various soil constituents as discussed above. Its value is seen to increase at a decreasing rate, as a function of the logarithm of age (Fig. 17.3f); it increases rapidly during the first 500 years of pedogenesis and then begins to stabilise. It may approach a steady state after about 1000 years, but if so, translocation apparently continues, with the mineral horizons continuing to alter in their concentration of translocated constituents, if not appreciably in their thickness. This is suggested by other studies of Scandinavian podzols (Ellis & Matthews 1984), and is of considerable importance in the context of the changing hydraulic conductivity of the podzol profile, and the developing impediment to drainage.

The Ulvådalen chronosequence may be examined in the context of other Scandinavian studies which indicate the time required to produce a podzolic profile. In northern Norway, it has been suggested that a visually and chemically distinguishable podzolic profile develops in 250–1000 years (Ellis 1980a). It appears that in Sweden a minimum period of around 100 years is necessary for the formation of visually recognisable profiles (Tamm 1950), while in Finland, chemically and visually distinguishable profiles can develop in 200–300 years and 400–500 years, respectively (Jauhiainen 1973). A profile both visually and chemically podzolic can apparently develop in only 55 years in Ulvådalen (site B). Although much shorter than the other time estimates, this period is similar to the 75 year period suggested by chronosequence investigations in southeastern Alaska (Crocker & Dickson 1957). Environmental contrasts influence the differences in the time required for podzol development (Ellis 1980a), and slower translocation rates and podzolisation than at Ulvådalen might be expected in the more severe climate of northern Norway and the drier conditions experienced in Sweden and Finland.

Interpretation of the palaeosol site

Several difficulties associated with ^{14}C dating of soil organic horizons (Matthews 1980, this volume) must be evaluated prior to interpretation of the palaeosol ^{14}C dates (Table 17.1) in the contexts of the chronosequence, and the timing of slope instability and profile burial. Removal of modern rootlets from the palaeosol surface organic-rich horizon prior to ^{14}C dating minimised the possibility of contamination by living organic material, and since the palaeosol is developed in and buried by acidic material, the dates are considered to be uncontaminated by minerogenic carbon. Although the palaeosol is buried by material in which a podzolic profile has subsequently developed, the ^{14}C-dated horizon is situated below the maximum depth of the Bs horizon of the podzol, and is therefore unlikely to be seriously contaminated by illuvial organic matter. The older of the two dates obtained (Table 17.1) was for a sample pretreated with HCl; the younger date was for an untreated sample. It is therefore likely that the 'fulvic acid' fraction was removed by the HCl pretreatment, and since this fraction is generally considered representative of more mobile organic matter, its removal should cause the remaining acid-insoluble material to possess a slightly greater age (Matthews 1980). The older date is thus the more accurate on which to base interpretation of the timing of profile burial.

However, the 'apparent mean residence time' (AMRT), which represents the mean time over which organic material has resided in the horizon prior to profile burial (Scharpenseel & Schiffmann 1977), must be subtracted from the ^{14}C date to obtain the time since burial. Recent investigations suggest that the AMRT of organic matter in Norwegian soils may exceed 1000 years (Matthews 1980, 1981), which would imply a much more recent date of burial of the palaeosol than the 1920 years BP calibrated ^{14}C date. Since AMRT appears to increase with depth in a horizon (Matthews 1980), the ^{14}C dates from the uppermost parts most closely reflect the time since burial. Unfortunately, it was necessary to sample over the entire 5 cm thickness of the palaeosol organic-rich horizon in order to obtain sufficient organic matter for dating, and depth-controlled sampling was impossible. Thus the AMRT may represent a significant component of the ^{14}C dates obtained. However, some assessment of the age of the podzolic soil developed in the overlying material (site E) is possible by comparison with the chronosequence discussed above, and this permits approximate division of the ^{14}C age into AMRT and time since burial.

In terms of visual horizon differentiation, this profile appears more fully developed than the site C podzol but much less so than that at site D (Table 17.2). An age intermediate between these is supported by the values of solum thickness, B horizon organic C content, and Ea : B horizon ratios of clay (0.30) and combined Fe$_p$ and Al$_p$ (0.15) contents (Table 17.2 & Fig. 17.3). However, *precise* insertion in the chronosequence is unrealistic because local environmental conditions of altitude and slope are different at site E, and as a result some properties lie outside the range of values from sites C and D, e.g. surface organic C and the Ea : B ratio of Fe$_d$ + Al$_d$ (Fig. 17.3). Since the podzol profile at the palaeosol site appears significantly older than the soil at site C, it is considered unlikely to be younger than about 1000 years old. This accords roughly with the difference between the ^{14}C age of the buried soil organic horizon (1920 years BP) and a mean residence time of 1000 years (Matthews 1980, 1981). If logarithmic regression relationships are fitted to the chronosequence data for solum thickness and organic C in the B horizon and the site E values are inserted in the equations, age estimates are obtained which range from 1290–1470 years, with an average of 1380 years. This predicts a mean residence time of only 540 years, which may be acceptable given that environmental conditions at this site differ from those at the sites investigated by Matthews.

The palaeosol profile itself is not podzolic, but resembles the Brown Soil described in northern Norway (Ellis 1979). This is considered to be genetically related to the podzol, but lacks a bleached eluvial horizon since environmental conditions are less conducive to podzolisation (Ellis 1979, 1980b). The palaeosol may therefore have developed beneath a tundra vegetation type incapable of preventing creep or slope wash on this 18° slope, with the result that formation of visually distinct horizons including the bleached eluvial zone was inhibited. The podzol profile overlying the palaeosol may, in contrast, have developed under conditions of slope stability caused by birch tree cover. If so, the altitudinal tree limit must have risen after the episode of slope instability approximately 1000 years ago. The site is only some 50 m below the present tree limit, so could conceivably have been in the tundra vegetation zone prior to the slope instability. The tree line history of the Romsdal area is unknown, but there is evidence for Flandrian fluctuations elsewhere in Scandinavia (e.g. Kullman 1981).

The mechanics of slope instability in Ulvådalen

The south-facing valley side of Ulvådalen is characterised by a smooth, gneissic free face above 1100 m, an upper debris-covered slope of 30−35° which is vegetated but covered with patches of scree, and which decreases to 17−23° above a structurally controlled mid-slope convexity, and a lower concavity decreasing from 25−28° to 5−11° near the River Ulvåa (see Figs 17.2 & 17.5). The 1960 debris slides were generally initiated on the steep slopes beneath the free face as shallow translational slides. Eye-witness accounts summarised by Rapp (1963) suggest downslope propagation of the initial slides followed by conversion to debris flows which accelerated into avalanche forms over the mid-slope convexity. High moisture content and the fine sand (20−22%) and silt-clay (9−21%) contents probably aided fluidisation and transport of the coarser clasts. The tracks generally widen downslope and are associated with marginal levees of coarser debris; basal lobes are often absent, however, where the tracks reached Ulvådalsvatn or the River Ulvåa.

The till and the regolith above the probable ice limit are gravelly sands with 20−50% by weight of gravel and median sand fraction diameters of 130−300 µm. Geotechnical testing of such heterogeneous non-cohesive material inevitably involves disturbed samples, which were obtained from the head of the easternmost slide track, the bCu (parent material) horizon of the adjacent buried soil, and the Bs horizon of the overlying podzol. Direct shear tests were performed in a 6 cm shear box after consolidation under normal stresses of 20−60 KN m^{-2}, similar to the stresses experienced in the field. The tests performed were on dry samples sheared at 0.0007 mm s^{-1}, and repetitive testing was effected after screening particles coarser than 2, 4 and 6.73 mm, to give ratios of shear box length to maximum grain diameter of 30, 15 and 8.9. Chandler (1973) reports that ratios >12 are required to inhibit development of unrepresentative stresses around large grains. However, in these tests the small number of coarser particles produced no systematic effect on the measured angle of shearing resistance, which varied from 38.4° to 40.9°. The bCu horizon of the buried soil, with a gravel content of 37−49%, had $\varphi' = 38.8°$ to 39.6°, while for the less gravelly slope deposit forming the Bs horizon of the overlying podzol, φ' was 38.4°. At the head of the slide, $\varphi' = 40.9°$, possibly because the gravel content tended to be rougher

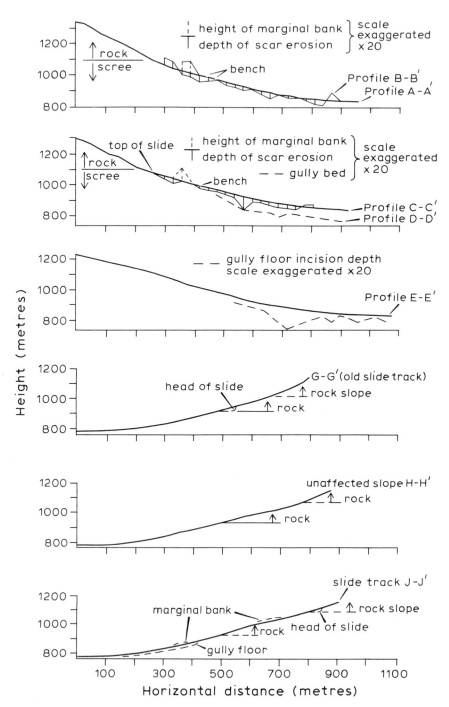

Figure 17.5 Typical slope profiles along and between slide tracks on the south-facing valley side of Ulvådalen (see Fig. 17.1 for profile locations).

in surface texture as a result of prolonged weathering above the ice limit (cf. Sollid & Sørbel 1979).

On slopes of 30−35° these friction angles suggest that the regolith is stable when dry, because the threshold angle for shallow sliding under such conditions equals the angle of shearing resistance. However, application of the infinite slope, shallow slide stability analysis, assuming groundwater flow parallel to the surface (Skempton & De Lory 1957), indicates that if the water table runs parallel to the surface, it needs to rise 0.2−0.7 of the distance from the shear plane to the surface to reduce the factor of safety (F) to unity and cause instability. Since

$$F = \frac{(\gamma_s - m\gamma_w) \tan \varphi'}{\gamma_s \tan \beta} \tag{17.1}$$

the proportional height of the water table above the shear plane (m) is found by inserting appropriate values for the slope angle β, friction angle φ', and saturated sediment (γ_s) and water (γ_w) unit weights in Equation 17.1 and setting F to unity.

Such rises in the water table are not extreme, but may be unusual on these slopes because downslope drainage is rapid. However, several factors could have contributed to increased pore-water pressures in the 1960 event. High-intensity thunderstorm rainfall was augmented by instantaneous runoff from the bedrock slopes above and so supplied moisture at rates exceeding the permeability of the regolith both vertically, because the B horizon represents a drainage impediment particularly below the tree line where podzols are developed, and laterally, because the gradient was reducing down slope onto the rock ledge underlying the mid-slope convexity. In addition, scars on the rock face above some slides suggest that rockfalls may have triggered translational slides by suddenly loading the slope and forcing a sharp, undrained rise in pore-water pressure (Hutchinson & Bhandari 1971), thereby reducing the effective shear strength of the material. Subsequent translation of debris and loading downslope may then have produced a progressive failure.

The main pedological influence on slope stability is thought to lie in the temporally-increasing drainage impediment which arises because translocation rates into the B horizon are maintained while the increase of solum thickness becomes slower. Thus the devastation wrought on the Ulvådalen hillslope reflects the occurrence of an extreme meteorological event at a propitious moment in soil evolution. Nevertheless, shear surfaces would be expected at the base of the Ea horizon at depths of 15−20 cm, whereas the mean erosion depth of 0.4 m in the slide tracks approximates to the total solum thickness of a mature (10^3 − 10^4 years) podzol. This may be because of the subsequent debris flow and later gully erosion, rather than the initial slide, however. Despite these complexities, the Ulvådalen study illustrates the points that geomorphological work is not simply the product of event magnitude and frequency, and that the relation of slope failure to 'forcing' meteorological events (heavy rain or rapid snowmelt) is non-linear and possibly cyclic. Walker (1963) provides further support for this episodic form of erosion: distinctive soils on Quaternary debris avalanche deposits in Australia suggest that talus deposits have twice built up beneath the Triassic Hawkesbury sandstone scarp, and have then failed. Although failures were attributed to wetter climatic phases, critical stages in the accumulation of sediment supply may also be important. On a shorter timescale, Newson (1980) showed that a storm in mid-Wales in August 1973 produced significant slope failure, while a similar event in 1977 mainly involved channel changes, because the potentially unstable slopes had failed in 1973. The interaction between intrinsic failure thresholds generated by weathering, pedogenesis and debris accumulation, and extrinsic thresholds caused by meteorological or hydrological events therefore considerably complicates the magnitude−frequency relations of slope erosional processes.

Temporal implications

Assessment of the magnitude and frequency of slope processes and their relationship to slope profile morphological development is thus highly problematic. Debris slide frequencies based on dating lobes on debris cones (Rapp & Nyberg 1981) may be overestimates of frequencies acting on *profiles* because each depositional event may derive from a different part of an avalanche gully system, rather than being successively generated from one profile. Normally, the probability of occurrence of a particular type of event on a profile line decreases as the meteorological trigger becomes spatially localised and as the necessary prior period of weathering and pedogenesis increases. Furthermore, one process type may redistribute regolith downslope and therefore alter the probability of occurrence of other process types. Thus in Ulvådalen there is evidence of slope instability in 1960, and about 55, 500 and 1000−1400 years BP. This is not indicative of secular environmental change, because extreme events may occur on a slope at any time, although regional incidence may be enhanced during periods of climatic deterioration (Grove 1972). The buried soil is, however, suggestive of a change of tree limit and a possible climatic amelioration since 1000−1400 years BP. These earlier, dated, events also fail to define a return period for slope failure because each involves a different type of process; the slope profile is thus best viewed as a palimpsest of a range of morphologically and sedimentologically effective processes. Although such conclusions may seem negative, indicative as they are of the considerable problems of interpretation that arise in assessing temporal aspects of slope development, the investigation of Ulvådalen soils and slope deposits illustrates the potential of soil studies in the analysis of the history and mechanisms of alpine slope processes.

Acknowledgements

Financial assistance for fieldwork was provided by the British Geomorphological Research Group and the Sir Philip Reckitt Educational Trust. Radiocarbon dating was funded by the Natural Environment Research Council and performed at the NERC Radiocarbon Laboratory under the direction of Dr D. D. Harkness. Laboratory analyses were conducted, and illustrative material prepared, by members of the technical staff of the Geography and Geology Departments at Hull University. To all these individuals and organisations we offer our sincere thanks.

References

Andersen, B. G. 1980. The deglaciation of Norway after 10,000 B.P. *Boreas* **9**, 211−16.

Anderson, H. A., M. L. Berrow, V. C. Farmer, A. Hepburn, J. D. Russell and A. D. Walker 1982. A reassessment of podzol formation processes. *J. Soil Sci.* **33**, 125−36.

Avery, B. W. 1980. *Soil classification for England and Wales.* Soil Survey Tech. Monogr. No. 14.

Bascomb, C. L. 1974. Physical and chemical analyses of ‹2 mm samples. In *Soil Survey laboratory methods*, B. W. Avery and C. L. Bascomb (eds), 14−41. Soil Survey Tech. Monogr. No.6.

Beaty, C. B. 1974. Debris flows, alluvial fans, and a revitalised catastrophism. *Zeitschr. Geomorph. Suppl.* **21**, 39−51.

Bockheim, J. G. 1980. Solution and use of chronofunctions in studying soil development. *Geoderma* **24**, 71−85.

Brunsden, D. and J. B. Thornes 1979. Landscape sensitivity and change. *Trans Inst. Brit. Geogs New Ser.* **4**, 463−84.

Butler, B. E. 1967. Soil periodicity in relation to landform development in south-eastern Australia. In *Landform studies from Australia and New Guinea*, J. Jennings and J. A. Mabbutt (eds), 231−55. Cambridge: Cambridge University Press.

Caine, N. 1980. The rainfall intensity-duration control of shallow landslides and debris flows. *Geogr. Ann.* **62A**, 23−7.

Chandler, R. J. 1973. The inclination of talus, arctic talus terraces, and other slopes composed of granular materials. *J. Geol.* **81**, 1−14.

Childs, C. W., R. L. Parfitt and R. Lee 1983. Movement of aluminium as an inorganic complex in some podzolised soils, New Zealand. *Geoderma* **29**, 139−55.

Clark, R. M. 1975. A calibration curve for radiocarbon dates. *Antiquity* **49**, 251−66.

Crocker, R. L. and B. A. Dickson 1957. Soil development on the recessional moraines of the Herbert and Mendenhall Glaciers, south-eastern Alaska. *J. Ecol.* **45**, 169−85.

Curtis, L. 1975. Landscape periodicity and soil development. In *Processes in physical and human geography*, R. F. Peel, M. Chisholm and P. Haggett (eds), 249−65. London: Heinemann.

Dahl, E. 1975. Flora and plant sociology in Fennoscandian tundra areas. In *Fennoscandian tundra ecosystems. Part 1. Plants and microorganisms*, F. E. Wielgolaski (ed.), 62−7. Berlin: Springer.

Ellis, S. 1979. The identification of some Norwegian mountain soil types. *Norsk Geogr. Tidsskr.* **33**, 205−11.

Ellis, S. 1980a. Physical and chemical characteristics of a podzolic soil formed in Neoglacial till, Okstindan, northern Norway. *Arctic Alpine Res.* **12**, 65−72.

Ellis, S. 1980b. Soil−environmental relationships in the Okstindan Mountains, north Norway. *Norsk Geogr. Tidsskr.* **34**, 167−76.

Ellis, S. 1983. Micromorphological aspects of arctic-alpine pedogenesis in the Okstindan Mountains, Norway. *Catena* **10**, 133−48.

Ellis, S. and J. A. Matthews 1984. Pedogenic implications of a ^{14}C-dated paleopodzolic soil at Haugabreen, southern Norway. *Arctic Alpine Res.* **16**, 77−91.

Evans, L. J. 1978. Quantification and pedological processes. In *Quaternary soils*, W. C. Mahaney (ed.), 361−78. Norwich: Geo Abstracts.

FitzPatrick, E. A. 1970. A technique for the preparation of large thin sections of soils and unconsolidated materials. In *Micromorphological techniques and applications*, D. A. Osmond and P. Bullock (eds), 3–13. Soil Survey Tech. Monogr. No. 2.

Grove, J. M. 1972. The incidence of landslides, avalanches and floods in western Norway during the Little Ice Age. *Arctic Alpine Res.* **4**, 131–8.

Hutchinson, J. N. and R. K. Bhandari 1971. Undrained loading, a fundamental mechanism of mudflows and other mass movements. *Geotechnique* **21**, 353–8.

Innes, J. L. 1983. Stratigraphic evidence of episodic talus accumulation on the Isle of Skye, Scotland. *Earth Surf. Proc. Landforms* **8**, 399–403.

Jacobson, G. L. and H. J. B. Birks 1980. Soil development on recent end moraines of the Klutlan Glacier, Yukon Territory, Canada. *Quat. Res.* **14**, 87–100.

Jauhiainen, E. 1973. Age and degree of podzolization of sand soils on the coastal plain of northwest Finland. *Commentat. Biol.* **68**, 1–32.

Jenny, H. 1946. Arrangement of soil series and soil types according to functions of soil-forming factors. *Soil Sci.* **61**, 375–91.

Kihl, R. 1975. Physical preparation of organic matter samples for submission to a radiocarbon dating laboratory. *Quat. News.* **16**, 4–6.

Kullman, L. 1981. Some aspects of the ecology of the Scandinavian subalpine birch forest belt. *Wahlenbergia* **7**, 99–112.

Larsson, S. 1982. Geomorphological effects on the slopes of Longyear valley, Spitsbergen, after a heavy rainstorm in July 1972. *Geogr. Ann.* **64A**, 105–25.

Mahaney, W. C. 1974. Soil stratigraphy and genesis of Neoglacial deposits in the Arapaho and Henderson cirques, central Colorado Front Range. In *Quaternary environments*, W. C. Mahaney (ed.), 197–240. Geogr. Monogr. 5. Toronto: York University.

Matthews, J. A. 1980. Some problems and implications of ^{14}C dates from a podzol buried beneath an end moraine at Haugabreen, southern Norway. *Geogr. Ann.* **62A**, 185–208.

Matthews, J. A. 1981. Natural ^{14}C age/depth gradient in a buried soil. *Naturwissenschaften* **68**, 472–4.

Moore, T. R. 1976. Sesquioxide-cemented soil horizons in northern Quebec: their distribution, properties and genesis. *Can. J. Soil Sci.* **56**, 333–44.

Newson, M. 1980. The geomorphological effectiveness of floods – a contribution stimulated by two recent events in mid-Wales. *Earth Surf. Proc.* **5**, 1–16.

Rapp, A. 1963. The debris slides at Ulvådal, western Norway. An example of catastrophic slope processes in Scandinavia. *Nach. Akad. Wissen. Gottingen, Math. Physik. Klasse* **13**, 195–210.

Rapp, A. and R. Nyberg 1981. Alpine debris flows in northern Scandinavia. *Geogr. Ann.* **63A**, 183–96.

Rapp, A. and L. Strömquist 1976. Slope erosion due to extreme rainfall in the Scandinavian mountains. *Geogr. Ann.* **58A**, 193–200.

Scharpenseel, H. W. and H. Schiffmann 1977. Radiocarbon dating of soils, a review. *Zeitschr. Pflanzenernähr. Düng. Bodenk.* **140**, 159–74.

Schumm, S. A. 1975. Episodic erosion: a modification of the geomorphic cycle. In *Theories of landform development*, W. N. Melhorn and R. C. Flemal (eds), 69–85. 6th Ann. Binghamton Symposium.

Selby, M. J. 1976. Slope erosion due to extreme rainfall: a case study from New Zealand. *Geogr. Ann.* **58A**, 131–8.

Simmons, I. G. and P. R. Cundill 1974. Late Quaternary vegetational history of the North York Moors. II Pollen analyses of landslip bogs. *J. Biogeog.* **1**, 253–61.

Skempton, A. W. and F. A. De Lory 1957. Stability of natural slopes in London Clay. In *Proceedings of the 4th International Conference on Soil Mechanics Foundation Engineering*, Vol. 2, 378–81.

Sollid, J. L. and L. Sørbel 1979. Deglaciation of western central Norway. *Boreas* **8**, 233–9.

Starkel, L. 1976. The role of extreme (catastrophic) meteorological events in contemporary evolution of slopes. In *Geomorphology and climate*, E. Derbyshire (ed.), 203–46. London: Wiley.

Tallis, J. H. and R. H. Johnson 1980. The dating of landslides in Longdendale, north Derbyshire, using pollen-analytical techniques. In *Timescales in geomorphology*, R. A. Cullingford, D. A. Davidson and J. Lewin (eds), 189–205. Chichester: Wiley.

Tamm, O. 1950. *Northern coniferous forest soils*. Oxford: Scrivener.

Walker, P. H. 1963. Soil history and debris-avalanche deposits along the Illawarra scarpland. *Austral. J. Soil Res.* **1**, 223–30.

Wolman, M. G. and R. Gerson 1978. Relative scales of time and effectiveness of climate in watershed geomorphology. *Earth Surf. Proc.* **3**, 189–208.

Yaalon, D. H. 1971. Soil-forming processes in time and space. In *Paleopedology*, D. H. Yaalon (ed.), 29–39. Jerusalem: International Society of Soil Science and Israel Universities Press.

18

Palaeosols and the interpretation of the British Quaternary stratigraphy

J. Rose, J. Boardman, R. A. Kemp and C. A. Whiteman

Introduction

When the Geological Society of London issued its Special Report *Correlation of Quaternary deposits in the British Isles* (Mitchell *et al.* 1973) the stratigraphic units were identified on lithological and biological evidence, and correlations were made by radiocarbon dating and diagnostic spectra of plant, animal and rock assemblages. Soils had no role in this compilation. Similarly, the systematic review of recent advances in *British Quaternary studies* (Shotton 1977) produced for the Xth INQUA Congress in Birmingham, UK, failed to consider Quaternary soils, and Bowen's *Quaternary geology* (1978, p. 31) gives just one reference to palaeosols within the British Quaternary stratigraphy. Since then, Catt (1979, 1983, 1985) has reviewed the place and significance of soils in British Quaternary studies, but soils continue to have only a minor, albeit an increasing, role in the determination and correlation of British Quaternary environments (Rose & Allen 1977, Boardman 1985, Kemp 1985, Rose *et al.* 1985).

Hitherto, the most effective and widespread use of soil properties in a stratigraphic context has been the use of periglacial soil structures such as ice-wedge casts and involutions as evidence for permafrost climates (West 1969, Williams 1969, 1975, Watson 1977). The development of the *paleo-argillic* concept (Avery 1985) which implies humid, temperate, soil-forming processes prior to the Devensian Glaciation has taken place in an open-ended stratigraphic framework and has, so far, not made any contribution of significance to the interpretation of the British Quaternary stratigraphy, although there is no doubt that the potential exists (Bullock & Murphy 1979, Sturdy *et al.* 1979).

The situation in Britain contrasts markedly with that in other parts of the world such as Belgium, Czechoslovakia and North America, where soil stratigraphy plays an important part in the identification and correlation of interglacial and interstadial episodes of the Middle and Upper Pleistocene (Paepe 1971,

Kukla 1975, Follmer 1978, Morrison 1978). For instance in Belgium, the last two temperate stages are represented by the Rocourt and Profondville Soils and in Czechoslovakia the last 900 000 years of Quaternary time are subdivided and the events spatially correlated on the basis of soil properties and terrace development. In North America, where fewer palynological studies of interglacial deposits have been made, the last two interglacial stages are mainly represented by the Sangamon and Yarmouth Soils.

Definitions and types of palaeosols

Despite the apparent simplicity of the definition that 'palaeosols (are) soils of an environment of the past, formed either by burial under geological materials or by a natural change of climate or topographic conditions of soil formation' (Catt 1979), many opinions exist about both the emphasis that must be placed on particular types of soil evidence (Fenwick 1985) and the length of time over which the soil has been fossilised (Catt 1979). For the purpose of this chapter and associated work (Rose et al. 1985) the definition given above is considered satisfactory. Palaeosols are identifiable by any evidence that indicates the presence of a former land surface that has undergone some form of alteration in response to in situ surface processes. This implies the use of physical as well as chemical and biological criteria, and is important for the identification of periglacial soils where polar desert conditions existed. In such regions vegetation and chemical alteration is minimal or non-existent, whereas certain physical processes are extreme (Péwé 1974, Tedrow 1978). This means that active layer processes and atmospherically induced thermal contraction are considered as evidence for soil formation, but equally it does not include features formed below the permafrost table, such as pingos. In terms of duration, the definition has no limits, although very short-lived surfaces, such as bars in a river channel which can be identified by sedimentary structures, are eliminated unless they show evidence of some form of surface alteration such as root development (Rose et al. 1980) or thermal contraction (Bryant 1983).

Soil stratigraphic units (American Commission on Stratigraphic Nomenclature 1961) must comprise an assemblage of soil properties from a range of localities, which have a similar stratigraphic relationship to one another, derived independently from lithological, biological, geomorphological or absolute dating evidence. Any individual soil stratigraphic unit may show a range of development and a range of soil features caused by changes in the soil-forming environment such as parent material, topography, drainage, vegetation and climate. The terms 'soil stratigraphic unit' (American Commission on Stratigraphic Nomenclature 1961) and 'geosol' (Morrison 1978) have been used to refer to soils in a stratigraphic context, although as far as Britain is concerned only three soils: the Valley Farm Rubified Sol Lessivé, the Barham Arctic Structure Soil (Rose & Allen 1977), and the Troutbeck Palaeosol (Boardman 1981), have been given such a status.

Palaeosols can be defined as *buried* or *relict* depending upon their history in relation to the present land surface (Valentine & Dalrymple 1976). As implied by the definition, buried soils are covered by an overburden that impedes soil-forming processes and causes fossilisation. The Sangamon Soil in Illinois is a buried palaeosol developed on Illinoian till and loess and buried beneath Wisconsin till and loess (Leverett 1898). Relict palaeosols are those that have remained exposed at the land surface through at least two distinct soil-forming environments. The soil characteristics older than those currently forming are the relict component. For instance, periglacial structures formed during the Devensian can be a relict soil component of a soil on the present land surface currently experiencing temperate climate pedogenesis. Relict soils can provide a summation of surface processes but their usefulness for palaeoenvironmental interpretation depends on the earlier events being represented by stronger processes, such as cryoturbation, clay illuviation or carbonate accumulation, and not being obscured by subsequent pedogenesis. In a time context, relict soils have been used most effectively to analyse land surfaces that have experienced soil formation over different lengths of time, such as a sequence of moraines (Alexander 1982) or alluvial deposits (Chartres 1980).

From a stratigraphic point of view, buried soils are of more value because upper and lower boundaries may be defined by datable sediment bodies, and any inferences derived from their soil properties can be related to a specific period of time, although the actual value of the interpretation will depend on the length of the interval and its range and history of environmental changes. Relict soils are defined only by a lower boundary with their upper age extending through to the present day. This is not a problem when the time span is short and the interval is characterised by relatively simple environmental history, but when it is long enough to cover several episodes of climatic and environmental change, interpretation may be difficult although not impossible (Bullock & Murphy 1979, Chartres 1980). For the purpose of this study both types of palaeosol will be considered, but inevitably buried palaeosols will form the basis of interpretation and relict palaeosols will be restricted to those that began forming in the Late Devensian, although it is hoped that in the future it may be possible to identify paleo-argillic features of Ipswichian age.

Objectives

In this chapter British palaeosols will be classified and discussed according to their environmental significance and stratigraphic position. Three groups have been identified: stadial soils and relict features, formed during the coldest parts of Quaternary time; interstadial soils, formed in periods of limited or temporary climatic amelioration; and temperate soils, formed usually in interglacial conditions, but also in intervals of relative warmth that were not bounded by glacial events. The occurrences of palaeosols that can be allocated to a particular stratigraphic event, or period of time, are shown in Figure 18.1, which also gives a

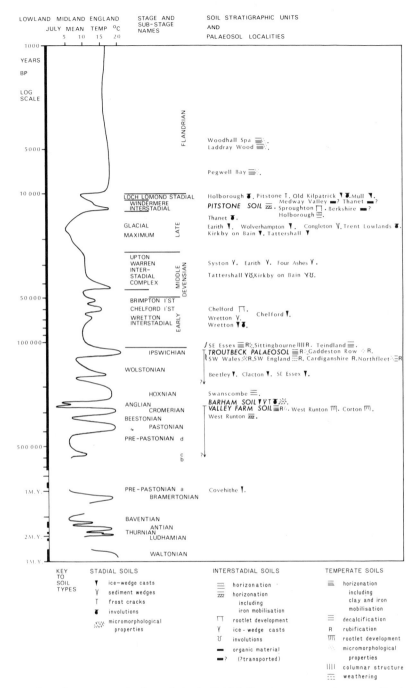

Figure 18.1 Occurrences of palaeosols in the British Quaternary succession, according to age, location and soil characteristics. The scheme of climatic change is based on Zagwijn (1975) with modifications for the Devensian on Coope (1975) and Bryant *et al.* (1983), for the 'Ipswichian/ Wolstonian' interval on Shotton (1983) and for the Bramertonian on Funnell *et al.* (1979). The positioning of the Bramertonian within the hiatus is largely speculative being correlated tentatively with the main climatic amelioration of the early Waalian. It is expected that this scheme will change in the near future.

summary of the British Pleistocene stratigraphy as currently accepted, although almost certainly this will change in the near future. Using this information this chapter aims to review the soil characteristics associated with each of the three categories and to examine the environments that can be reconstructed for the particular stratigraphic events. Additionally, wherever it is possible, the soil evidence is evaluated in terms of the understanding it gives to the interpretation of stratigraphic problems and the British Quaternary history.

Stadial soils and relict features

Evidence for soil formation during the most severe climatic conditions of the Pleistocene usually takes the form of ground-ice structures and thermal contraction cracks (West 1969, Williams 1969, 1975), although recently reference has been made to micromorphological features and organic content, and an attempt has been made to evaluate these soils in pedological terms (Rose et al. 1985). This type of evidence has been used either to demonstrate the climate during part, or all, of a depositional sequence, or to reconstruct, on a regional basis, the environment during formation of an established soil stratigraphic unit.

In the first case evidence generally takes the form of isolated or discontinuous ground-ice structures such as involutions, frost cracks and ice-wedge casts in an aggrading depositional sequence, and as such reflects the effects of periglacial soil-forming processes on ephemeral land surfaces such as alluvial fans or braided river floodplains. It was partly on these criteria that permafrost conditions were recognised in the Beestonian Stage in East Anglia (West & Wilson 1966, West 1980) and during deposition of the Middle Pleistocene Kesgrave Sands and Gravels in East Anglia and the lower Thames valley (Rose et al. 1976, Rose & Allen 1977). Similarly, permafrost conditions have been attributed to the Early Devensian on the basis of periglacial structures below sediments from Wretton in Norfolk (West et al. 1974) and Chelford in Cheshire (Worsley 1977). Similar evidence from Ouse river deposits at Earith in Cambridgeshire (Bell 1970) and Trent system river gravels from Four Ashes in Staffordshire (Morgan 1973) has also been used to indicate permafrost in the Middle Devensian.

In the second case, where widely distributed cold climate soils can be shown to represent a particular stratigraphic event, it has been possible to reconstruct environmental conditions across a region. This technique can be applied effectively to three episodes of the Pleistocene in Britain: the Loch Lomond Stadial between 11 000 and 10 000 years BP; the time of maximum glacier expansion in the Late Devensian between about 25 000 and 14 000 years BP; and the early part of the Anglian Glaciation.

Stratigraphically well-defined evidence from the Loch Lomond Stadial gives a picture of ice-wedge casts in lowland western and central Scotland (Sissons 1974, Rose 1975), the Isle of Man and Wales (Watson 1977), frost cracks in Buckinghamshire (Evans 1966) (Fig. 18.6), involutions in Cumbria (Johnson 1975) and south-east England (Kerney 1963) and an absence of evidence for any form of ground-ice development in the South-West (Brown 1977). This suggests a discernible thermal gradient from active, if discontinuous, permafrost in Scotland, the Isle of Man and Wales, seasonally frozen ground in the Midlands and eastern England and cool temperate conditions in the South-West. The evidence for the severity of the climate in eastern England is supported by persistent disruption of existing horizonation in chalkland soils (Kerney 1963, Evans 1966, 1968) and the local development of polar desert with associated wind-blown sands in the Vale of York (Matthews 1970) and Lincolnshire (Straw 1963), equivalent to conditions that existed at the same time in the Low Countries (Van der Hammen 1957, Paepe & Vanhoorne 1967). The soil evidence for the Loch Lomond Stadial complements that derived from fossil Coleoptera (Coope 1975) by providing additional information on distribution and temperature conditions for parts of the year other than the summer.

Williams (1969, 1975) and Watson (1977) have used periglacial soil structures to reconstruct the environment during the maximum of the Late Devensian Glaciation. Unfortunately, in the majority of cases, evidence for this episode takes the form of relict structures in soils on the present land surface, and only rarely is it preserved in a sedimentary or geomorphic sequence that can be used to define the lower and upper boundaries of the palaeosol (Galloway 1961). Typically, a maximum age is provided by the age of the parent material such as Late Devensian tills in Cheshire and Staffordshire (Worsley 1966, Morgan 1971) and Late Devensian tills and gravels throughout Scotland (Sissons 1974). Beyond the limit of Late Devensian glacigenic deposits, a maximum age can be provided by river sediments such as those of the Avon No.2 Terrace (Shotton 1968) and the Ouse Low Terrace (Bell 1970), or development in loess of Late Devensian age (Catt 1978). Evidence for the minimal age is usually absent, being assumed from knowledge of the subsequent climatic history. Thus, a degree of ambiguity is inevitable as some of the features ascribed to the period of maximum glaciation could equally have formed during the Loch Lomond Stadial. Figure 18.2 shows the distribution of periglacial soil structures that are believed to have formed between about 26 000 and 13 000 BP.

Briefly, interpretation of periglacial soil structures suggests that at the maximum of glaciation, about 18 000 years BP, mean annual temperatures were of the order of 25 °C below the present and July mean temperatures were about 10 °C lower (Watson 1977). Active permafrost was extensive throughout Britain beyond the ice margin, although of a discontinuous nature in the South-West owing to either snow cover in the cooling season or maritime influences from the Atlantic. A temperature gradient of about +1.5 °C per 100 km in a southwest direction across England is estimated (Williams 1969, 1975). The presence of wind-blown sediment (Catt 1978) and the absence of contemporaneous organic material suggests that polar desert conditions were prevalent in northern and eastern parts of the country at least, and the development of ice and sediment wedges in Late Devensian glacigenic sediments in Wales, northern England and Scotland indicates that active permafrost followed deglacierisation and persisted until most of the country was ice-free. Estimations of active layer

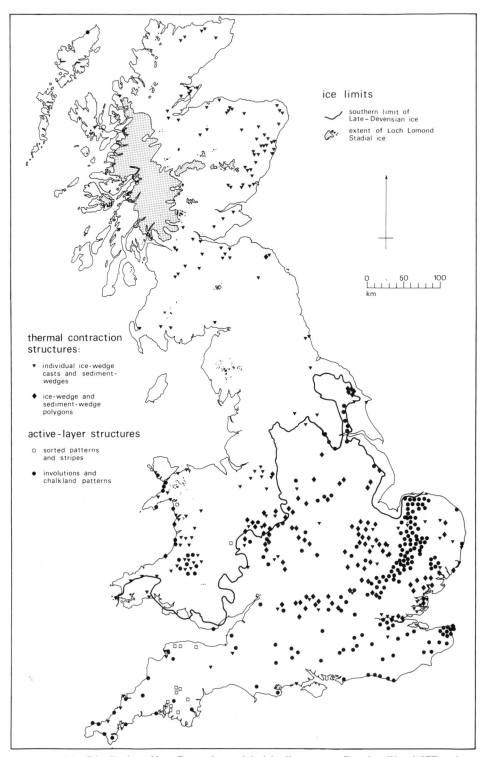

Figure 18.2 Distribution of Late Devensian periglacial soil structures. Based on West (1977) and Watson (1977) with additional information from Gaunt (1976), Girling (1977), Greig (1981), Douglas (1982), personal communication from Dr D. Sutherland and personal observations.

thickness from the depth of involutions suggests that towards the end of the stadial, when July mean temperatures were some 7 °C lower than at present, accumulated heat was similar to that of the boreal zone at present (Williams 1969, 1975, Watson 1977).

Unlike the periglacial soils developed in the Late Devensian, the early Anglian soil is a buried palaeosol. It has been recognised with clearly defined upper and lower boundaries throughout East Anglia and the lower Thames valley, and has been given the status of a soil stratigraphic unit (Rose & Allen 1977, Rose *et al.* 1985). It is known as the Barham Soil from its type site in the Gipping Valley near Ipswich. It can be traced across 6500 km² developed in Cromerian soil material or early Anglian freshwater sediments, and it is buried throughout by tills and sands and gravels of the Anglian Glaciation.

The soil properties observed consist of involutions, frost cracks, ice-wedge casts, sand wedges, silt accumulations, disrupted argillans, banded fabric and interstitial organic material (Rose *et al.* 1985). In addition it is associated with wind-blown sand and silt derived from sediments transported by the encroaching Anglian ice sheet. A typical development and stratigraphic position is given in Figure 18.3 and the sites at which it has been observed are shown in Figure 18.4. Throughout Essex, Suffolk and parts of Hertfordshire the soil shows complex development with ice-wedge casts up to 8 m deep and disruption sequences involving macrostructures such as involutions and sand wedges, and microscale features such as *in situ* and translocated papules (Rose *et al.* 1985). In northeast Norfolk, however, where the soil is developed on short-lived land surfaces in an aggrading sediment body, the macrostructures are smaller and simple and can be ascribed with considerable precision to different parts of the late Cromerian and early Anglian and to the vegetation that developed during this period of time (West 1980).

Using soil and vegetational evidence together, it is possible to reconstruct the environment of this period (Rose *et al.* 1985). Permafrost first developed during the last part of the Cromerian Interglacial (Cr IVc) while the landscape maintained a cover of birch woodland. With the onset of the Anglian Glaciation and the increasing severity of climate, grass and herb tundra replaced woodland vegetation and the periglacial soils became more complex. Large, non-sorted polygons, caused by extensive thermal contraction, were widespread, as was the development of vein-ice. In the active layer, cryostatic and hydrostatic stresses produced involutions, disrupted clay argillans and caused the migration of clay-size particles to form sorted lamellae (Van Vliet-Lanoë 1976). In some localities organic matter gives the soil a very dark greyish brown colour (10YR 3/2) and the amorphous distribution, without any distinct horizonation, combined with the ground-ice structures, suggests an affinity with arctic brown and meadow tundra soils, depending upon the drainage at the site (Tedrow 1978, Bockheim 1978). At the same time, silt- and clay-size materials were moved downwards forming cappings on the top of the larger particles. With a further deterioration of climate, glaciers reached the vicinity of eastern England contributing disaggregated sediment capable of being transported by the wind. The development

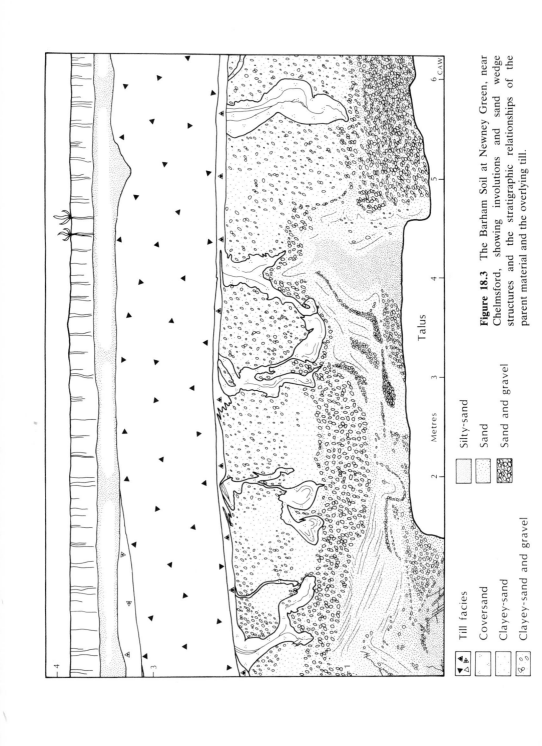

Figure 18.3 The Barham Soil at Newney Green, near Chelmsford, showing involutions and sand wedge structures and the stratigraphic relationships of the parent material and the overlying till.

Till facies

Coversand

Clayey-sand

Clayey-sand and gravel

Silty-sand

Sand

Sand and gravel

Talus

Metres

CAW

Figure 18.4 Location of sites of Valley Farm and Barham Soils in eastern England.

of sand wedges demonstrates the existence of dry permafrost and perhaps regional aridity, although this last interpretation must be viewed with some caution as the frequent development of sand wedges and arctic brown soil material indicates a preference for well-drained sites and the possibility of topographically controlled desiccation (Rose *et al.* 1978). At this time vegetation was virtually absent, polar desert soils were prevalent and mean annual temperatures were of the order of −12 °C or less. Eventually, glaciers overrode the periglacial soil, preserving it in some places and eroding it in others.

In terms of landscape evolution and stratigraphy, the Barham Soil is important because it marks the major change in the landscape of eastern England, and in conjunction with the aeolian sediments, it forms a distinctive stratigraphic mar-

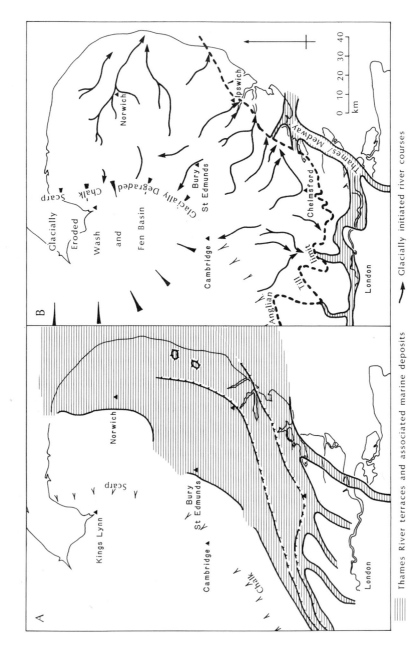

Figure 18.5 The landscape of eastern England (a) at the time of the Barham Soil and (b) after the Anglian glaciation. These figures are compiled from the work of the authors and information from Dr P. Allen, Dr D. R. Bridgeland and Dr R. W. Hey.

‖‖‖ Thames River terraces and associated marine deposits

⟶ Glacially initiated river courses

ker. When the Barham Soil formed, the landscape of East Anglia was dominated by river terraces formed when the Thames flowed across the region. After the Anglian glaciation the topography was dominated by a radial valley pattern determined by the hydrology of the ice sheet (Fig. 18.5). As a stratigraphic marker it identifies perhaps the most severe climatic deterioration in the British Quaternary and is an important component of many relict palaeosols beyond the Anglian ice limit (Rose *et al.* 1976, Kemp 1985).

Interstadial soils

Although interstadials must have occurred on numerous occasions during the cold stages of the British Quaternary, with the exception of an early Beestonian ferric podzol (Valentine & Dalrymple 1975), only those within the Devensian are sufficiently well known to be the subject of systematic analysis (Fig. 18.1). Interstadial soils have been described from the Middle Devensian in Scotland (FitzPatrick 1965, Edwards *et al.* 1976, Romans 1977, Connell *et al.* 1982) but their age and status is, as yet, far from clear owing to profile superimposition and ambiguous radiocarbon dates. Worsley (1977) has reported a rootlet horizon from the Chelford Interstadial at Chelford, and Straw (1980) has described an interstadial palaeosol profile from the Middle Devensian in Norfolk. Periglacial structures such as ice-wedge casts and involutions in aggrading sediment bodies have been ascribed to the Middle Devensian interstadial (Coope *et al.* 1961, Bell 1970, Bell *et al.* 1972, Morgan 1973).

The best understood interstadial soil in Britain is that developed during the Windermere Interstadial during the Devensian Lateglacial (Pennington 1977, Coope 1977). Soils of this age have been recognised widely from chalk dry valley sites (Dalrymple 1958, Kerney 1963, Kerney *et al.* 1964, Evans 1966, 1968, Paterson 1971), loess accumulations (Kerney 1965) and floodplain deposits (Rose *et al.* 1980) in southern England. Organic accumulations of equivalent age are known even more widely from lake basins in all parts of Britain (Sissons 1979).

During this time summer temperatures reached levels higher than the present, although winter temperatures remained low (Coope 1975). Vegetation changed from open tundra to closed birchwood in southern and midland England, and patchy birchwood and scrub in northern and western Britain (Pennington 1977) before reverting once more to birch scrub and tundra. Buried soils dating from this interstadial are preserved in low-lying locations because high-energy slope and river activity during the succeeding Loch Lomond Stadial caused erosion of the hillside slopes and sedimentation in the valley bottoms.

Most descriptions record the soil as a grey mud (Paterson 1971), grey chalk mud (Kerney 1963) or a humic mud (Evans 1968) on the basis of colour with little reference to horizonation, structure, catenary position, or whether the soil material is *in situ* or transported. At a chalk dry valley site at Holborough in Kent, Dalrymple (1958) described a rendzina buried beneath 1.8 m of 'solifluc-

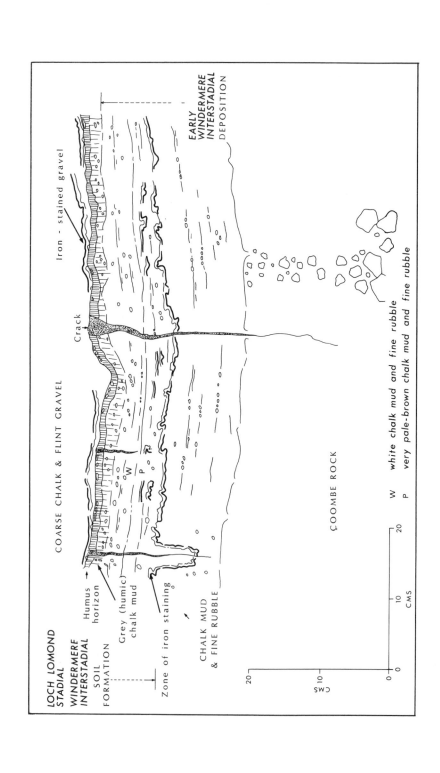

LOCH LOMOND
STADIAL

WINDERMERE
INTERSTADIAL

SOIL
FORMATION

COARSE CHALK & FLINT GRAVEL

Iron - stained gravel

EARLY
WINDERMERE
INTERSTADIAL
DEPOSITION

Crack

Humus
horizon

Grey (humic)
chalk mud

Zone of iron staining

CHALK MUD
& FINE RUBBLE

COOMBE ROCK

W   white chalk mud and fine rubble
P   very pale-brown chalk mud and fine rubble

W
P

0   10   20
         CMS

20

10
CMS

0

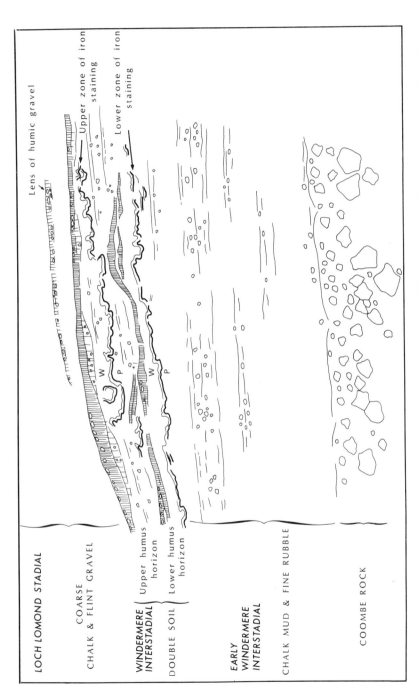

Figure 18.6 The Pitstone Soil at Pitstone, Buckinghamshire (after Evans 1966 reproduced, with permission, from Proc. Geol. Assoc., 1966, 1977).

tion material' in terms of a 5 cm A horizon composed of black mull humus developed on chalk 'solifluction' parent material with carbonate concretions. The most complete description of a soil from this interstadial is that from a chalk dry valley at Pitstone in Buckinghamshire where transverse and longitudinal exposures show catenary relationships, and profile characteristics are given by detailed drawings and written descriptions (Evans 1966) (Fig. 18.6). On the evidence available to date, this can be taken as a characteristic profile, and it is suggested that the soil of the Windermere Interstadial be known as the Pitstone Soil.

At its fullest development this soil shows two profiles with distinct horizonation (Fig. 18.6). The unaltered parent material is white (10YR 8/2) chalk rubble and mud. This is followed by 5 cm of pale-brown (10YR 8/4) chalk mud with iron staining, 5 cm of white (2.5Y 9/2) chalk mud, 2.5 cm of light-grey (10YR 7/1) humic chalk mud and 2.5 cm of black (10YR 2/1) humus with chalk fragments. As at other locations (Kerney 1963, 1965, Kerney et al. 1964, Evans 1968), the soils contain a land snail fauna indicative of locally occurring open ground and a broken substratum. Where the two profiles exist one above the other the lower is less well developed, with humic chalk mud absent, the horizonation disrupted and the snail population more limited in numbers and range of species.

This evidence indicates two periods when physical processes were diminished and ground-surface stability was sufficient for humus to accumulate and horizonation to develop by mixing and iron mobilisation. The two periods are separated by an episode in which a chalk mud and fine rubble were deposited on the lower soil, probably because of enhanced soil erosion in the upper part of the catchment. Environmental conditions appear to have been more severe during the first of the two phases of soil formation. Without absolute dates the precise age of the two parts of the soil cannot be determined, but the pattern has a parallel in the Low Countries where the Stabroek Soil of Belgium formed between about 13 000 and 12 000 years BP (Paepe & Vanhoorne 1967, Paepe 1971) and the Roksem Soil of Belgium (Paepe 1971) and the Usselo Soil of the Netherlands (Van der Hammen 1957) formed between 11 800 and 11 000 years BP.

Soil formation during the Windermere Interstadial, as recorded at Pitstone, may have wider implications. The presence of two soils and the greater development of the upper, rather than the lower, brings our attention to the problem of climatic change during the Devensian Lateglacial. In recent years Coleopteran evidence has led to the view that the acme of Lateglacial climate was about 13 000 to 12 300 years BP rather than during classical Allerød time between 11 800 and 11 000 years BP, and that the Older Dryas climatic deterioration was of little significance (Coope 1975, 1977). The more weakly developed lower soil may be explained by the history of plant and animal colonisation rather than the effects of climate as suggested above, but consideration should also be given to proximity of the palaeosol sites to mainland Europe and a more continental climatic regime. In this case, the climatic response to changes in the position of the North Atlantic Polar Front (Ruddiman et al. 1977) may vary with proximity to the maritime influence (Watts 1980) so that a different pattern and magnitude of climatic change was experienced in southeastern, compared with western Britain.

Finally, attention must be drawn to the relative scarcity of Lateglacial buried palaeosols. Although to some extent this may reflect the amount of scientific investigation in Britain, the striking contrast between this situation and the frequency of occurrence of Neoglacial palaeosols in Scandinavia (Griffey & Worsley 1978) or Lateglacial palaeosols in the Low Countries (Vink & Sevink 1971) suggests a more fundamental explanation. It is therefore proposed that their relative absence is due to extensive soil removal during the Loch Lomond Stadial (between 11 000 and 10 000 years BP), when severe cold and a maritime location caused conditions of exceptional ground ice activity, slope instability and snowmelt runoff (Rose *et al.* 1980).

Temperate soils

In those parts of the world where soil stratigraphy is important, emphasis is placed on soils formed in the warmer intervals of the Quaternary (see the introduction to this chapter). Similarly, in Britain, particular attention has been given to soils formed during the temperate stages. This is partly because of the familiarity with soil characteristics developed on stable temperate land surfaces like those of the present interglacial, but it also reflects the influence of overseas experience (MacClintock 1933, Zeuner 1959). With the many changes that have occurred in the formulation of the British stratigraphic succession, particular climatic intervals have been given many names. For convenience, the stage names given in Mitchell *et al.* (1973) will be used in the following review even though the term may not have been in existence at the time at which the palaeosol was described. Where correlation cannot be made, a palaeosol is of little stratigraphic value and will therefore not be discussed.

The majority of references to temperate soils concern either reasonably well-dated buried fragments of only local significance or spatially extensive relict material in a poorly dated context. Only two temperate palaeosols are sufficiently extensive to deserve the status of soil stratigraphic units: the Valley Farm Soil in East Anglia (Rose & Allen 1977, Kemp 1985) and the Troutbeck Palaeosol in the Lake District (Boardman 1981, 1985), and even these cover a wide stratigraphic range.

Of the dated buried fragments, soils and soil material from Northfleet (Zeuner 1959, Kerney & Sieveking 1977) and Sittingbourne (Tilley 1964) in Kent, the Southend area of Essex (Gruhn & Bryan 1969, Gruhn *et al.* 1974), south-west England (Stephens 1970) and south and west Wales (Ball 1960, Mitchell 1960, Bowen 1966) have been allocated to the Ipswichian Stage. Palaeosols or palaeosol material of Hoxnian age have been described from Swanscombe, Kent (Conway & Waechter 1977) and Slindon in Sussex (Dalrymple 1957), and a Cromerian palaeosol has been described from north Norfolk (West & Wilson 1966, West 1980). Temperate soils from East Anglia (MacClintock 1933) and

Hampshire (Clarke & Fisher 1983) are in too vague a stratigraphic context to be dated.

Paleo-argillic horizons contain relict components of temperate soils. They are recognised by deep decalcification (>1 m) and oxidation, bright brown and red colours (matrix with chroma >4 in hues redder than 10YR, and mottles equal to or redder than 5YR) not inherited from the parent material, high illuvial clay content (>10%) (Sturdy *et al.* 1979) and pronounced orientation of matrix clays with strong birefringence patterns when viewed in thin section (Catt 1979). The extensiveness of this soil material can be seen by the fact that 32 Soil Series mapped by the Soil Survey of England and Wales contain a paleo-argillic component (Catt 1979). Unfortunately, however, its stratigraphic significance is minimal as the parent material is predominantly of pre-Quaternary age, such as Carboniferous Limestone in south Wales, Derbyshire and Staffordshire, Jurassic rocks in south Wales and Yorkshire, Cretaceous rocks in Devon and Kent and Tertiary rocks in south-east and central southern England (Catt 1979, 1983). Of equally limited stratigraphic significance are the soil materials containing pedogenic haematite, kaolinite and gibbsite from Palaeozoic rocks in Scotland (Stevens & Wilson 1970, Wilson & Brown 1976, Lawson 1983) and Wales (Ball 1964). At best, the presence of Devensian loess in these types of soils means that they can be shown to predate the maximum of the Last Glaciation (Catt 1978).

Where relict temperate soils are developed in parent material of Quaternary age, a finer stratigraphic resolution is possible, although their value, hitherto, has been more in the form of evaluating patterns of soil development than in deciphering Quaternary stratigraphy (Wooldridge & Cornwall 1964, Bullock & & Mackney 1970, Bullock & Murphy 1979, Sturdy *et al.* 1979). The greatest potential for stratigraphic correlation using relict soils exists where they are developed in a chronological sequence of lithologically similar material. Chartres (1980) has adopted this method on the terraces of the river Kennet in Berkshire and shown a range of soil properties, particularly degree of illuviation and argillan colour, to be related to age. However, at the present time the age of the terraces is so poorly understood that the wider implications of these results have yet to be realised.

The Troutbeck Palaeosol is developed at a number of localities in the Mosedale and Thornsgill valleys of the northeastern Lake District, Cumbria (Boardman 1981, 1985) where it covers an area of about 8 km^2. The valleys are cut into a succession of glacial deposits with the palaeosol developed on glacial and glaciofluvial sediments of the Thornsgill Formation. It is buried by the Threlkeld Till which was deposited by the Late Devensian ice sheet and covers most of the region (Fig. 18.7). Part of the soil has survived erosion during the Late Devensian owing to its location in buried valleys and probably the cold-based nature of the ice sheet.

Palaeosol characteristics consist of oxidation features and rock weathering. Oxidation takes the form of a colour change from dark grey (N 3/0) and dark bluish grey (10BG 4/1) to yellowish brown (10YR 5/6). Weathering extends to

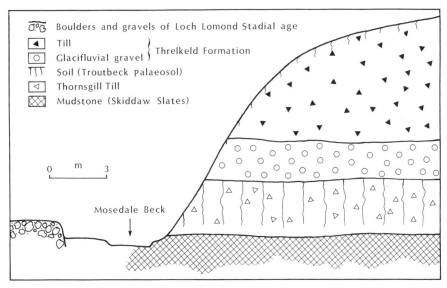

Figure 18.7   Stratigraphic position of the Troutbeck Palaeosol in the northeastern Lake District.

a depth of 15 m and is expressed by the rotting of andesite lavas and micro-granites to a soft and friable condition and a yellowish brown colour. At two sites, the zone of severe weathering grades into relatively unweathered parent material which may be till or Skiddaw Group mudstones. Micromorphological evidence from Thornsgill shows that *in situ* weathering resulted from a fluctuating water table producing gleying (Boardman 1983). The upper horizons of the palaeosol have been lost through erosion; sheared blocks are seen in the overlying Threlkeld Till. Scanning electron microscope evidence suggests that quartz grains in the severely weathered zone of the palaeosol have been subjected to a high-energy chemical environment, whilst X-ray diffractometry shows that the alteration products consist of illite, kaolinite and vermiculite. In contrast, Flandrian weathering profiles in the Threlkeld Till amount to 2 to 3 m of oxidation, clasts remain fresh and unweathered, and matrix and clast colour changes are minimal.

The depth and severity of weathering of the Troutbeck Palaeosol implies a long period of humid temperate conditions, considerable chemical weathering, and a stable land surface. Incision of streams into the landscape controlled the general decline of the groundwater levels and therefore the advance of a weathering front through the sediments. Stratigraphically, the period or periods of weathering must predate the Late Devensian glacial event, but succeed an earlier glacial episode (Boardman 1980), and are interglacial rather than interstadial in character. The duration of the interval of weathering is unknown but the contrast between the Troutbeck and Flandrian soil profiles suggests at least 100 000 years of temperate conditions, and as such indicates an interval or intervals of humid temperate conditions between episodes of glaciation in northern England which have hitherto not been recognised (Boardman 1985).

The Valley Farm Soil is a major soil stratigraphic unit extending over large parts of southern East Anglia (Rose *et al.* 1976, Kemp 1985) (Fig. 18.4). It is developed in the Kesgrave Sands and Gravels, a fluvial deposit laid down by a proto-Thames in a periglacial environment (Rose *et al.* 1976). Superimposed upon the truncated remnants of this soil is the early Anglian Barham Soil, and both are buried beneath Anglian glacigenic deposits. Beyond the limits of the Anglian ice margins, the Valley Farm Soil was never buried at any stage by sufficient thicknesses of sediments to isolate it from post-Anglian pedogenic processes. Consequently, in this region at least, some of the relict features of the stagnogleyic paleo-argillic brown earths (Tendring Association) mapped by the Soil Survey of England and Wales are derived from the Valley Farm Soil (Kemp 1985).

Typical properties of the Valley Farm Soil are summarised in Table 18.1 along with details of the Kesgrave and Barham Sands and Gravels for comparison. The soil is red or reddish brown with hues in the range 10R to 7.5YR, colours that reflect the significant amounts of haematite present in the palaeosol (Kemp 1985). Grey mottles are frequently associated with the soil and indicate surface-water gleying. This process is a consequence of the high clay contents resulting from extensive illuviation. This illuvial clay is easily identifiable in thin sections as well-oriented, undisturbed, fragmented or deformed grain and void coatings (Kemp 1985).

The evidence of the reddening, clay illuviation and surface water gleying processes is sufficient to indicate that the Valley Farm Soil developed at least partly during the temperate Cromerian Stage prior to the Anglian Glaciation. As the Kesgrave Sands and Gravels have been shown to comprise a series of terrace levels of different ages (Hey 1980, Allen 1983), it is probable that the Valley Farm Soil is a complex soil stratigraphic unit representing different periods of time extending back to the beginning of the Cromerian, or to the Beestonian, Pastonian, or even earlier, depending upon the age of the terrace surface upon which it has developed (Rose 1983, Kemp 1985).

In terms of the stratigraphy of East Anglia, the Valley Farm Soil represents a major break in the depositional history of the region. Its presence as a stratigraphic marker has been important in demonstrating, along with mineralogical and sedimentary evidence, the existence of the Thames river deposits (Kesgrave Sands and Gravels) as separate from the Anglian glaciofluvial outwash (Barham Sands and Gravels). Previously, all the sands and gravels beneath the Anglian till had been considered to be of a glaciofluvial origin (Clayton 1957, Bristow & Cox 1973, Lake *et al.* 1977). For the future, the presence of soil evidence means that it may be possible to elucidate further the environmental conditions in this part of East Anglia for a period of time in which evidence was previously not available (Kemp 1985).

Conclusions

Palaeosols have been identified at a wide range of occurrences in the British Quaternary stratigraphy (Fig. 18.1), but in the majority of cases the evidence is

Table 18.1 Typical properties of the Kesgrave Sands and Gravels, Valley Farm Soil and Barham Sands and Gravels.

	Colours	Fe$_d$[1] % (<2 mm)	% Clay (<2 mm)	Silt Silt + Sand ratio	Non-opaque heavy sand mineralogy	Clay mineralogy	Micromorphology
Barham Sands and Gravels	Yellow to yellowish brown 10YR hues	0.1–0.5	0.5–3.0	0.01–0.03	Stable minerals, particularly zircon, tourmaline and rutile, in addition to small quantities of weatherables such as collophane and apatite.	Interstratified smectites with subsidiary amounts of mica and kaolinite. Rarely goethite.	Illuvial clay absent.
Valley Farm Soil	Red to reddish yellow 2.5YR, 5YR and 7.5YR hues. Grey mottles with 2.5Y hues.	0.3–1.5	15.0–30.0	0.01–0.03	Stable minerals, particularly zircon, tourmaline and rutile.	Interstratified smectites with subsidiary amounts of mica kaolinite, goethite and haematite. Rarely lepidocrocite.	All clay is in the form of well-oriented undisturbed, fragmented or deformed coatings.
Kesgrave Sands and Gravels	Yellowish brown to white. 7.5YR, 10YR and 2.5Y hues.	0.0–0.5	0.5–3.0	0.01–0.03	Stable minerals, particularly zircon, tourmaline and rutile.	Interstratified smectites with subsidiary amounts of mica and kaolinite. Rarely goethite.	Illuvial clay absent.

[1] Fe $_d$: dithionite-extractable iron.

fragmentary, profiles are disrupted and truncated, and the stratigraphic resolution is crude.

On only six occasions in the British Quaternary succession is soil information adequate to justify the recognition of soil stratigraphic units, and in only three of these cases can the soil properties justifiably be related to a single stratigraphic episode: that which formed in Loch Lomond Stadial, the Pitstone Soil of the Windermere Interstadial and the Barham Soil of the early Anglian (Table 18.2). Even in these cases, boundaries may not be precise and the episodes may not be uniform. For instance, features of the Barham Soil date from the late Cromerian and the Pitstone Soil may represent a relatively complex sequence of climatic change.

Where palaeosols do occur, they may provide evidence of stratigraphic and palaeoenvironmental significance. The soil developed during the Windermere Interstadial may possibly support the view that the pattern of climatic change during this interval showed two thermal optima separated by a temporary deterioration rather than a single early amelioration as indicated by Coleopteran evidence. The Troutbeck Palaeosol provides evidence for a long period of temperate weathering between two glacial episodes in northern England. The Barham Soil provides new and more detailed evidence of the early Anglian stadial in eastern England. The Valley Farm Soil gives a history of environmental change in a part of the British Quaternary succession that has hitherto been difficult to resolve. The Loch Lomond and Late Devensian Stadial soils and relict features permit a reconstruction of the contemporary climate and environment. In a different way, the presence of periglacial soil structures in the Middle Devensian interstadial, similar to those formed in the Late Devensian and other stadials, emphasises the crudeness of some soil properties as environmental indicators when they are not close to the developmental thresholds, and consequently the need to use a range of evidence when classifying stratigraphic episodes.

In terms of the history of the British landscape, the Valley Farm and Barham soils define the topography of part of eastern England before the relief and drainage of that region was significantly altered by the erosion and deposition of Anglian glaciers. Likewise, the Troutbeck Palaeosol shows that the glacial troughs of part of the northeastern Lake District are older than the Ipswichian Interglacial and were not formed by erosion during the Last Glaciation, as has previously been suggested (King 1976). Equally, the distribution of the Pitstone Soil in chalkland dry valleys shows the extent of erosion and deposition within a catchment by river and slope processes during the succeeding Loch Lomond Stadial.

In eastern England, the widespread occurrence of the Valley Farm and Barham Soils permits the reconstruction of synchronous buried land surfaces. In this case, the buried land surface comprises a series of terraces formed when the river Thames drained north-east and eastwards across the region. Prior to the recognition of these palaeosols, such a reconstruction would not have been possible as the upper elevation of the terrace gravels could have been determined by

Table 18.2 Soil stratigraphic units in the British Quaternary.

Soil stratigraphic Unit	Age	Environment of formation	Location	Comments
no name	Loch Lomond Stadial	periglacial/cool temperate	Britain	
Pitstone Soil	Windermere Interstadial	cool temperate	southern England	
no name	main Late Devensian Stadial	periglacial	Britain	mainly relict and may include soil of Loch Lomond Stadial age
Troutbeck Palaeosol	Ipswichian and older	temperate	northeastern Lake District	
Barham Soil	early Anglian	periglacial	eastern England	
Valley Farm Soil	Cromerian to Pastonian and possibly earlier	warm temperate and cold	eastern England	complex environmental history, probably began forming at different times

subsequent erosion rather than represent the original surface of the river flood-plain. The survival of relict features formed during the Late Devensian glacial maximum and the Loch Lomond Stadial in soils at the present-day land surface demonstrates clearly that large areas of Britain have changed little in the inter-vening time. As would be expected, their particular distribution shows that the most stable areas are those with low relief, but it brings attention particularly to chalkland plateaux (with Devensian loess in the topsoils), river terraces and, perhaps rather more unexpectedly, till surfaces in Wales, Scotland and west midland England.

The survival of palaeosols beneath tills, coversand, loess, river and slope deposits indicates the activity of these transporting processes on the shaping of the British landscape at particular times. Burial by wind-blown sand and silt, or beneath colluvial diamictons is widely reported in many parts of the world and reflects the relatively low energy of these particular processes during stadial conditions. Burial beneath fluvial sediments is far less common as the likelihood exists that soils will be destroyed by river scour. The survival of the Pitstone Soil in the base of chalkland valleys, and the periglacial structures on braided river bars almost certainly reflects the high sediment load of the periglacial rivers which caused the burial, and their lack of additional capacity for further erosion. Similarly, the burial of lowland soils by glacial deposits almost certainly reflects an excess of easily deformable bed material, such as the clay-rich lodgement and melt-out tills of the Anglian ice sheet in East Anglia (Perrin *et al.* 1979). In this case the ice appears to have flowed over its own far-travelled deposit, eroding its bed only locally, and generally preserving the land surface. Survival beneath glacial deposits is far less easily comprehended in the relatively high-relief Lake District, and in this case preservation may be attributable to cold-based conditions during at least part of the glacial event.

For the future, the possibility exists that soils may provide diagnostic spectra of properties that will permit a stratigraphic signature, but as yet this has only been achieved with the relict soil features of the Kennet terraces in Berkshire and with the Valley Farm Soil in East Anglia. The full stratigraphic value of Quaternary palaeosols will only be realised when they can be dated by indepen-dent chronological methods. At present this seems most likely to be achieved by amino acid ratios (Limmer & Wilson 1980) or thermoluminescence dating.

Acknowledgements

The authors wish to thank Dr J. A. Catt for helpful criticism of the original manuscript and to Dr P. Allen, Dr D. R. Bridgeland, Dr R. W. Hey, Professor F. W. Shotton, Dr D. Sutherland and Professor R. G. West for information leading to the compilation of some of the figures. The help of Mr G. Reeve and Miss Loraine Rutt in drawing the figures is also greatly appreciated.

References

Alexander, M. J. 1982. Soil development at Engabredal, Holandsfjord, North Norway. In *Okstindan Preliminary Report for 1980 and 1981*, J. Rose (ed.), 1–23. London: Birkbeck College.

Allen, P. 1983. *Middle Pleistocene stratigraphy and landform development of south-east Suffolk*. Unpublished Ph.D. thesis, University of London.

American Commission on Stratigraphic Nomenclature 1961. Code of stratigraphic nomenclature. *Am. Assoc. Petrolm Geol. Bull.* **45**, 645–65.

Avery, B. W. 1985. Argillic horizons and their significance in England and Wales. In *Soils and Quaternary landscape evolution*, J. Boardman (ed.), 69–86. Chichester: Wiley.

Ball, D. F. 1960. Relict-soil on limestone in South Wales. *Nature* **187**, 497–8.

Ball, D. F. 1964. Gibbsite in altered rock in north Wales. *Nature* **204**, 673–4.

Bell, F. G. 1970. Late Pleistocene floras from Earith, Huntingdonshire. *Phil Trans R. Soc. Lond. Ser. B* **258**, 347–78.

Bell, F. G., G. R. Coope, R. J. Rice and T. H. Riley 1972. Mid-Weichselian fossil bearing deposits at Syston, Leicestershire. *Proc. Geol. Assoc.* **83**, 197–211.

Boardman, J. 1980. Evidence for pre-Devensian glaciation in the northeastern Lake District. *Nature* **286**, 599–600.

Boardman, J. 1981. *Quaternary geomorphology of the northeastern Lake District*. Unpublished Ph.D. thesis, University of London.

Boardman, J. 1983. The role of micromorphological analysis in an investigation of the Troutbeck Paleosol, Cumbria, England. In *Soil micromorphology (1): techniques and applications*, P. Bullock and C. P. Murphy (eds), 281-8. Berkamsted: ABA.

Boardman, J. 1985. The Troutbeck Paleosol, Cumbria, England. In *Soils and Quaternary landscape evolution*, J. Boardman (ed.), 231–60. Chichester: Wiley.

Bockheim, J. G. 1978. A comparison of the morphology and genesis of Arctic Brown and Alpine Brown Soils in North America. In *Quaternary soils*, W. C. Mahaney (ed.), 427–53. Norwich: Geo Abstracts.

Bowen, D. Q. 1966. Dating Pleistocene events in south-west Wales. *Nature* **211**, 475–6.

Bowen, D. Q. 1978. *Quaternary geology*. Oxford: Pergamon.

Bristow, C. R. and F. C. Cox 1973. The Gipping Till: a reappraisal of East Anglian glacial stratigraphy. *J. Geol Soc. Lond.* **129**, 1–37.

Brown, A. 1977. Late Devensian and Flandrian vegetational history of Bodmin Moor, Cornwall. *Phil Trans R. Soc. Lond. Ser. B* **276**, 251–320.

Bryant, I. D. 1983. Facies sequences associated with some braided river deposits of late Pleistocene age from southern Britain. In *Modern and ancient fluvial systems*, J. D. Collinson and J. Lewin (eds), 267–75. Oxford: Blackwell.

Bryant, I. D., D. T. Holyoak and K. A. Moseley 1983. Late Pleistocene deposits at Brimpton, England. *Proc. Geol Assoc.* **94**, 321–43.

Bullock, P. and D. Mackney 1970. Micromorphology of strata in the Boyn Hill terrace deposits, Buckinghamshire. In *Micromorphological techniques and applications*, D. A. Osmond and P. Bullock (eds), 97–106. Soil Survey Tech. Monogr. No. 2.

Bullock, P. and C. P. Murphy 1979. Evolution of a paleo-argillic brown earth (paleudalf) from Oxfordshire, England. *Geoderma* **22**, 225–53.

Catt, J. A. 1978. The contribution of loess to soils in lowland Britain. *Counc. Brit. Arch. Res. Rep.* No. 21, 12–20.

Catt, J. A. 1979. Soils and Quaternary geology in Britain. *J. Soil Sci.* **30**, 607–42.

Catt, J. A. 1983. Cenozoic pedogenesis and landform development in southeast England. In *Residual deposits: surface related weathering processes and materials*, R. C. L. Wilson (ed.), 251–8. Oxford: Blackwell.

Catt, J. A. 1985. Soils and Quaternary stratigraphy in the United Kingdom. In *Soils and Quaternary landscape evolution*, J. Boardman (ed.), 161–78. Chichester: Wiley.

Chartres, C. J. 1980. A Quaternary soil sequence in the Kennet Valley, central southern England. *Geoderma* **23**, 125−46.

Clarke, M. R. and P. E. Fisher 1983. The Caesar's Camp Gravel − an early Pleistocene fluvial periglacial deposit, southern England. *Proc. Geol Assoc.* **94**, 345−57.

Clayton, K. M. 1957. Some aspects of the glacial deposits of Essex. *Proc. Geol Assoc.* **68**, 1−21.

Connell, E. R., K. J. Edwards and A. M. Hall 1982. Evidence for two pre-Flandrian palaeosols in Buchan, northeast Scotland. *Nature* **297**, 570−2.

Conway, B. W. and J. de A. Waechter 1977. Lower Thames and Medway Valleys − Barnfield Pit, Swanscombe. In *Southeast England and the Thames Valley, INQUA Guidebook*, E. R. Shephard-Thorn and J. J. Wymer (eds), 38−44. Norwich: Geo Abstracts.

Coope, G. R. 1975. Climate fluctuations in north-west Europe since the last interglacial, indicated by fossil assemblages of Coleoptera. In *Ice ages ancient and modern*, A. E. Wright and F. Moseley (eds), 153−68. *Geol. J.* Spec. Issue No. 6.

Coope, G. R. 1977. Fossil coleopteran assemblages as sensitive indicators of climatic changes during the Devensian (last) cold stage. *Phil Trans R. Soc. Lond. Ser. B* **286**, 313−40.

Coope, G. R., F. W. Shotton and I. Strachan 1961. A Late Pleistocene fauna and flora from Upton Warren, Worcestershire. *Phil Trans R. Soc. Lond. Ser. B* **244**, 380−417.

Dalrymple, J. B. 1957. The Pleistocene deposits of Penfolds Pit, Slindon, Sussex and their chronology. *Proc. Geol Assoc.* **68**, 294−303.

Dalrymple, J. B. 1958. The application of soil micromorphology to fossil soils and other deposits from archaeological sites. *J. Soil Sci.* **9**, 199−209.

Douglas, T. D. 1982. Periglacial involutions and the evidence for coversands in the English midlands. *Proc. Yorks Geol Soc.* **44**, 131−43.

Edwards, K. J., C. J. Caseldine and D. K. Chester 1976. Possible interstadial and interglacial pollen floras from Teindland, Scotland. *Nature* **264**, 742−4.

Evans, J. G. 1966. Late-glacial and Post-glacial subaerial deposits at Pitstone, Buckinghamshire. *Proc. Geol Assoc.* **77**, 347−64.

Evans, J. G. 1968. Periglacial deposits on the chalk of Wiltshire. *Wilts. Arch. Nat. Hist. Mag.* **63**, 12−26.

Fenwick, I. 1985. Palaeosols − problems of recognition and interpretation. In *Soils and Quaternary landscape evolution*, J. Boardman (ed.), 3−21. Chichester: Wiley.

FitzPatrick, E. A. 1965. An interglacial soil at Teindland, Morayshire. *Nature* **207**, 621−2.

Follmer, L. R. 1978. The Sangamon Soil in its type area − a review. In *Quaternary soils*, W. C. Mahaney (ed.), 125−66. Norwich: Geo Abstracts.

Funnell, B. M., P. E. P. Norton and R. G. West 1979. The crag at Bramerton, near Norwich, Norfolk. *Phil Trans R. Soc. Lond. Ser. B* **287**, 490−534.

Galloway, R. W. 1961. Ice wedges and involutions in Scotland. *Biul. Peryglac.* **10**, 169−93.

Gaunt, G. D. 1976. The Devensian maximum ice limit in the Vale of York. *Proc. Yorks. Geol Soc.* **40**, 631−7.

Girling, M. A. 1977. Tattershall and Kirkby-on-Bain. In *Yorkshire and Lincolnshire, INQUA Guidebook*, J. A. Catt (ed.), 19−21. Norwich: Geo Abstracts.

Greig, D. G. 1981. Ice wedge cast network in eastern Berwickshire. *Scot. J. Geol.* **17**, 119−22.

Griffey, N. J. and P. Worsley 1978. The pattern of Neoglacial glacier variations in the Okstindan region of northern Norway during the last three millenia. *Boreas* **7**, 1−18.

Gruhn, R. and A. L. Bryan 1969. Fossil ice wedge polygons in southeast Essex, England. In *The periglacial environment: past and present*, T. L. Péwé (ed.), 351−63. Montreal: McGill-Queen's University Press.

Gruhn, R., A. L. Bryan and A. J. Moss 1974. A contribution to the Pleistocene chronology in southeast Essex, England. *Quat. Res.* **4**, 53−71.

Hey, R. W. 1980. Equivalents of the Westland Green Gravels in Essex and East Anglia. *Proc. Geol Assoc.* **91**, 279−90.

Johnson, R. H. 1975. Some late Pleistocene involutions at Dalton-in-Furness, northern England. *Geol. J.* **10**, 23−34.

Kemp, R. A. 1985. The Valley Farm Soil in southern East Anglia. In *Soils and Quaternary landscape evolution*. J. Boardman (ed.), 179−96. Chichester: Wiley.

Kerney, M. P. 1963. Late-glacial deposits on the chalk of south-east England. *Phil Trans R. Soc. Lond. Ser. B* **246**, 203−54.

Kerney, M. P. 1965. Weichselian deposits in the Isle of Thanet, east Kent. *Proc. Geol Assoc.* **76**, 269−74.

Kerney, M. P. and G. de G. Sieveking 1977. Lower Thames and Medway Valleys − Northfleet. In *Southeast England and the Thames Valley, INQUA Guidebook*, E. R. Shephard-Thorn and J. J. Wymer (eds), 44−9. Norwich: Geo Abstracts.

Kerney, M. P., E. H. Brown and T. J. Chandler 1964. The Late-glacial and Post-glacial history of the chalk escarpment near Brook, Kent. *Phil Trans R. Soc. Lond. Ser. B* **248**, 134−204.

King, C. A. M. 1976. *Northern England*. London: Methuen.

Kukla, G. J. 1975. Loess stratigraphy of central Europe. In *After the Australopithecines*, K. W. Butzer and G. L. Isaac (eds), 99−188. The Hague: Mouton.

Lake, R. D., R. A. Ellison and B. S. P. Moorlock 1977. Middle Pleistocene stratigraphy in southern East Anglia. *Nature* **265**, 663.

Lawson, T. J. 1983. A note on the significance of a red soil on dolomite in north-west Scotland. *Quat. News.* **40**, 10−12.

Leverett, J. 1898. The weathered zone (Sangamon) between the Towan loess and the Illinoian till sheet. *J. Geol.* **6**, 171−81.

Limmer, A. J. and A. T. Wilson. 1980. Amino-acids in buried paleosols. *J. Soil Sci.* **31**, 147−53.

MacClintock, P. 1933. Interglacial soils and the drift sheets of eastern England. *Report XVI International Geology Congress*, Washington, 1041−53.

Matthews, B. 1970. Age and origin of aeolian sand in the Vale of York. *Nature* **227**, 1234−6.

Mitchell, G. F. 1960. The Pleistocene history of the Irish Sea. *Adv. Sci.* **17**, 313−25.

Mitchell, G. F., L. F. Penny, F. W. Shotton and R. G. West 1973. *A correlation of Quaternary deposits in the British Isles*. Geol Soc. Lond. Spec. Rep.

Morgan, A. V. 1971. Polygonal patterned ground of late Weichselian age in the area north and west of Wolverhampton, England. *Geogr. Ann.* **53A**, 146−56.

Morgan, A. V. 1973. The Pleistocene geology of the area north and west of Wolverhampton, Staffordshire. *Phil Trans R. Soc. Lond. Ser. B* **265**, 233−97.

Morrison, R. B. 1978. Quaternary soil stratigraphy − concepts, methods and problems. In *Quaternary soils*, W. C. Mahaney (ed.), 77−108. Norwich: Geo Abstracts.

Paepe, R. 1971. Dating and position of fossil soils in the Belgian Pleistocene stratigraphy. In *Paleopedology: origin, nature and dating of paleosols*, D. H. Yaalon (ed.), 261−9. Jerusalem: International Society of Soil Scientists and Israel Universities Press.

Paepe, R. and R. Vanhoorne 1967. *The stratigraphy and palaeobotany of the Late Pleistocene in Belgium*. Geol Surv. Belgium, Mem., No. 8.

Paterson, K. 1971. Weichselian deposits and fossil periglacial structures in north Berkshire. *Proc. Geol Assoc.* **82**, 455−68.

Pennington, W. 1977. The late Devensian flora and vegetation of Britain. *Phil Trans R. Soc. Lond. Ser. B* **280**, 247−70.

Perrin, R. S. M., J. Rose and H. Davies 1979. The distribution, variation and origins of pre-Devensian tills in eastern England. *Phil Trans R. Soc. Lond. Ser. B* **287**, 535−70.

Péwé, T. L. 1974. Geomorphic processes in polar deserts. In *Polar deserts and modern man*, T. L. Smiley and J. H. Zumberge (eds), 33−52. Tuscon: University of Arizona Press.

Romans, J. C. C. 1977. Stratigraphy of a buried soil at Teindland Forest, Scotland. *Nature* **268**, 622−8.

Rose, J. 1975. Raised beach gravels and ice wedge casts at Old Kilpatrick, near Glasgow. *Scot. J. Geol.* **11**, 15−21.

Rose, J. 1983. Early Middle Pleistocene sediments and palaeosols in west and central Essex. In *Diversion of the Thames, QRA field guide*, J. Rose (ed.), 135–9. Leighton Buzzard: Gemprint.

Rose, J. and P. Allen 1977. Middle Pleistocene stratigraphy in south-east Suffolk. *J. Geol Soc. Lond.* **133**, 83–102.

Rose, J., P. Allen and R. W. Hey 1976. Middle Pleistocene stratigraphy in southern East Anglia. *Nature* **263**, 492–4.

Rose, J., R. A. Kemp, C. A. Whiteman, P. Allen and N. Owen 1985. The early Anglian Barham Soil of Eastern England. In *Soils and Quaternary landscape evolution*, J. Boardman (ed.). 197–229. Chichester: Wiley.

Rose, J., R. G. Sturdy, P. Allen and C. A. Whiteman 1978. Middle Pleistocene sediments and palaeosols near Chelmsford, Essex. *Proc. Geol Assoc.* **89**, 91–6.

Rose, J., C. Turner, G. R. Coope and M. D. Bryan 1980. Channel changes in a lowland river catchment over the last 13 000 years. In *Timescales in geomorphology*, R. A. Cullingford, D. A. Davidson and J. Lewin (eds), 159–75. London: Wiley.

Ruddiman, W. F., C. D. Sancetta and A. McIntyre 1977. Glacial/interglacial response rate of subpolar North Atlantic waters to climatic change: the record in ocean sediments. *Phil Trans R. Soc. Lond. Ser. B* **280**, 119–41.

Shotton, F. W. 1968. The Pleistocene succession around Brandon, Warwickshire. *Phil Trans R. Soc. Lond. Ser. B* **254**, 387–400.

Shotton, F. W. 1977. *British Quaternary studies, recent advances*. Oxford: Clarendon Press.

Shotton, F. W. 1983. United Kingdom contribution to the International Geological Correlation Programme, Project 24, Quaternary Glaciations of the Northern Hemisphere. *Quat. News.* **39**, 19–25.

Sissons, J. B. 1974. The Quaternary of Scotland: a review. *Scot. J. Geol.* **10**, 311–37.

Sissons, J. B. 1979. The Loch Lomond Stadial of the British Isles. *Nature* **280**, 199–203.

Stephens, N. 1970. The west country and southern Ireland. In *The glaciations of Wales and adjoining regions*, C. A. Lewis (ed.), 267–314. London: Longman.

Stevens, J. H. and M. J. Wilson 1970. Alpine podzols on the Ben Lawers Massif, Perthshire. *J. Soil Sci.* **21**, 85–95.

Straw, A. 1963. Some observations on the 'cover sands' of north Lincolnshire. *Lincs. Nats. Union* **15**, 260–9.

Straw, A. 1980. The age and geomorphological context of a Norfolk paleosol. In *Timescales in geomorphology*, R. A. Cullingford, D. A. Davidson and J. Lewin (eds), 305–15. London: Wiley.

Sturdy, R. G., R. H. Allen, P. Bullock, J. A. Catt and S. Greenfield 1979. Paleosols developed on chalky boulder clay in Essex. *J. Soil Sci.* **30**, 117–37.

Tedrow, J. C. F. 1978. Development of Polar Desert Soils. In *Quaternary soils*, W. C. Mahaney (ed.), 413–25. Norwich: Geo Abstracts.

Tilley, P. D. 1964. The significance of loess in southeast England. In *Report on the VIth INQUA Congress, Warsaw, 1961*, Vol. 4, 591–6.

Valentine, K. W. G. and J. B. Dalrymple 1975. The identification, lateral variation, and chronology of two buried paleocatenas at Woodhall Spa and West Runton, England. *Quat. Res.* **5**, 551–90.

Valentine, K. W. G. and J. B. Dalrymple 1976. Quaternary buried paleosols: a critical review. *Quat. Res.* **6**, 209–22.

Van der Hammen, T. 1957. The stratigraphy of the Late Glacial. *Geol. Mijb.* **19**, 250–4.

Van Vliet-Lanoë, B. 1976. Traces de ségrégation de glace en lentilles associées aux sols et phénomènes périglaciaires fossiles. *Bíul. Peryglac.* **26**, 41–56.

Vink, A. P. A. and Sevink, J. 1971. Soils and paleosols in the Lutterzand. *Med. Rijks. Geol. Dienst. N.S.* **22**, 165–85.

Watson, E. 1977. The periglacial environment of Great Britain during the Devensian. *Phil Trans R. Soc. Lond. Ser. B* **280**, 183–97.

Watts, W. A. 1980. Regional variation in the response of vegetation to Lateglacial climatic events in Europe. In *Studies in the Lateglacial of northwest Europe*, J. J. Lowe, J. M. Gray and J. E. Robinson (eds), 1–21. Oxford: Pergamon.

West, R. G. 1969. Stratigraphy of periglacial features in East Anglia and adjacent areas. In *The periglacial environment*, T. L. Péwé (ed.), 411–15. Montreal: McGill-Queen's University Press.

West, R. G. 1977. *Pleistocene geology and biology*, 2nd edn. London: Longman.

West, R. G. 1980. *The pre-glacial Pleistocene of the Norfolk and Suffolk coasts.* Cambridge: Cambridge University Press.

West, R. G. and D. G. Wilson 1966. Cromer Forest Bed Series. *Nature* **209**, 497–8.

West, R. G., C. A. Dickson, J. A. Catt, A. H. Weir, and B. W. Sparks 1974. Late Pleistocene deposits at Wretton, Norfolk II. Devensian deposits. *Phil Trans R. Soc. Lond. Ser. B* **267**, 337–420.

Williams, R. G. B. 1969. Permafrost and temperature conditions in England during the last glacial period. In *The periglacial environment*, T. L. Péwé (ed.), 399–410. Montreal: McGill-Queen's University Press.

Williams, R. G. B. 1975. The British climate during the last glaciation; an interpretation based upon periglacial phenomena. In *Ice ages ancient and modern*, A. E. Wright and F. Moseley (eds), 95–120. *Geol. J.* Spec. Issue.

Wilson, M. J. and C. J. Brown 1976. The pedogenesis of some gibbsitic soils from the Southern Uplands of Scotland. *J. Soil Sci.* **27**, 513–22.

Wooldridge, S. W. and I. W. Cornwall 1964. A contribution to a new datum to the prehistory of the Thames valley. *Bull. Inst. Arch. Lond. Univ.* **4**, 223–32.

Worsley, P. 1966. Some Weichselian fossil frost wedges from East Cheshire. *Mercian Geol.* **1**, 357–65.

Worsley, P. 1977. The Cheshire–Shropshire plain. In *Wales and the Cheshire–Shropshire lowland, Xth INQUA Guidebook*, D. Q. Bowen (ed.), 53–60. Norwich: Geo Abstracts.

Zagwijn, W. H. 1975. Variations in climate as shown by pollen analysis, especially in the Lower Pleistocene of Europe. In *Ice ages ancient and modern*, A. E. Wright and F. Moseley (eds), 137–52. *Geol. J.* Spec. Issue.

Zeuner, F. E. 1959. *The Pleistocene period: its climate, chronology and faunal successions.* London: Hutchinson.

Part V

SOIL–GEOMORPHIC APPLICATIONS

19

Soil degradation and erosion as a result of agricultural practice

R. P. C. Morgan

Introduction

Removal of forest or grassland vegetation so that land can be used for agriculture results in disturbance of the soil. The natural vegetation is replaced by agricultural plant assemblages whose survival depends on mechanical cultivation of the soil, the application of herbicides, pesticides and fertilisers, and, in many parts of the world, the implementation of irrigation and drainage. Tillage is used to destroy the residue of previous crops, to produce a tilth which will allow easy placement and germination of seeds and emergence of the crop, and to control weeds. Over the years a reasonable standard system of tillage, involving ploughing and one or more disc harrowings, followed by planting, has been found suitable for a wide range of soils. Certain soils, however, are susceptible to damage as a result of compaction, smearing, excessive working, and in extreme cases, pulverisation. These effects on the soil, known collectively as degradation, reduce its value as a growing medium. In recent years, greater farm mechanisation has increased the impact of wheeled traffic on the land, while the omission of grass leys from arable farming systems has brought about a decline in the organic content of the soil. These changes in farming practice have extended the range of soil types which are prone to degradation.

As well as reducing crop yield directly, soil degradation can result in loss of soil aggregate structure and breakdown of the soil into its primary particles so that it is less resistant to erosion by raindrop impact, surface runoff and wind. Compaction and smearing of the soil reduce its infiltration capacity and promote greater runoff which, in turn, also leads to greater erosion. Thus, soil erosion is an extreme form of soil degradation in which natural geomorphological processes are accelerated so that soil is removed at rates ten and sometimes several thousand times faster than is the case under conditions of natural vegetation, and much faster than rates at which new soil forms.

Mechanics of soil erosion

Soil erosion is a two-phase process comprising detachment of soil particles from the soil mass and their transport down slope, down stream or down wind. A framework for understanding soil erosion in this way was provided by Meyer and Wischmeier (1969) for water erosion on a hillslope. On a segment of a hillside, detachment occurs as a result of raindrop impact on the soil surface and the action of runoff. The detached particles combine with an input of sediment from up slope to provide a reservoir of material for transport. Rainfall and runoff are also the agents of transport. If more particles are detached than the transporting agents have the capacity to remove, erosion equals the transport capacity rate and is 'transport-limited'. If fewer particles are detached than can be transported, erosion equals the detachment rate and is 'detachment-limited'. This basic framework has been simplified by considering only raindrop detachment and runoff transport to form the sediment phase of a soil erosion model, to which has been added a water phase covering rainfall energy and runoff volume (Fig.

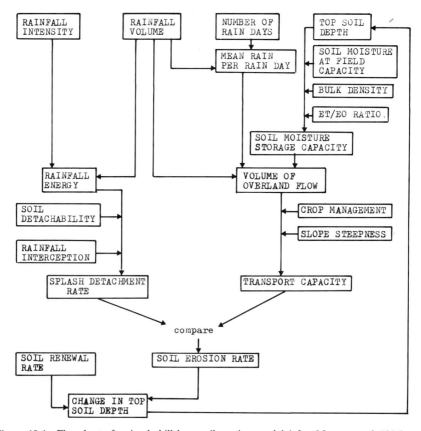

Figure 19.1 Flowchart of a simple hillslope soil erosion model (after Morgan *et al.* 1984).

19.1). Operating equations have been taken from the published researches of geomorphologists and agricultural engineers (Morgan *et al.* 1984).

In many recent soil erosion models (Foster 1982) a distinction is made between interrill erosion, which results from a combination of raindrop impact and overland flow, and rill erosion, where the runoff becomes channelled. Interrill erosion is usually viewed as transport-limited because, while raindrops strike the soil surface at velocities approaching 9 m s^{-1}, which is sufficient to detach soil particles, overland flow velocities are generally around 0.01 m s^{-1}, which is too low for detachment. In contrast, flow velocities in rills are around 0.5 − 1.0 m s^{-1}, sufficient to detach and entrain soil particles, and erosion is usually detachment-limited with the rate depending upon the detachment rate in the rill itself and the rate of supply of material to the rill from the interrill areas. In practice, the distinction between rill and interrill erosion may not be easy to make because Moss *et al.* (1982) and Merritt (1984) have shown that lines of concentrated flow develop within thin, supercritical sheet flow before any channels are formed, and that these flow lines are closer to rills in their behavioural properties than to overland flow. If rills effectively extend headwards along flow lines, separation of rill and interrill erosion according to the presence or absence of channels is likely to be misleading. Such extension, however, could help to explain why hillslope erosion by unchannelled flow has been found in some instances to be detachment-limited (Morgan 1977, van Asch 1980).

The detachability of soil particles is dependent on grain size, but the relationship is not linear. Soil has an inherent resistance to erosion and the erosive agent must attain a critical or threshold condition for detachment to occur. For water erosion the critical velocity increases with increasing grain size for particle sizes above 0.20 mm because of the greater force required to dislodge the larger and heavier particles. Below this grain size, critical velocities increase with decreasing particle size for cohesive materials but decrease slightly for cohesionless particles (Fig. 19.2). Once a particle has been detached it can be transported until the flow velocity drops below a threshold fall velocity which increases with particle size throughout the grain size range.

A similar relationship has been established for the detachment of soil particles by wind except that the most detachable particles are around 0.05 − 0.10 mm in size, smaller than for water erosion. Usually, two threshold velocity curves are recognised for wind erosion: the static or fluid threshold relates to the action of moving air alone and the dynamic or impact threshold relates to the action of moving air laden with sediment. Impact threshold velocities are about 80% of the fluid threshold velocities.

Plots of the kinetic energy of rainfall required to detach a unit weight of sediment (Savat 1982) show the same inverse parabolic relationship with grain size, with the most detachable particles being about 0.12 − 0.15 mm in diameter. The reasons why the most erodible particles are smallest for wind erosion and increase in size through splash erosion to erosion by turbulent running water are not fully understood. In part, they must relate to the considerable difference between the densities of the fluids. A sand grain in the air is some 2000 times

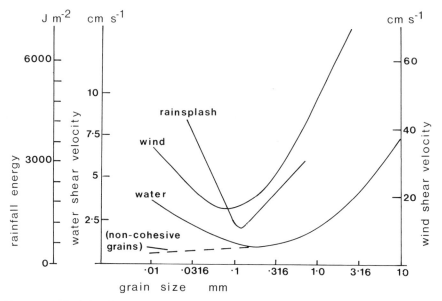

Figure 19.2 Relationship between grain size and critical shear velocities of wind and water and critical rainfall energy for movement of soil particles (after Savat 1982).

more massive than the surrounding fluid, whereas it is only about 2.6 times more massive than water (Bagnold 1979). This difference in relative densities also explains why the critical shear velocity required to move a particle of 0.10 mm diameter is about 10 mm s^{-1} for water whereas it is about 120 mm s^{-1} for wind.

Soil erodibility

Erodibility defines the resistance of the soil to erosion. It is dependent upon the texture and chemical composition of the soil and the way these affect its shear strength, aggregate stability and tendency to surface crusting. The relationships described above between grain size and erosion support the use of the particle-size distribution as an indicator of soil erodibility, but this is inadequate under many circumstances because it does not take into account the propensity of soils to flocculate and deflocculate, their capacity for volume change through swelling, the ability of the clay particles to hold or replace surface ions, and the soil fabric (i.e. the pattern of orientation of the particles).

Shear strength

The shear strength of a soil is a measure of its resistance to shear stresses exerted by gravity, moving fluids and mechanical loads. A soil derives its shear strength from the frictional resistance met by particles when they are forced to slide over one another or to move out of interlocking positions, the extent to which stress or force can be absorbed by greater solid-to-solid contact between the particles, the cohesive forces related to interparticle chemical bonding of the clay minerals, and surface tension forces within the moisture films in unsaturated soils. Although these components are understood qualitatively, they cannot readily be combined in a physically based interpretation, so that shear strength is still expressed for practical purposes by an empirical equation:

$$s = c + \sigma_n \tan \phi \qquad (19.1)$$

where s is the shear stress required for failure to occur, c is cohesion σ_n, is the stress normal to the shear plane (all in units of force per unit area) and ϕ is the angle of shearing resistance. Both c and ϕ should be regarded as empirical parameters rather than as physical properties of the soil, and their values are normally obtained by plotting data from shear tests carried out under different normal stresses (Marshall & Holmes 1979). In granular soils, shear strength (s) is mainly due to friction and is proportional to the normal stress (σ_n). There may be some apparent cohesion due to granular interlocking. With moist or wet clay soils, ϕ can often be taken as zero and s is equal to c.

As noted above, cohesion arises from chemical forces acting between the particles, surface tension in the moisture film and bonding by cementing agents. Clay minerals are silicates of positively charged atoms, most commonly silicon (Si^{4+}), aluminium (Al^{3+}) and magnesium (Mg^{2+}), surrounded by negatively charged atoms of oxygen (O^{2-}). Clay particles are attracted to one another at short range by Van der Waals' forces which are sufficiently strong to overcome the repelling forces resulting from the similar (negative) electric charges present on the particle surfaces. This means that, in simple terms, shear strength arising from cohesion depends upon factors influencing the distance apart of the clay particles. These are the ion content of the soil water, the size of the ions and the moisture content of the soil.

When a soil adsorbs water, the cations in the solution are attracted to the negative charges on the clay surfaces and an electric double layer is formed comprising the particle surface and the surrounding adsorbed cations. Hydrogen ions in the solution are exchanged for cations adsorbed to the surfaces and the resulting hydrogen bonding between the clay surfaces and the water molecules adsorbed to them is an attractive force contributing to cohesion. In addition, the smaller cations adsorbed to the surfaces are replaced by larger ones, thereby reducing the distance between the clay particles.

As the moisture content of a soil increases, its shear strength decreases and its behaviour under stress changes. At low moisture contents the soil behaves as a solid and fractures under stress, but as the moisture content increases, the soil becomes plastic and yields without fracture. The moisture content at which this change in behaviour occurs is defined as the plastic limit. With further increases in wetting, the soil will eventually reach its liquid limit when it will start to flow under its own weight. The values of shear strength at these limits depend upon the mineralogy and percentage of clay in the soil.

Initially, an increase in the water content of a soil can be contained within its structure on a microscopic scale. Water molecules enter the spacings in the lattice structure of the clay minerals causing the spacings therein to increase from about 0.9 to 2.0 nm. The repelling effects of this increase in spacing are offset by hydrogen bonding. At this stage in the wetting process, the increase in soil volume occurs at a rate lower than the rate of water uptake and there is no visible swelling. Once the plastic limit has been reached, continued wetting and hydration of the cations result in an increase in volume at the same rate as water uptake, and visible swelling occurs. If the new water solution has a higher electrolyte concentration than the solution already surrounding the clay, the water will be attracted in osmotically. The cations associated with the clay surfaces are then unable to move freely into the new solution but are held as if surrounded by a semipermeable membrane. In this stage of wetting the thickness of the double layer and the distances between the surfaces of adjacent particles increase. This two-phase wetting process is paralleled in the drying stage when first normal shrinkage occurs where volume decreases at approximately the same rate as water loss, and then residual shrinkage takes place as the clay loses its plasticity and volume decreases less rapidly than the water content.

The application of double layer theory to swelling has been criticised because of inadequacies when applied to divalent and trivalent cations which do not hydrate fully, and to clays which do not have an expanding crystal lattice with regular arrangement of parallel plates. It also ignores the effects of cementing agents, the bonding of particles by crystal growth, and reorientation of the fabric. Nevertheless, whilst the mechanism of shrinking and swelling clays remains imperfectly understood, the theory provides a helpful qualitative appreciation of the process (Marshall & Holmes 1979).

The effects of electrochemical changes on the resistance of the soil to erosion vary over time. This is partly the result of relative rates of cation exchange in relation to the residence time of the water (Thornes 1980). In general, chemical reactions are almost instantaneous in kaolinites, take a few hours in illites, but require much longer with smectites (montmorillonites). This is because clays with a high content of calcium ions develop little or no separation between the layers so that most of the reaction takes place within the lattice structure, whereas with smectitic clays most of the reaction occurs within the double layer. It is these latter clays, with high sodium contents, that are most subject to swelling on wetting and to loss of shear strength.

The interactions between the moisture content of the soil and the chemical composition of the clay particles and the soil water make the behaviour of clays, particularly swelling clays, difficult to predict. The identical treatment of different types of clay can have totally different effects on their strengths (Thornes 1980). Most clays lose strength when first wetted because the free water releases the bonds between the particles. Under moist, unsaturated conditions, however, some clays may regain their strength over time, a process known as thixotropic behaviour, as the hydration of clay minerals and adsorption of free water promote hydrogen bonding (Grissinger & Asmussen 1963). Strength may also be regained if, following swelling, small rotations of the particles bring about slight changes to the fabric, for instance, a change from strong particle alignments parallel to the eroding water to a more random orientation (Grissinger 1966). The shear strength of smectitic clays is largely dependent upon the sodium adsorption ratio. As this increases, i.e. the replacement of calcium or magnesium ions or both by sodium increases, so does water uptake, and the shear strength decreases. Under these conditions the soil behaviour is also affected by the salt concentration in the water. When this is high, the initial resistance of the soil to erosion is maintained at much higher sodium adsorption ratios, probably reflecting a tendency of the soil to form strong flocculated structures with the salts in the water (Arulanandan et al. 1975).

The behaviour of compressible soils on saturation and when subjected to further loading depends upon whether the water can drain from the soil. If drainage cannot take place, pressure will increase in the water and the compaction load will not be supported by the particles. The soil will deform, behaving as a plastic material. If the soil can drain, more of the load will be carried by the particles, under greater intergranular stress, and it is more likely to remain below the plastic limit and retain a higher shear strength.

Although shear strength was shown by Chorley (1959) and Eyles (1968) to affect the broad pattern of landforms, for example by influencing differences in altitude in the landscape, it has been little used as an index of soil erodibility. Shear strength can be used as a basis for understanding the detachment and entrainment of soil particles by running water, but the concept of a critical velocity is more commonly used as a surrogate (Thornes 1979). One reason for the limited use of shear strength is that valid measurements are difficult to obtain both in the laboratory and in the field so that theory cannot be readily verified experimentally. Valid data apply to a surface layer of soil only a few millimetres thick with resistance to shear at zero confining stresses and low velocity impacts reaching a maximum of $4.5 - 6.0$ m s^{-1} with raindrops. Shear failure is usually static, i.e. the result of overcoming the frictional resistance required to initiate motion rather than the dynamic force needed to keep one particle sliding over another.

Recently, however, attempts have been made to understand splash erosion in terms of shear. Ghadiri and Payne (1977) have examined the impact forces involved in raindrops of different sizes striking a target and Huang et al. (1982) have analysed the forces involved in the lateral movements of water away from and back to the point of impact, which are set up immediately following the raindrop hitting the soil surface. Local velocities in these impinging and Rayleigh jets are nearly double those of the raindrop impact and the resulting shear stresses peak around the circumference of the impact crater. Since the soils are often saturated and this process is virtually instantaneous, there is no time for drainage and undrained failure occurs (Al-Durrah & Bradford 1982).

Aggregate stability

The chemical bonding of clay and organic particles results in the development of aggregates so that most soils in the field comprise primary particles of silts,

sands and occasionally gravel, and aggregates of fine particles. The spaces between the aggregates and primary particles constitute the pores. The resistance
of the soil to erosion by wind and water depends upon the stability of the aggregates and, therefore, of the pore spaces. Large, stable aggregates resist raindrop
impact and allow infiltration of rainwater to occur. Wetting of the soil decreases
its strength because of reduced cohesion, softening of cements and swelling
during water adsorption. Rapid wetting can also entrap air in the pores, and
under compression by the advance of the wetting front through the soil profile,
the air pressure may exceed the tensile strength of the soil before escaping ex-
plosively. This process of aggregate collapse is known as slaking. In swelling soils,
wetting may cause the pores to close so that the infiltration rate decreases and
surface runoff becomes more likely. The freezing of water in the pore spaces of
fine-textured soils may promote aggregation by compression of the soil as ice
lenses grow. In large cracks, however, freezing and thawing can cause aggregates
to crumble. Thus the effects of freeze—thaw are ambivalent and may well depend
upon the moisture content at the time of freezing and the texture of the soil.

Assessments of aggregate stability based on dry sieving of soil samples for
wind erosion (Chepil 1942) and wet sieving for water erosion (Bryan 1968) have
been suggested as indexes of soil erodibility. Although the soil particles are agi-
tated during sieving, the direct action of wind, running water and raindrop
impact is not simulated. A more dynamic test for water erosion is to examine
the response of small pats of soil to the impact of simulated individual raindrops
of a standard size (5.5 mm diameter and weighing about 0.1 g) and velocity (using
a fall height of 1 m). Results of this water-drop test may show little correlation
with assessments of erodibility based on soil texture, particularly where clay soils
are concerned. This could be because aggregate stability is determined more by
the nature of the parent material of the soil through its contribution to the stabilising ion content and the type of clay mineral than by particle size and organic
content (De Meester & Jungerius 1978).

Surface crusting
Following the breakdown of aggregates by raindrop impact, the disaggregated
particles are re-sorted to form a surface cover of closely packed individual grains,
resulting in a surface skin or seal which prevents entry of water and air. As the
skin dries, it becomes a crust which may also hinder the emergence of seedlings.
Thus the crust reduces the productivity of the soil as well as seriously reducing
the infiltration rate and, in turn, promoting greater surface runoff and more
erosion. Detailed studies of crust development (Farres 1978) show that raindrop
impact is the critical process, not saturation of the soil. After rainfall, aggregates
can be observed intact beneath the 2 − 4 mm thick crust even though they are
completely saturated; tapping these aggregates, however, causes their instant
breakdown. From this evidence, it appears that saturation reduces the internal
strength of the aggregate but it does not disintegrate until struck by raindrops.

A consistency index (C_{5-10}) has been developed by De Ploey (1981) to assess
the susceptibility of soils to crusting. The index is defined as $w_5 - w_{10}$ where w_5

and w_{10} are the water contents as a percentage of dry weight at which the two sections of a pat of soil in the Casagrande cup touch each other over a distance of 10 mm, after 5 and 10 blows respectively. Index values greater than 3 denote stable soils and values less than 2.5, unstable soils. The index has been tested on soils from Brabant and Flanders in Belgium, defining stable soils as those showing crusting in the field over less than 50% of the surface and no evidence of surface wash at the end of the winter rains. These soils are generally well aggregated but they still break down under raindrop impact. According to De Ploey and Mücher (1981), they are more prone to detachment by raindrops than are unstable soils but the detachment is of stable aggregates from soil crumbs and clods, so that crusts do not form. Detachment of unstable soils, however, produces compaction and liquefaction of the surface, resulting in the collapse of their internal structure. Crustability decreases with increasing clay and organic matter, which provide greater strength to the soil, so that loams and sandy loams with low clay and organic contents are the most vulnerable.

Agricultural practices

In terms of its effects on soil structure, mechanised agriculture produces the following generalised sequence: compaction during seedbed preparation, interactive effects of the crops during the growing season, loosening processes after harvest and weather influences in the non-productive season. The response of the soil to these events determines its susceptibility to degradation and erosion. Thus the way in which a soil is managed is critical to the maintenance of its fertility and to its conservation. Relevant management practices relate to the accumulation of organic matter, the role of tillage, the use of soil conditioners and the influence of the crop cover.

Organic content

The addition of organic material to the soil is widely accepted as good farming practice improving the agronomic behaviour of mineral soils (Hamblin & Davies 1977). With the trend towards larger, mechanised, arable farms there is no demand to grow grass, clover or alfalfa and organic matter can only be added by uneconomic methods (Davies 1982).

Greenland et al. (1975) show that soils with less than 2% organic carbon, equivalent to about 3.5% organic content, have unstable aggregates. Most soils contain less than 15% organic content and a large number, especially sands and sandy loams, have less than 2%. In a review of trends in the organic content of arable soils, Newbould (1982) cites studies from Sweden showing a fall in organic carbon from 3.5 to 2.5% in 60 years of cereal production, except where straw residue was returned to the soil and nitrogen regularly added. In the Great Plains of Canada the organic content has been virtually halved since 1900 from 1.7 to 0.9% under continuous cereal growing until 1935 and alternating cereals and fallow since then. Thus there is some evidence to indicate that continuous cropping can lead, on some soils, to increasing vulnerability to degradation and erosion.

Voroney *et al.* (1981) suggest that soil erodibility decreases linearly with increasing organic content over the range from 0 to 10%. A linear relationship is also implicit in the procedure used to determine the K-factor values of soil erodibility in the Universal Soil Loss Equation (Wischmeier & Smith 1978). The research base for supporting a linear relationship is slight, however, for few studies have explicitly examined erodibility as a function of organic content. Clearly, the linear relation cannot be extrapolated to very high organic contents because peat soils are highly erodible by wind and water; nor does it apply to some soils of low organic content which become excessively hard under dry conditions. Although theoretically increasing organic matter is a means of promoting structural stability and erosion resistance, too much organic material can be a problem, as is the case with Scottish hill peats where organic carbon contents of $9 - 30\%$ are associated with poor drainage, a slow rate of breakdown of the material into humus and immobilisation of the plant nutrients (Newbould 1982).

Adding organic matter to a soil in the quantities required to affect erodibility substantially can pose difficulties in terms of supply. Jones (1971) shows that, in Nigeria, ploughing in maize residue at $0.5 - 1.0$ kg m^{-2} increases organic carbon content in absolute terms by only $0.004 - 0.017\%$ whilst maintaining an existing level of organic content using farmyard manure requires an annual application of 1 kg m^{-2}. A three-year ley, however, is equivalent to an annual application of farmyard manure of 1.2 kg m^{-2}.

The effects of organic matter on structural stability of a soil depend upon the type of material and its humification coefficient, i.e. on the quantity of humus it produces on breakdown. Living organics, such as green manures, provide an immediate improvement in stability, but because they yield only limited amounts of humus, the effect only lasts a few months. Partially decomposed material, like farmyard manure, takes longer to have an effect but the results are longer lasting. Farmyard manure equivalents of various organic materials have been determined by comparing their effects after ten annual applications to soil at a rate equal to 1% of the weight of the tilled layer (Table 19.1; Kolenbrander 1974).

Numerous field experiments on soil organic matter show that the addition of straw, green manures and farmyard manure will lessen the rate of decrease in structural stability compared with no return of crop residues to the soil, but that grass leys are the only effective way of building up the organic content. Whether this is necessary depends on the soil type. On the silty clay loam at Rothamsted Experimental Station, organic carbon contents have remained virtually unchanged in trials started in 1852 on unmanured and chemically fertilised plots and have increased slightly on plots treated annually with 3.5 kg m^{-2} of farmyard manure. In contrast, on the sandy loam soils at Woburn Experimental Farm, organic carbon contents have decreased from 1.5 to 0.76% over 100 years and rotational cropping has had little effect in preventing this decline. The growing of three-year leys in a five-year rotation with the addition of farmyard manure to the first arable crop every five years was the only treatment to produce an

Table 19.1 Farmyard manure equivalents (FYM) of some organic materials (after Kolenbrander 1974).

Material	Humification coefficient	FYM equivalent
plant foliage	0.20	0.25
green manures	0.25	0.35
cereal straw	0.30	0.45
roots of crops	0.35	0.55
farmyard manure	0.50	1.00
deciduous tree litter	0.60	1.40
coniferous tree litter	0.65	1.60
peat moss	0.85	2.50

increase in organic carbon, from 1.02 to 1.44% over 33 years (Johnston 1982). Since ley farming is uneconomic, the effects of continuous arable cropping on sandy and sandy loam soils give rise for concern. These are the soils most susceptible to crusting and on which soil erosion rates of $2 - 45$ t ha^{-1} yr^{-1} have been recorded in the eastern midlands of England (Morgan 1984) and $1 - 82$ t ha^{-1} yr^{-1} in Hesbaye, Belgium (Bollinne 1978), considerably in excess of the probable soil loss tolerances of $1 - 2$ t ha^{-1} yr^{-1}.

Tillage

The response of a soil to the vertical and horizontal forces resulting from wheeled traffic and tillage implements depends upon its shear strength, the nature of the confining stresses and the direction in which the force is applied.

The main effect of driving a tractor across a field is to apply force from above and compact the soil. Although this may increase shear strength by increasing bulk density, a sandy soil may lose its aggregate structure almost entirely under compaction, promoting in turn surface crusting and reduced infiltration. Ten passes of a 3 t tractor on a sandy loam soil in Bedfordshire increased bulk density from 1.40 to 1.56 Mg m^{-3} and reduced infiltration capacity from 420 to 83 mm h^{-1} (Martin 1979). Such soils are then highly susceptible to water erosion as evidenced above from studies in lowland England and northern Belgium. The pattern of compaction in a soil depends upon tyre pressure, the width of the wheels and the speed of the vehicle, the speed controlling the contact time between wheel and soil. Compaction generally extends to the depth of previous tillage, up to 300 mm for deep ploughing, 180 mm for normal ploughing and 60 mm with zero tillage (Pidgeon & Soane 1978).

Many tillage tools drawn by the tractor are designed to apply upward forces to cut and loosen the compacted soil, sometimes to invert and mix it, and smooth and shape its surface. When the moisture content of the soil is below the plastic limit the soil fails by cracking with the soil aggregates sliding over one another but remaining unbroken. The soil ahead of loosening tools moves forwards and upwards over the entire working depth with a distinct shear plane being formed from the base of the tool, crescentic in the case of tines but modified where the tool turns and inverts the soil, as with a ploughshare. Effective soil loosening occurs where the confining stress resisting the upward movement is less than that resisting sideways movement. Clearly, vertical confining stress is zero at the surface and increases with depth until, at a critical depth, it equals lateral confining stresses and crescentic failure ceases to occur. Below this depth the soil moves only forwards and sideways, no distinct shear plane is formed (Godwin & Spoor 1977) and lateral failure occurs with a risk of compaction.

The application to sandy soils of high forces, as from a Rotavator or combine harvester, can produce such a large number of failure planes that the soil is pulverised; it is then readily eroded by water and, on drying into a fine dust, also by wind. Thus, whilst coarse structures on heavy soils can be improved by tillage, the structures of less cohesive soils may be destroyed. They can then only be cured by natural processes such as weathering. Within a month of the completion of the compaction experiment in Bedfordshire referred to above, the bulk density of the soil had decreased to 1.48 Mg m^{-3} and the infiltration capacity had increased to 337 mm h^{-1} (Martin 1979).

Soils with a moisture content above the plastic limit will fail compressively during tillage, produce few or no fissures and become smeared (Spoor & Godwin 1979). The type of clay mineral in the soil will determine its behaviour under these conditions (Spoor et al. 1982). Unconfined swelling is likely to occur following tillage of smectitic clays and this may result in either the disappearance of aggregates or the formation of new and less stable ones. If drainage cannot occur and undrained failure takes place, an already smeared soil may become puddled or even turn into a slurry.

The shaping of the soil surface by tillage produces a given level of roughness depending upon the implement used. Ploughing produces a rough, cloddy surface with local variations in height of 120 − 160 mm. Secondary cultivation reduces the variation to 30 − 40 mm and drilling and rolling can decrease the roughness further (Evans 1980). Tillage-imparted roughness is greater for clay soils than for sands and sandy loams. A rough surface with high storage volume and a dense network of macropores will reduce surface runoff and increase a soil's resistance to water and wind erosion. These conditions are unsuitable for seed germination and crop establishment, however, so that secondary cultivation is required to produce a finer tilth. The stability of the soil clods comprising this tilth is a critical factor influencing soil erodibility.

Stuttard (1984) shows that clods break down through raindrop impact and slumping and that the stability of individual clods depends upon the conditions at the time of their formation. Clod behaviour with respect to formation conditions is determined by differences in density at low moisture conditions but density has little effect when moisture content is high. If, as suggested by this preliminary study, the effects of formation density and moisture persist throughout the period of clod breakdown, it may be feasible, through greater attention to operating schedules, to use tillage to produce an erosion-resistant surface that is compatible with seedbed requirements. This approach could considerably extend the role of tillage in soil conservation beyond the minimum and zero tillage practices currently recommended. Some tillage systems are already being tried experimentally along these lines. Dixon and Simanton (1980) review various forms of land pressing and imprinting for water erosion control whilst Selman (1981) describes a method of rolling and pressing of the soil to form weather-resistant clods for controlling wind erosion.

Soil conditioners

Various synthetic polymer emulsions, oil- or rubber-based, are commercially available as soil stabilisers to control wind erosion and prevent surface capping. The molecules making up the polymer form bridges across adjacent clay mineral particles, bonding them into soil aggregates which remain stable for periods of two weeks to six months. The emulsions are best applied to a dry soil to prevent raindrops breaking down the natural soil aggregates and washing out the additive before it can take effect.

When soil conditioners are applied to very small aggregates they may prevent infiltration of water. This effect is more marked if the conditioners have been modified by the addition of asphalt to make them hydrophobic. Although asphalt emulsions create increased runoff, they are effective in stabilising the soil surface and preventing crusting and erosion. Subsequent discing to a depth of 200 mm may break up the soil, however, and promote aggregate destruction (Gabriels & De Boodt 1978). This problem could be alleviated by incorporating the emulsion in the top 100 − 200 mm of the soil but, unless the newly formed aggregates are at least 2 mm in size, and ideally greater than 5 mm (Pla 1977), infiltration rates are generally too low. Experiments with polyacrylamide conditioners suggest that high infiltration rates may be achieved regardless of the size of the aggregates.

Bitumen, Portland cement, hydrated lime, animal slurry and sewage sludge have all been tried as soil conditioners in agriculture but they are either, like the polymers, too expensive to use except for high value crops such as vegetables, or in the cases of slurries and sludges, are unpleasant to handle, create odours and, without specialised soil-incorporation equipment, can only be applied in small quantities.

Plant cover effects

Plant covers reduce the volume and energy of the rainfall reaching the ground surface and would therefore be expected to minimise the effects of raindrop impact, prevent surface crusting and reduce erosion. It is generally accepted that the rate of erosion decreases as the canopy of the plant cover increases.

Field studies under winter wheat, winter oats and spring barley showed that rates of soil detachment by rainfall were reduced to 10 − 60% of those in open ground but that, at these lower rates of detachment, were unexpectedly inversely related to rainfall energy (Morgan 1982). This negative relationship, also reported in laboratory experiments with Brussels sprouts by Noble and Morgan

(1983) and with Brussels sprouts, sugar-beet and potatoes by Finney (1984), implies that plant covers provide better protection against erosion from high than from low intensity rains. Noble and Morgan (1983) found that as the percentage canopy cover of Brussels sprouts increased so did the detachment rate per unit of rainfall energy, indicating that the rainfall reaching the ground surface was a more effective detaching agent. One reason for this may be the increase in median volume drop size from 3.2 mm in the design storm to 6.0 mm under leaf drips. Finney (1984) recorded similar increases for all three species of plants studied. The effects on soil detachment are ambivalent, however, with detachment rates being reduced in the early stages of canopy growth but increasing under potatoes and Brussels sprouts to rates close to those in open ground when the canopy cover exceeds 20 − 25% (Morgan *et al.* 1983).

Although much more research is required into rainfall interception and soil detachment by rainfall under plants, the results of these studies have important implications for soil management. The increases in median volume drop size of the transformed raindrops beneath the plants compared with the natural rainfall are most marked for low intensities, explaining the poorer protection afforded by plants in such rains. The drop size increase appears to offset the effects of lower impact velocities of drops falling from the crop canopy. Whilst some crops may be effective in protecting the soil, the high rates of soil detaehment under large-leaved crops such as Brussels sprouts imply that others may result in as much destruction of soil aggregates, soil degradation and soil erosion as rain in open ground.

Towards an index of soil erodibility

Various indexes of soil erodibility are reviewed by Bryan (1968) and summarised in Morgan (1979). The most commonly used is the *K*-factor which expresses erodibility as the mean annual soil loss per unit of a rainfall erosion index for a standard set of slope and crop management conditions. The *K*-factor has proved valuable in the Universal Soil Loss Equation (Wischmeier & Smith 1978) and its numerical value can be estimated for a given soil from readily measurable properties of grain size, organic content, permeability and structure, with grain size as the dominant parameter. Although the relationships between grain size and erodibility are, as shown earlier, broadly understood, *K*-factor values relate largely to erosion as a mechanical process and do not take account of chemical bonding of the clay particles. They tend to express rather poorly the true erodibility of swelling clay soils. Nor does the *K*-factor consider the differences which exist in the size of the most erodible particles between splash, runoff and wind erosion, so that a value for a particular soil is not necessarily valid as a measure of, for instance, both splashability and rillability.

Advances in our understanding of erosion mechanics make the development of an index based on shear strength and aggregate stability more feasible. However, soil behaviour is difficult to predict, especially with cohesive soils.

because of the effects of different clay minerals and changes in soil moisture and bulk density. These changes occur over short time periods, thereby invalidating the use of an average value to describe erodibility. Agricultural practices can seriously weaken certain problem soils through reduction in their organic content, the development of a surface crust, and compaction, smearing and pulverisation during tillage. There is concern that recent farming methods have extended the range of problem soils. The growing of certain crops, especially large-leaved vegetables, on these soils may enhance degradation and erosion rather than protect the soil, because the transformation of raindrop properties during interception offsets the effects of reductions in rainfall volume and energy. When this occurs, a plant cover can only decrease erosion through a reduction in surface runoff but this becomes less likely if the transformed raindrops promote surface crusting.

An erodibility index is required for use in the new generation of soil erosion models which takes account of short-term variations in soil strength and the nature of the eroding agent. Application of these models will also require a better distinction between the domains of each process than recent research indicates is presently possible, for example, between rill and interrill erosion.

Acknowledgement

I am grateful to Mr G. Spoor (Professor of Applied Soil Physics, Department of Agricultural Engineering, Silsoe College) for comments and discussion on an earlier version of this chapter.

References

Al-Durrah, M. M. and J. M. Bradford 1982. Parameters for describing soil detachment due to single water drop impact. *J. Soil Sci. Soc. Am.* **46**, 836–40.

Arulanandan, K., P. Loganathan and R. B. Krone 1975. Pore and eroding fluid influences on the surface erosion of a soil. *J. Geotech. Engng Div. Am. Soc. Civ. Engnrs* **101**, 53–66.

Bagnold, R. A. 1979. Sediment transport by wind and water. *Nordic Hydrol.* **10**, 309–22.

Bollinne, A. 1978. Study of the importance of splash and wash on cultivated loamy soils of Hesbaye (Belgium). *Earth Surf. Proc.* **3**, 71–84.

Bryan, R. B. 1968. The development, use and efficiency of indices of soil erodibility. *Geoderma* **2**, 5–26.

Chepil, W. S. 1942. Measurement of wind erosiveness of soils by dry sieving procedure. *Sci. Agr.* **23**, 154–60.

Chorley, R. J. 1959. The geomorphic significance of some Oxford soils. *Am. J. Sci.* **257**, 503–15.

Davies, D. B. 1982. Soil degradation and soil management in Britain. In *Soil degradation*, D. Boels, D. B. Davies and A. E. Johnston (eds), 19–26. Rotterdam: Balkema.

De Meester, T. and P. D. Jungerius 1978. The relationship between the soil erodibility factor *K* (Universal Soil Loss Equation), aggregate stability and micromorphological properties of soils in the Hornos area, S. Spain. *Earth Surf. Proc.* **3**, 379–91.

De Ploey, J. 1981 Crusting and time-dependent rainwash mechanisms on loamy soil. In *Soil conservation: problems and prospects*, R. P. C. Morgan (ed.), 139−52. Chichester: Wiley.

De Ploey, J. and H. J. Mücher 1981. A consistency index and rainwash mechanisms on Belgian loamy soils. *Earth Surf. Proc. Landforms* **6**, 319−30.

Dixon, R. M. and J. R. Simanton 1980. Land imprinting for better watershed management. In *Symposium on watershed management*, 809−26. New York: Am. Soc. Civ. Eng.

Evans, R. 1980. Mechanics of water erosion and their spatial and temporal controls: an empirical viewpoint. In *Soil erosion*, M. J. Kirkby and R. P. C. Morgan (eds), 109−28. Chichester: Wiley.

Eyles, R. J. 1968. Morphometric explanation: a case study. *Geographica (Univ. Malaya)* **4**, 17−23.

Farres, P. 1978. The role of time and aggregate size in the crusting process. *Earth Surf. Proc.* **3**, 243−54.

Finney, H. J. 1984. *The effect of crop covers on rainfall characteristics and splash detachment.* J. Agric. Engng. Res. **29**, 337−43.

Foster, G. R. 1982. Modeling the erosion process. In *Hydrologic modeling of small watersheds*, C. T. Haan, H. P. Johnson and D. L. Brakensiek (eds), 297−380. Am. Soc. Agric. Engrs Monogr. No. 5.

Gabriels, D. and M. De Boodt 1978. Evaluation of soil conditioners for water erosion control and sand stabilization. In *Modification of soil structure*, W. W. Emerson, R. D. Bond and A. R. Dexter (eds), 341−8. Chichester: Wiley.

Ghadiri, H. and D. Payne 1977. Raindrop impact stress and the breakdown of soil crumbs. *J. Soil Sci.* **28**, 247−58.

Godwin, R. J. and G. Spoor 1977. Soil failure with narrow tines. *J. Agric. Engng Res.* **22**, 213−28.

Greenland, D. J., D. Rimmer and D. Payne 1975. Determination of the structural stability class of English and Welsh soils using a water coherence test. *J. Soil Sci.* **26**, 294−303.

Grissinger, E. H. 1966. Resistance of selected clay systems to erosion by water. *Water Resour. Res.* **2**, 131−8.

Grissinger, E. H. and L. E. Asmussen 1963. Discussion of channel stability in undisturbed cohesive soils by E. M. Flaxman. *J. Hydraul. Div. Am. Soc. Civ. Engrs* **89**, 259−64.

Hamblin, A. P. and D. B. Davies 1977. Influence of soil organic matter on the physical properties of some East Anglian soils of high silt content. *J. Soil Sci.* **28**, 11−23.

Huang, C., J. M. Bradford and J. H. Cushman 1982. A numerical study of raindrop impact phenomena: the rigid case. *J. Soil Sci. Soc. Am.* **46**, 14−19.

Johnston, A. E. 1982. The effects of farming systems on the amount of soil organic matter and its effect on yield at Rothamsted and Woburn. In *Soil degradation*, D. Boels, D. B. Davies and A. E. Johnston (eds), 187−202. Rotterdam: Balkema.

Jones, M. J. 1971. The maintenance of soil organic matter under continuous cultivation at Samaru, Nigeria. *J. Agric. Sci. Cambridge* **77**, 473−82.

Kolenbrander, G. J. 1974. Efficiency of organic manure in increasing soil organic matter content. In *Transactions of the 10th international congress on soil science*, Vol. II, 129−36.

Marshall, T. J. and J. W. Holmes 1979. *Soil physics*. Cambridge: Cambridge University Press.

Martin, L. 1979. Accelerated soil erosion from tractor wheelings: a case study in mid-Bedfordshire. In *Comptes-Rendus, Coll. Erosion agricole des sols en milieu tempérée non-Mediterranéen*, 157−61. Strasbourg: Université Louis Pasteur.

Merritt, E. 1984. *The identification of four stages during micro-rill development.* Paper presented to Earth Surf. Proc. Landform **9**, 493−6.

Meyer, L. D. and W. H. Wischmeier 1969. Mathematical simulation of the process of soil erosion by water. *Trans Am. Soc. Agric. Engrs* **12**, 754−8, 762.

Morgan, R. P. C. 1977. *Soil erosion in the United Kingdom: field studies in the Silsoe area, 1973−75.* Nat. Coll. Agric. Engng Silsoe, Occasional Paper No. 4.

Morgan, R. P. C. 1979. *Soil erosion*. London: Longman.

Morgan, R. P. C. 1982. Splash detachment under plant covers: results and implications of a field study. *Trans Am. Soc. Agric. Engrs* **25**, 987−91.

Morgan, R. P. C. 1985. Soil erosion measurement and soil conservation research in cultivated areas of the UK. *Geogr. J.*

Morgan, R. P. C., H. J. Finney, D. D. V. Morgan and C. A. Noble 1983. *Predicting hillslope runoff and erosion using CREAMS model.* Final Report, Natural Environment Research Council, Research Grant No. GR/4282.

Morgan, R. P. C., D. D. V. Morgan and H. J. Finney 1984. *A predictive model for the assessment of soil erosion risk.* J. Agric. Engng. Res. **30**, 245–53.

Moss, A. J., P. Green and J. Hutka 1982. Small channels: their experimental formation, nature and significance. *Earth Surf. Proc. Landforms* **7**, 401–15.

Newbould, P. 1982. Losses and accumulation of organic matter in soils. In *Soil degradation*, D. Boels, D. B. Davies and A. E. Johnston (eds), 107–31. Rotterdam: Balkema.

Noble, C. A. and R. P. C. Morgan 1983. Rainfall interception and splash detachment with a Brussels sprout plant: a laboratory simulation. *Earth Surf. Proc. Landforms* **8**, 569–77.

Pidgeon, J. D. and D. B. Soane 1978. Soil structure and strength relations following tillage, zero tillage and wheel traffic in Scotland. In *Modification of soil structure*, W. W. Emerson, R. D. Bond and A. R. Dexter (eds), 371–8. Chichester: Wiley.

Pla, I. 1977. Aggregate size and erosion control on sloping land treated with bitumen emulsion. In *Soil conservation and management in the humid tropics*, D. J. Greenland and R. Lal (eds), 109–15. Chichester: Wiley.

Savat, J. 1982. Common and uncommon selectivity in the process of fluid transportation: field observations and laboratory experiments on bare surfaces. In *Aridic soils and geomorphic processes*, D. H. Yaalon (ed.), 139–60. *Catena* Suppl. No. 1.

Selman, M. 1981. The control of wind erosion on sandlands. In *Proceedings of the SAWMA Conference. Soil and crop loss: developments in erosion control.* National Agricultural Centre, Stoneleigh.

Spoor, G. and R. J. Godwin 1979. Soil deformation and shear strength characteristics of some clay soils at different moisture contents. *J. Soil Sci.* **30**, 483–98.

Spoor, G., P. B. Leeds-Harrison and R. J. Godwin 1982. Potential role of soil density and clay mineralogy in assessing the suitability of soils for mole drainage. *J. Soil Sci.* **33**, 427–41.

Stuttard, M. J. 1984. *Effect of clod properties on clod stability to rainfall: laboratory simulation.* J. Agric. Engng. Res. **30**, 141–7.

Thornes, J. B. 1979. Fluvial processes. In *Process in geomorphology*, C. Embleton and J. B. Thornes (eds), 213–71. London: Edward Arnold.

Thornes, J. B. 1980. Erosional processes of running water and their spatial and temporal controls: a theoretical viewpoint. In *Soil erosion*, M. J. Kirkby and R. P. C. Morgan (eds), 129–82. Chichester: Wiley.

van Asch, Th. W. J. 1980. *Water erosion on slopes and landsliding in a Mediterranean landscape.* Utrechtse Geografische Studies No. 20, Geografisch Instituut, Rijkuniversiteit Utrecht.

Voroney, R. P., J. A. van Veen and E. A. Paul 1981. Organic carbon dynamics in grassland soils. II Model validation and simulation of the long-term effects of cultivation and rainfall erosion. *Canadian J. Soil Sci.* **61**, 211–24.

Wischmeier, W. H. and D. D. Smith 1978. *Predicting rainfall erosion losses.* USDA Agric. Res. Serv. Handbook No. 282.

20

Forecasting the trafficability of soils

M. G. Anderson

The need for trafficability models

Models for off-road vehicle behaviour are required for three principal purposes. Firstly, in the context of general agricultural practice, the use of machinery increases the problem of soil compaction. It therefore becomes a requirement to be able to forecast the degree of compaction a given type of vehicle may cause on a particular soil. Empirical approaches have been adopted in this field, with the key variable being soil dry density (γ_d). Raghavan and McKyes (1978) selected γ_d as the dependent variable since it is altered whenever there is a change in the moisture content, or the external forces acting on it. They were then able to establish, for four soil types, eight statistical models predicting γ_d and taking the form:

$$\gamma_d = f(\text{moisture content, tyre slip, contact pressure,}$$
$$\text{number of passes, depth and distance})$$

Schemes of this type enable the broad factors resulting in soil compaction following vehicle activity to be identified. It is relatively easy, if somewhat demanding of data, to extend the analysis into examination of traffic−soil−plant relationships in order to make assessments of plant (root) damage caused by the operation of agricultural machinery (see Rhagavan *et al.* 1979, Huck *et al.* 1975, Taylor 1971).

A second area at which trafficability studies are directed relates to dynamic forecasting requirements. Since the strength of a soil is of major importance to vehicle mobility, and the effective strength varies significantly with soil moisture, a means of forecasting soil moisture variation could form the basis of a real-time soil trafficability model (Carlson *et al.* 1970). The requirement and principal applications are, in this case, primarily related to military needs, and work has been undertaken by the US Army Corps of Engineers (e.g. Collins 1967, Moltham 1967). This work based the trafficability predictions on a rating cone index (RCI), defined by Collins (1971) as:

M. G. ANDERSON

397

$$\ln \text{RCI} = 4.60 + \frac{2.123 + 0.008\,C - 0.693\ln M}{0.149 + 0.002\,C} \qquad (20.1)$$

where M is the moisture content (percentage dry weight) and C is the percentage of clay. RCI values were found below which specified vehicles could not complete more than 40–50 passes (only tracked vehicles can operate when the RCI is less than 50, and four-wheel drive trucks can operate when the RCI is in the range 60–80). Smith and Meyer (1973) used empirical relationships of the form of Equation 20.1 with empirically derived soil moisture prediction equations to establish a dynamic model for trafficability. The concept of index-related performance (see Table 20.1) has certain obvious deficiencies. If the complete terrain–operator–vehicle system is examined, it becomes desirable to consider a physically based analytical model to relate soil strength to vehicle performance, since such a scheme would facilitate an analysis of terrain–track interaction which could incorporate steering and shearing forces.

The third area of trafficability modelling is the utilisation of the fundamental mechanical properties of the soil (such as the angle of internal friction ϕ', cohesion c' and density γ). The cone index has been successfully used as a description of soil strength in establishing empirical soil–vehicle relations, as has already been observed. In order to match recent developments applying the Coulomb equation to the substantial data bases available for cone index measures, it is necessary to attempt to relate the cone index to the engineering properties of the soil. Rohani and Baladi (1981) have successfully developed a mathematical

Table 20.1 Terrain factors affecting vehicle performance (after Shamburger 1967).

Terrain factor	Unit of measure	Range	No. of classes
surface composition			
soil mass strength	RCI	0–100	5
soil surface strength	kg cm^{-2}	0–0.28	9
surface geometry			
slope	deg	0–>45	7
spacing of vertical obstacles	m	0–>45	5
terrain approach angle	deg	<100–220	8
step height	cm	0–>210	8
vegetation			
spacing of stems ≤5, 13, 23 and 127 cm	m	0–>9	4
spacing of stems ≥2.5, 8, 15 and 25 cm	m	0–>9	4
hydrological geometry			
contact approach angle	deg	<145–180	4
step height	cm	<30–122	5
water depth	m	1–>1.4	2

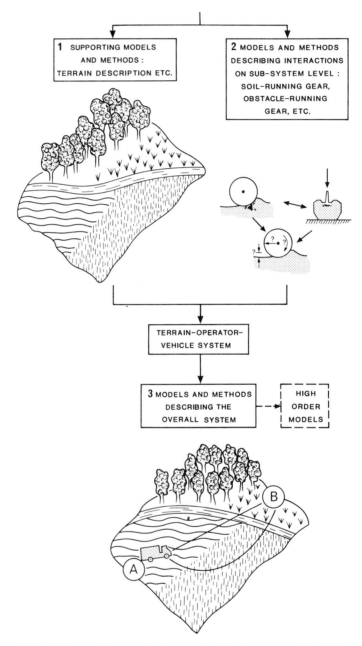

Figure 20.1   Basic model types related to trafficability forecasting (after Melzer 1982).

model for the cone penetration process and thereby achieved very high correlations between measured and predicted cone index values, using the engineering properties of the soil.

The three areas of analysis briefly outlined here, each of which is concerned with the predictive performance of off-road vehicular traffic models, cover the spectrum from basic terrain description, through models describing the detailed interaction between track and terrain, to models describing the overall system (Fig. 20.1). The salient elements can be identified more precisely in the context of their components, as shown in Figure 20.2. The multiplicity of the information

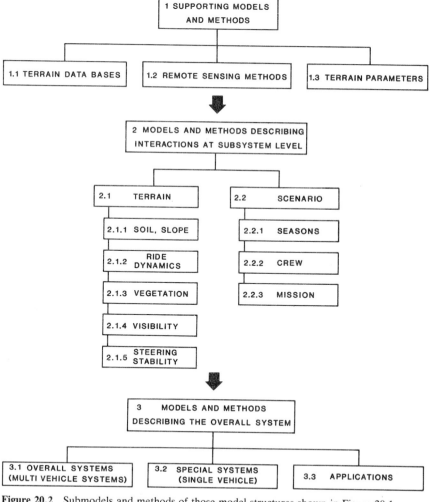

Figure 20.2 Submodels and methods of those model structures shown in Figure 20.1.

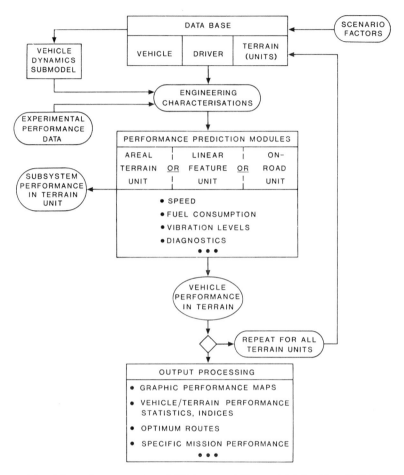

Figure 20.3   Schematic presentation of US Army mobility model (after Nuttall 1976).

field that is relevant has frequently necessitated resorting to a classificatory approach, producing for example, terrain-type maps (Shamburger 1967). Table 20.1 illustrates the commonly measured terrain factors, together with an indication of the number of classes into which each factor is categorised for data base summaries.

There are two inherent problems with describing terrain (as well as other factors) in a quantitative way. One is the enormous amount of data required, and the second is the need to make predictions of trafficability in areas in which the user may be unable to acquire a full suite of desired information.

There has therefore been a significant shift in modelling emphasis, with a move away from index-based analysis towards dynamic modelling of overall systems. Such a move should serve in the long term not only to improve forecasting capability but also simultaneously to render the predictive scheme parsimonious. The US Army mobility model is the only approach in the transport field that allows a comprehensive evaluation of the terrain–vehicle–operator system from the viewpoint of overall technical systems analysis (Melzer 1982). The general model scheme is shown in Figure 20.3. Nuttall (1976), in outlining this model, observes that the acquisition of detailed mobility-oriented terrain data from a variety of sources requires considerable effort, but need be done only once for any given area. In the context of the soil moisture dynamics, this has meant, as in the case of the soil moisture strength prediction (SMSP) model (Smith & Meyer 1973), the establishment of soil moisture recording sites at representative locations and the formulation of soil-water accretion and depletion curves. Such curves are empirically fitted and relate only to seasonal patterns in their parameterisation – the sole driver being total storm rainfall. Moisture content predicted in this manner is then the input to equations of the form of Equation 20.1.

Formulating an improved soil moisture submodel for trafficability models

Whilst state-of-the-art trafficability models now have the ability to model over-all systems dynamically, with an increasing emphasis on engineering material properties, an area that still requires predictive development is the soil moisture forecasting component; it is this potential that is now explored.

To circumvent the need for immense field monitoring programmes of soil moisture conditions for the purpose of establishing empirical rainfall–soil moisture relationships (e.g. Smith & Meyer 1973), it is suggested here that soil water finite difference models may be appropriate to reduce initial data require-ments and to yield greater accuracy in dynamic forecasting. The modelling scheme is outlined, together with an illustration of the output resolution. The questions of both field variability of the required input parameters, and the practical methods for acquisition of such input data, are discussed subsequently.

For the purposes of establishing a modelling framework for soil moisture, a basic computational cell structure can be considered (see Fig. 20.4a). Critical to the trafficability requirement is high resolution and accurate forecasting in the near-surface zone. The following criteria for the design of an initial model structure may therefore be set:

(a) incorporation of a non-isothermal evaporation submodel to act as a forcing function in the top computational cell;
(b) establishment of a multilayer model for the soil profile, in which all the media parameters may be varied for each layer;
(c) inclusion of field variability of parameters so that the sensitivity of the out-comes can be realistically assessed for field conditions; and
(d) retention of a parsimonious model as far as the input data requirements are concerned, the model being formulated such that no calibration is required, thus enhancing the operational potential of the scheme.

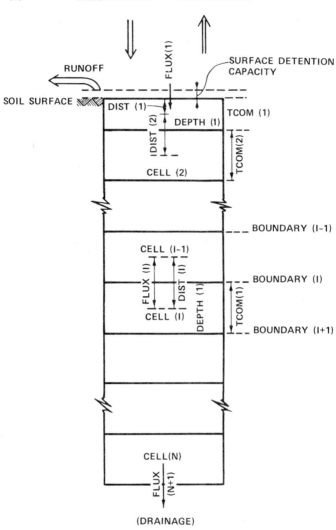

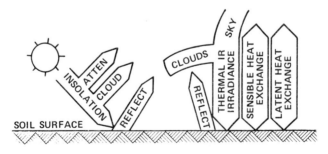

Figure 20.4 Basic element structure for (a) soil-water model, and (b) non-isothermal evaporation model (ATTEN = attenuation).

In the description of the model, each of these criteria and its associated components are outlined (see Table 20.2).

Non-isothermal evaporation estimation

Following Weisner (1970), a surface energy balance (see Fig. 20.4b) can be defined as

$$S_a - I = G + H + L \cdot E \qquad (20.2)$$

where S_a is the incoming radiation into a surface of albedo α_g, I is the thermal infra-red radiation, G is the heat flux from the ground, H is the heat flux from the air, L is the latent heat and E is the evaporation.

Table 20.2 Summary of (a) model governing equations, and (b) required input data. The mean and standard deviation are required for all soil properties listed, based on field variability tests (see Hillel 1980).

(a) Governing equations		
evaporation	Balick *et al.* (1981) model structure	Eqn 20.11
soil-water movement	Richard's equation (Bruce & Whisler 1973)	Eqn 20.12
unsaturated hydraulic conductivity	Millington−Quirk (1959) method	Eqn 20.13
rating cone index	Collins (1971)	Eqn 20.1
(b) Required input data		
location	latitude slope angle slope orientation	
time	Julian calendar day storm duration	
climatic data	atmospheric pressure cloud type met. instrument height	Hourly data for: cloud cover wind speed ground temperature relative humidity air temperature

soil properties	initial moisture content
	surface detention capacity
	saturated hydraulic
	   conductivity
	soil suction−moisture curves
	   (no hysteresis)
	clay percentage
	surface albedo	

Balick *et al.* (1981) provide a model structure that need only be marginally rearranged to estimate evaporation. An equation taken from Khale (1977) is used to compute the solar radiation absorbed at the ground with no cloud cover (see Balick *et al.* Eqn 20), and Haurwitz's (1948) empirical equation is then employed for overcast sky adjustments with respect to cloud cover and cloud type:

$$CA = (a/94.4)\exp\left[-m(b - 0.059)\right] \qquad (20.3)$$

where CA is the cloud adjustment factor, a and b are empirical coefficients dependent on cloud type and m is the secant of the solar zenith angle. Values for a and b are available for the following eight cloud genera, cirrus, cirro-stratus, altocumulus, altostratus, stratocumulus, stratus, nimbostratus, and fog (see Table 1 in Balick *et al.* 1981), and for intermediate cloud cover types (Pochop *et al.* 1968).

The empirical equation used to estimate atmospheric infra-red radiation on the surface ($I \downarrow_o$ is the Brunt equation (Sellers 1965):

$$I \downarrow_o = \varepsilon \delta\, T^4_a [c + b(e_a)^{0.5}] \qquad (20.4)$$

where ε is the emissivity which is assumed equal to unity, δ is the Stephan−Boltzman constant, T_a is the shelter air temperature (degrees kelvin), e_a is the water vapour pressure (millibars), $c = 0.61$ and $b = 0.050$. The cloud contributions to thermal infra-red irradiance $I \downarrow_t$ are treated with an empirical factor adapted from Sellers (1965):

$$I \downarrow_t = I \downarrow_o (1 + CIR \cdot C_c^2) \qquad (20.5)$$

where CIR is a coefficient dependent on cloud type (see Oke 1978) and C_c is cloud cover.

The surface is treated as a grey body emitter such that:

$$I \uparrow = \varepsilon_g \delta (T_g)^4 \qquad (20.6)$$

where $I \uparrow$ is the energy radiated from the surface, ε_g is the emissivity of the ground and T_g is the current surface temperature as predicted by the model. Thus I, the thermal infra-red input required by Equation 20.2, can be given as

$$I = I \downarrow_t - I \uparrow_t \qquad (20.7)$$

The conductive and convective sensible heat transfer H is estimated by an equation following Lamb (1974):

$$H = \left[-\rho C_p k^2 z^2 \quad \cdot \quad \frac{\partial \theta}{\partial z} \frac{\partial v}{\partial z} SCF \right] \qquad (20.8)$$

where

$$SCF = \begin{cases} 1.175 \ (1 - 15 \, Ri)^{0.75} & Ri \leqslant 0 \\ (1 - 5 \, Ri)^2 & 0 < Ri \leqslant 0.2 \\ 0 & Ri > 0.2 \end{cases}$$

and ρ is the air density, C_p is the specific heat of dry air at constant pressure, k is von Karman's constant (0.40), z is the observation height, $\partial\theta/\partial z$ and $\partial v/\partial z$ are the partial derivatives of potential temperature and wind speed, respectively, with respect to height z, and Ri denotes the Richardson number. Potential temperature θ is defined by the relation

$$\theta = T_a (1000/p)^{0.286} \tag{20.9}$$

where T_a and p are the air temperature and pressure, respectively. The Richardson number is defined by

$$Ri = (g/\theta) \ (\partial\theta/\partial z) \ (\partial v/\partial z)^2 \tag{20.10}$$

where g is the acceleration due to gravity and θ is the average potential temperature between the surface and the height z.

Neglecting G as a relatively minor element during a storm period in Equation 20.2, the evaporation determination can be made from

$$E = \frac{(S_a - I) - H}{L} \tag{20.11}$$

where S_a is the solar energy input modified by the albedo, I is the net thermal infra-red input as given by Equation 20.7, H is the sensible heat transfer given by Equation 20.8 and L is the latent heat of evaporation.

Multi-layer soil-water model

With Equation 20.11 being a forcing function in the top computation cell, the Richard's equation can be extended to include soil depth-dependent changes in physical properties (Bruce & Whisler 1973):

$$C \ (\psi,z) \ \frac{\partial\psi}{\partial t} = \frac{\partial(K \ (\psi,z)\partial\psi/\partial z)}{\partial z} + \frac{K(\psi,z)}{\partial z} \tag{20.12}$$

where K is the hydraulic conductivity, ψ is the soil water potential, z is the depth, and C is the volumetric water capacity.

In this analysis hysteresis is ignored (see Hillel & Van Bavel 1976, Gillham et al. 1979). To retain parsimony it is appropriate to estimate the unsaturated hydraulic conductivity (K_u) by the use of one of the established 'matching' procedures. It has been shown by Jackson (1972) that the Millington–Quirk

Table 20.3 Moment measures for soil parameter sampling distributions used in simulation.

Parameter	Mean	Standard deviation	Source	Used in
water content at OCM tension ($cm^3\ cm^{-3}$, θ_s)	0.525	0.040	Nielsen *et al.* (1973)	Eqn 20.13
water content at tensions >0 cm ($cm^3\ cm^{-3}$, θ)	not applicable	0.025	Gummaa (1978)	Eqn 20.13
initial water content for profile ($cm^3\ cm^{-3}$, θ_i)	0.28–0.50	0.020	McKim *et al.* (1980)	Eqn 20.12 (solutions)
saturated hydraulic conductivity (log, m s^{-1},K_s)	−6.0	0.20	Nielsen *et al.* (1973)	Eqn 20.13
surface detention capacity (m, Sdc)	0.01	0.001		
clay percentage (%)	49.0	6.0	Nielsen *et al.* (1973)	Eqn 20.1

(M−Q) method (Millington & Quirk 1959) provides reliable estimates of K_u from K_s and the appropriate suction−moisture curve, where K_u is estimated as follows

$$K_{u(i)} = K_s(\theta_i/\theta_s)\,p\ \frac{\displaystyle\sum_{j=1}^{m}[(2j + 1 - 2i)\ \psi_j^{-2}]}{\displaystyle\sum_{j=1}^{m}[(2j - 1)\ \psi_j^{-2}]} \qquad (20.13)$$

where $K_{u(i)}$ corresponds to θ_i, θ_s is volumetric water content at saturation, p is taken to be unity and the summations are made over the j increments into which the suction−moisture curve is divided.

Inclusion of parameter variability

There now exists a substantial amount of evidence on the variation of soil-water properties both within soil map units (e.g. Topp *et al.* 1980, Bonell *et al.* 1981) and at a site (e.g. Sharma *et al.* 1980). However, whilst the basic $\bar{x}$ and σ moment measures for a range of soil-water parameters have been established in different studies, spatial autocorrelation dependencies ($\bar{\alpha}$) are only rarely recorded (e.g. Webster & Cuanalo 1975). Freeze (1980) observes that measurements of hydraulic conductivity have been shown to exhibit correlation over distances that range from a few metres to several kilometres. The variability included in the model reported here is limited to the sampling of parameter values from distributions of the form $N(\bar{x}, \hat{\sigma})$, owing to the sparsity of data existing to specify spatial correlation structures. It is important to recognise at least the potential significance of $\bar{\alpha}$, but similarly to report the need for much more extensive estimations if realistic values are to be included in stochastic models which incorporate spatial autocorrelation elements for all relevant soil-water properties.

Parameter variability was included in elements of Equations 20.1, 20.12 and 20.13, as shown in Table 20.3. Sensitivity analysis of model structures of the type exemplified by Equation 20.12 are an additional key element in the application potential of coupled evaporation−soil water−soil strength models of the form proposed here. All parameters given in Table 20.3 were determined from a $N(\bar{x}, \hat{\sigma})$ distribution in the simulations which included parameter variability.

Illustrations of model potential for predicting RCI

Sensitivity to parameter input
The model scheme outlined here, although at this stage of development only coupled to empirical moisture−RCI relations (Eqn 20.1), requires that a sensitivity analysis be undertaken before initial assessments can be made of its potential ability. Anderson (1983) has reported the undertaking of such sensitivity tests of the model. Table 20.4 summarises the results of a sensitivity analysis

Table 20.4 Sensitivity of variables in trafficability prediction model (after Anderson 1983).

Very sensitive	Moderate sensitivity	Insensitive
start moisture content[1]	cloud cover	wind speed
precipitation	latitude	relative humidity
soil type	air temperature	atmospheric pressure
time of precipitation[2]	ground temperature suction−moisture curve	soil detention capacity surface slope surface permeability

[1]Moderate sensitivity following initial drainage.
[2]Evaporation can be very significant when storm occurs several daylight hours after simulation start time.

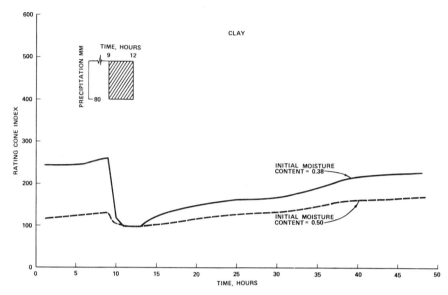

Figure 20.5　The impact of different initial volumetric water content on the RCI (after Anderson 1983).

with changes made to single parameters in each trial. Moisture content is a critical determinant, and Figure 20.5 illustrates the change in RCI effected by a change in prestorm moisture content of 0.12 cm^3 cm^{-3} in a clay soil (full data for the simulations given in Fig. 20.5 are presented by Anderson 1982, 1983). This finding reinforces the basis of the general thesis of this paper namely the need for incorporating a dynamic soil-water forecasting component in trafficability models. Of particular interest in Table 20.4 is the relative insensitivity of the suction—moisture curve. In the context of a typical soil catena, the implicit effect of changing permeability, suction—moisture curve and the remaining input variables shown in Table 20.2(b) under 'soil properties' can be illustrated. Figures 20.6a and b show a catena after Nye (1954). For the purposes of illustration here, profiles 2, 3 and 4 in Figure 20.6b were used to estimate trafficability performance during and after an 80 mm, 3 hour rainstorm. Initial volumetric water contents were set at 0.15, 0.20 and 0.35 for profiles 2, 3 and 4 respectively, in the uppermost 20 cm, with hydrostatic conditions assumed to be operative below 20 cm. Figure 20.6c illustrates the RCI values simulated using meteorological data otherwise similar to that used for the simulation shown in Figure 20.5. The post-storm recovery times to RCI values of 100 and 200 respectively show the complexity of response. Profile 4 (lower midslope) is the better performer in the former case, whilst profile 2 (upper slope) offers better trafficability if an RCI of 200 were a threshold criterion for a particular vehicle. The slope topography, depth to bedrock, and the form of the weathering front interrelate in a complex manner with the required model input parameters (Table 20.2). Generalisations are thus almost impossible to make, although the overall

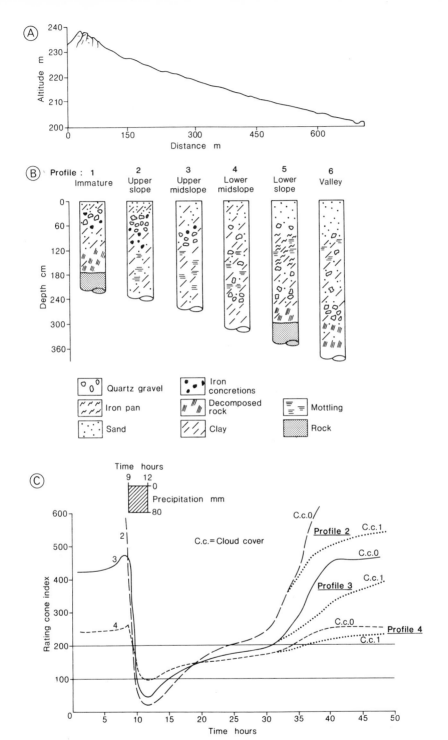

Figure 20.6 (A) Typical topographic section; (B) soil catena; and (C) associated RCI predictions for three soil profiles within the catena.

operation of the prediction scheme for RCI has been illustrated. A further element of RCI sensitivity is to be found in the determination of trafficability conditions along a specific route as different soil types are encountered.

Route selection

As depicted in Figure 20.1, models of overall systems are the required end product of trafficability forecasting. In the context of this model proposal, it becomes possible to make route selections on the basis of the RCI. Figure 20.7 illustrates just such an application for a straight line route of 30 km passing over three soil types. Rainfall of 80 mm between 0 and 3 hours was used in the simulation, and it is of interest to observe the changing relationship of soil type and RCI as drainage progresses. Initially, the clay soil provides the higher RCI, in the immediate post-storm period ($t = 3$ hours), but subsequently ($t > 6$ hours) sand and loam afford better trafficability conditions. Route selection can thus be simulated dynamically, with the scheme outlined above providing near-

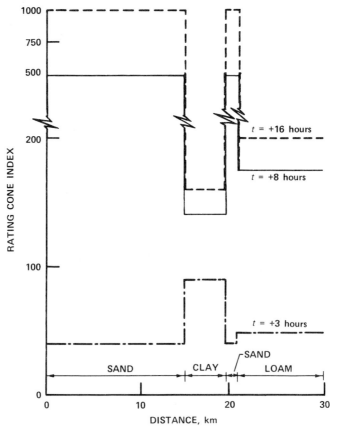

Figure 20.7   Changes in RCI with soil drainage on a 30 km route.

surface antecedent soil moisture values from which changing soil moisture conditions can be estimated (see Table 20.4).

Field variability effects

It has been stressed that model robustness requires the inclusion of field variability in line with the likely precision with which the required input variables (Table 20.2) can be estimated (Table 20.3). Figures 20.8 and 20.9 illustrate the simulated results of a 15 mm storm with soil hydrological properties stochastically determined within the $N(\bar{x}, \hat{o})$ distributions shown in Table 20.3. Figure 20.8 shows the bounds of the simulated RCI response for 25 runs. The bounds at the minimum RCI are 100 and 141, with the maximum and minimum differing by approximately 40 throughout the 14 hours shown. Of perhaps greater relevance is the fact that the standard deviation of the RCI simulations maintained a near-constant value of 12 in this example (the raw data for which are given in Anderson 1983). Undertaking this analysis for different initial soil relative saturation values facilitates the evaluation of probabilities of RCI less than 100 for each of these start conditions. Shown in Figure 20.9 are the results from two storms − 15 mm in 3 hours and 15 mm in 12 hours. The principal difference occurs for a range of higher relative saturation values, with similar probabilities occurring with initial relative saturations less than 0.9. The principal break point observed in this example is at a relative saturation of 0.78, probabilities dropping from 0.50 to 0.05 immediately either side of this value.

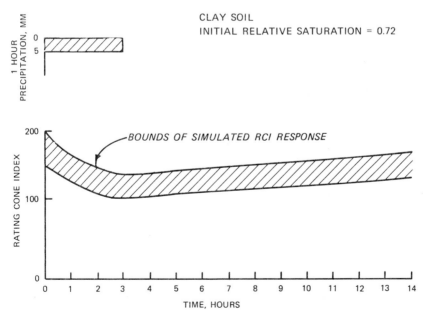

Figure 20.8 Bounds of RCI values generated under conditions of field variability, as given in Table 20.3 for a 15 mm storm.

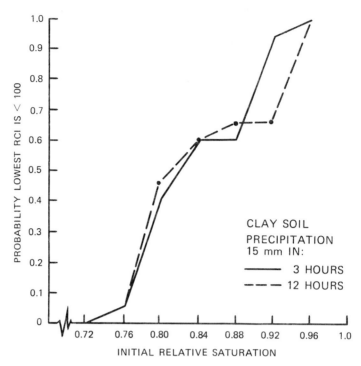

Figure 20.9   Impact of different antecedent soil moisture conditions and field variability (Table 20.3) upon RCI predictions.

There are too many variables involved in this procedure to illustrate all possible responses to differing input conditions, but with the variability given in Table 20.3 it is apparent that whilst initial moisture content is a key variable in prestorm conditions, cloud cover becomes a most significant variable in post-storm conditions in the context of RCI recovery. Inclusion of field variability as discussed does not mask this changing parameter dominance.

Discussion

The need for inclusion of soil moisture forecasting models in trafficability models has been stressed. Existing models used in this context are highly dependent upon a substantial empirical component to establish relatively coarse (often seasonal) moisture changes. The proposition here is that soil-water finite difference models eventually coupled to soil engineering property models should provide enhancement of trafficability forecasting. Whilst the overall robustness of the modelling strategy has been illustrated here (Tables 20.3 & 20.4 and Figs 20.8 & 20.9), of necessity this must be regarded as a first stage. The next requirement is to evaluate the performance of the scheme in the context of feasible data acquisition systems and methods. Given the tolerance of the model to

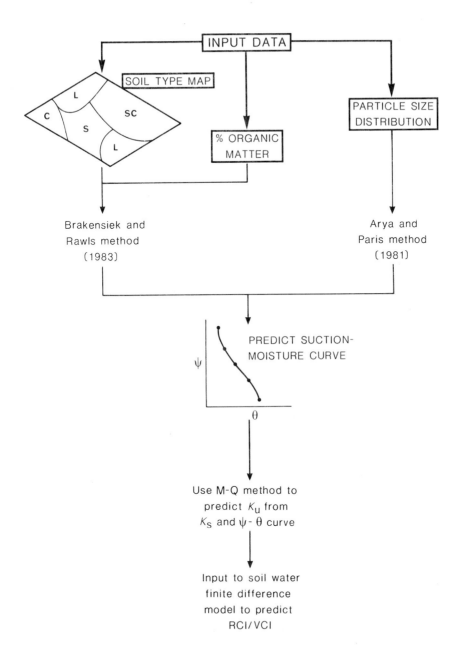

Figure 20.10   General scheme to enable the generation of suction–moisture curves from limited field data, and their incorporation into the model framework.

changes in the suction−moisture curve, it is possible to utilise estimation methods based upon either soil type maps or particle size distributions (if such information is available), to generate the suction−moisture curve (see Fig. 20.10). The requirement for initial moisture content, being of a higher resolution (Table 20.4), would necessitate remote sensing methods to achieve an adequate articulation of the model. With the establishment of a model structure of the type detailed above, but eventually coupled through the engineering properties of the soil (c', ϕ', etc.) by the model proposed by Rohani and Baladi (1981), it will be necessary to undertake a comprehensive survey of data acquisition methods. Such methods would not only need to supply the required input data (Table 20.2) within the field variability/model sensitivity limits which have been briefly illustrated here (Tables 20.3 & 20.4, Figs 20.8 & 20.9), but may also provide for real-time updating of model predictions given the necessary access to the field area of interest. Only when model coupling is complete and a deterministic overall structure is developed, including stochastic parameter variability, can an evaluation be made of the trade-offs between predictions in terms of spatial resolution, data acquisition systems, deterministic and stochastic model elements and field calibration needs. Despite a substantial research effort into the spatial variability of soil parameters (e.g. Topp *et al.* 1980), it is clear that for the purposes of trafficability forecasting this area is one of paramount significance to model selection, and to which much further work should now be directed in the context of available forecasting models.

Acknowledgements

Certain of the work reported here was undertaken by the author at the US Corps of Engineers Waterways Experiment Station, Vicksburg, Mississippi, USA, and the author is grateful to John Collins and Ed Link for their comments and advice. Dr C. Nuttall and Dr K. S. Richards made helpful comments on the initial draft of this chapter.

References

Anderson, M. G. 1982. *Assessment and implementation of drainage basin runoff models.* Final Report on Contract DAJA37−82−C−0092, US Army, 107.

Anderson, M. G. 1983. Forecasting off-road trafficability. *Appl. Geog.* **3**, 239−53.

Arya, L. M. and J. F. Paris 1981. A physico-empirical model to predict the soil moisture characteristic from particle-size distribution and bulk density. *Soil Sci. Soc. Am. J.* **45**, 1023−30.

Balick, M., L. E. Link, R. K. Scoggins and J. L. Solomon 1981. *Thermal modelling of terrain surface elements.* US Army Waterways Experiment Station, Technical Report EL−81−2.

Bonell, M., D. A. Gilmour and D. F. Sinclair 1981. Soil hydraulic properties and their effect on surface and subsurface water transfer in a tropical rainforest catchment. *Hydrol. Sci. Bull.* **26**, 1−18.

Brakensiek, D. L. and W. J. Rawls 1983. *Use of infiltration procedures for estimating runoff.* Paper presented to Soil Conservation Service, National Engineering Workshop, Tempe, Arizona, USA.

Bruce, R. R. and F. D. Whisler 1973. Infiltration of water into layered field soils. *Ecol. Stud.* **4**, 77–8.

Carlson, C. A., W. P. Bohnert and M. P. Meyer, 1970. *Trafficability predictions in tropical soils.* US Army Waterways Experiment Station Miscellaneous paper 4–355.

Collins, J. G. 1967. Influences of water tables on soil moisture and soil strength in *Report of Conference on Soil Trafficability Prediction* Appendix D. US Army Waterways Experiment Station.

Collins, J. G. 1971. Forecasting trafficability of soils. In *Technical Memorandum 3–331*, Report No. 10. US Army Waterways Experiment Station.

Freeze, R. A. 1980. A stochastic conceptual analysis of rainfall-runoff processes on a hillslope. *Water Res. Res.* **16**, 391–408.

Gillham, R. W., A. Klute and D. F. Heerman 1979. Measurement and numerical simulation of hysteretic flow in heterogeneous porous media. *Proc. Soil Soc. Am.* **43**, 1061–7.

Gummaa, G. S. 1978. *Spatial variability of* in situ *available water.* Unpublished Ph. D. thesis, University of Arizona.

Haurwitz, B. 1948. Insolation in relation to cloud types. *J. Met.* **5**, 1673–80.

Hillel, D. 1980. *Fundamentals of soil physics.* New York: Academic Press.

Hillel, D. and C. H. M. Van Bavel 1976. Simulation of profile water storage as related to soil hydraulic properties. *Proc. Soil. Sci. Soc. Am.* **40**, 807–15.

Huck, M. G., V. D. Browning and R. E. Young 1975. *Leaf water potential and moisture balance – field data.* Am. Soc. Agric. Eng., Paper No. 75–2582.

Jackson, R. D. 1972. On the calculation of hydraulic conductivity. *Soil Sci. Soc. Am. J.* **36**, 380–2.

Khale, A. B. 1977. A simple model of the earth's surface for geologic mapping by remote sensing. *J. Geophys. Res.* **82**, 1673–80.

Lamb, R. C. 1974. *The radiation and energy balance on a burned vs. an unburned natural surface.* Paper presented to the Conference on Fire and Forest Meteorology, Am. Met. Soc., & Soc. Am. Foresters, April 2–4, 1974.

McKim, H. L., J. E. Walsh and T. Pangburn 1980. Comparison of radio frequency, tensiometer and gravimetric soil moisture techniques. In *Proceedings of the 3rd Colloquium on Planetary Water*, New York, 129–35.

Melzer, K. J. 1982. Analytical methods of modelling: state-of-the-art report. *J. Terramechanics* **19**, 31–53.

Millington, R. J. and J. P. Quirk 1959. Permeability of porous media. *Nature* **183**, 387–8.

Moltham, H. D. 1967. Influence of soil variability on soil moisture and soil strength predictions. In *Report of Conference on Soil Trafficability Prediction,* Appendix E. US Army Waterways Experiment Station.

Nielsen, D. R., J. W. Biggar and K. T. Erh 1973. Spatial variability of field-measured soil-water properties. *Hilgardia* **42**, 215–59.

Nuttall, C. J. 1976. The user and terrain–machine system. *J. Terramechanics* **13**, 45–55.

Nye, P. H. 1954. Some soil forming processes in the humid tropics. *J. Soil Sci.* **5**, 7–21, 51–83.

Oke, T. R. 1978. *Boundary layer climates.* New York: Wiley.

Pochop, L. O., M. D. Shanklin and D. A. Horner 1968. Sky cover influence on total hemispheric radiation during daylight hours. *J. Appl. Met.* **7**, 484–9.

Raghavan, G. S. V. and E. McKyes 1978. Statistical models for predicting compaction generated by off-road vehicular traffic in different soils *J. Terramechanics* **15**, 1–14.

Raghavan, G. S. V., E. McKyes, R. Baxter and G. Gendran 1979. Traffic–soil–plant (maize) relations. *J. Terramechanics* **16**, 181–9.

Rohani, B. and G. Y. Baladi 1981. *Correlation of mobility cone index with fundamental engineering properties of soil.* US Army Waterways Experiment Station, Miscellaneous Paper SL–81–4.

Sellers, W. D. 1965. *Physical climatology.* Chicago: University of Chicago Press.

Shamburger, J. G. 1967. A quantitative method for describing terrain and ground mobility. *J. Terramechanics* **4**, 39–45.

Sharma, M. L., G. A. Gander and C. G. Hunt 1980. Spatial variability of infiltration in a watershed. *J. Hydrol.* **45**, 101−22.

Smith, M. H. and M. P. Meyer 1973. *Automation of a model for predicting soil moisture and soil strength (SMSP model).* US Army Waterways Experiment Station, Miscellaneous Paper SL−81−4.

Taylor, H. M. 1971. Effect of soil strength on seedling emergence, root growth and crop yield. In *Compaction of agricultural soils*, K. K. Barnes (ed.), 292−305. Am. Soc. Agric. Eng. Monogr.

Topp, G. C., W. D. Zebehuk and J. Dumanski 1980. The variation of *in situ* measured soil water properties within soil map units. *Canadian J. Soil Sci.* **60**, 497−509.

Webster, R. and H. E. Cuanalo, 1975. Soil transect correlograms of north Oxfordshire and their interpretation. *J. Soil Sci.* **26**, 176−94.

Weisner, C. J. 1970. *Hydrometeorology.* London: Chapman & Hall.

21

Geotechnical characteristics of weathering profiles in British overconsolidated clays (Carboniferous to Pleistocene)

D. J. Russell and N. Eyles

Introduction

The geological evolution of Great Britain has contrived to leave a considerable surface area underlain by overconsolidated clay soils. These soils (used in the engineering sense to refer to sediment, or weathered or unlithified bedrock) are of many ages and environments of deposition. However, common points are (a) their plasticity, caused by a significant clay mineral content, and (b) their overconsolidated nature which reflects their past loading by stresses far greater than those they experience at present. Because of their widespread occurrence, it is important to characterise the geotechnical response of overconsolidated clays to weathering, in particular, changes in properties such as strength. This is of fundamental importance given the influence of strength changes on the behaviour of foundations, natural and man-made slopes, and fill composed of these materials.

The speed and manner in which weathering occurs is controlled in part by mineralogy and structure. Although grouped together by virtue of the attributes stated above, there exists appreciable variation in mineralogical and textural characteristics between various types of overconsolidated clay soil. This is due to the varying modes and environments of deposition controlling composition, and varying stress histories controlling fabric and degree of overconsolidation. The most marked difference in origin within the overconsolidated clay group is perhaps that between the marine, dark-coloured mudrocks of Mesozoic and Cenozoic age, and the glacially-overconsolidated pebbly clays (lodgement tills) deposited by Quaternary glaciers. The contrasts in mineralogy and depositional and post-depositional history of these two argillaceous materials cause differences in the weathering they undergo, and are reflected in different com-

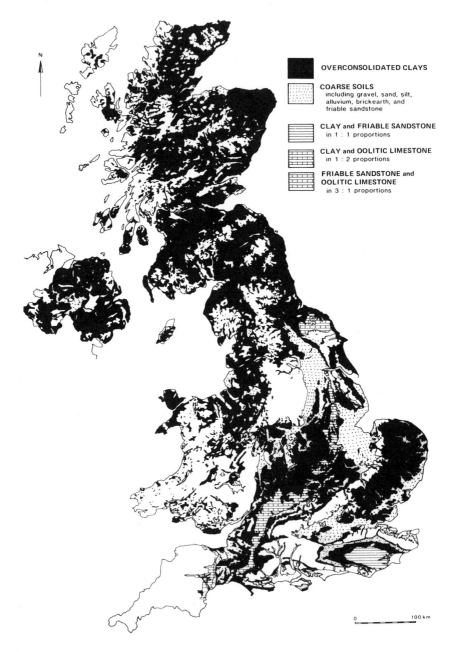

Figure 21.1 Distribution of overconsolidated clays (tills and argillaceous bedrock) in Great Britain. Modified from Dearman and Eyles (1982) and based on Bickmore and Shaw (1963).

positional and geotechnical characteristics of the weathered clay. Despite these differences, however, similar weathering processes can be identified for each lithology.

The key ingredients of clay soils are clay minerals. To provide the requisite clay mineral input to a depositional area, the source region either has to be subjected to appreciable chemical weathering, or be composed itself of argillaceous materials. From the Carboniferous period (about 350 ma) to at least Eocene times (about 40 ma), the former condition probably held for the areas close to Britain. Thus, many mudrocks of this age range were deposited in environments ranging from terrestrial (e.g. the Keuper Marl; Tucker 1978) to shallow shelf seas (e.g. Kimmeridge Clay; Morris 1979). The British land mass was better defined in Oligocene and later times (after 40 ma) and became a source of muddy detritus for surrounding areas (e.g. the North Sea). Still later, during multiple Quaternary glaciations, weathered bedrock, including mudrocks and their weathered products, became the source for extensive till sheets deposited by ice lobes. Thus, overconsolidated clay soils occur as surface material over a large part of the British land mass (Fig. 21.1).

Objectives

In the first two parts of this chapter, we present typical mineralogical and geotechnical profiles for both weathered Mesozoic mudrocks and tills of last glaciation (Devensian) age, and discuss the variation in properties with respect to the interaction of original lithology and subsequent processes. The third part reviews contrasts and similarities between the two groups. Environmental conditions (climate, hydrology) are as significant as original composition in their effect on weathering products (Jenny 1950). Thus, in the final section, the significance of the study of the evolution of weathering profiles in the widespread clay deposits to the reconstruction of Quaternary history in Britain is discussed.

Weathering of pre-Quaternary mudrocks

Initial clay properties

The major overconsolidated clays of post-Devonian age are listed in Table 21.1, with pertinent compositional and geotechnical data and references to studies of weathering of these units. A clear correlation can be seen in these data for pre-Pleistocene units between original burial depth and unweathered moisture content. The behaviour of clay as it is buried and exhumed is well known (cf. Cripps & Taylor 1981 and references therein). The physical response to uplift and unloading of a clay depends largely on the extent of diagenetic bonding undergone before unloading (Bjerrum 1967), and this is closely related to depth and duration of burial. For example, the matrix of a strongly bonded Carboniferous mudrock will tend to relax, and therefore increase in moisture content,

Table 21.1 Summary of depositional origin and geotechnical weathering characteristics of overconsolidated clays in Britain. Cu is undrained strength, and clay and moisture contents are percentages by weight.

Age	Name	Environment	Clay mineralogy	Clay content	Max. previous load (m of sed)	Cu, un'wd (kPa)	w'd (kPa)	M.C. (%) unw'd	w'd	P.I.(%)	Publication
Late Pleistocene about 20 000 gBP	lodgement till	sub glacial grounded ice	C.I.	23–50	ice sheet (17–20 000 kN m^{-2}) max.	150–300	75–250	10–17	12–30	14–40	Gilroy (1980); Eyles and Sladen (1981); Sladen and Wrigley (1983)
Eocene	London Clay	marine shelf	I, ML, K	40–72	152–396	100–400	100–175	19–28	23–49	40–65	
Cretaceous	Gault Clay	basinal marine	I, K, ML	38–62	425–520	300–550	60	18–30	32–42 3	27–80	
Jurassic	Weald	mudswamp	I, K, ML (varied)	17–71	1220–1370			25–34		28–32	
	Kimmeridge	anaerobic shallow marine	I, K, ML	57	1070–1220	130–470		18–22		24–59	
	Oxford	epeiric sea	I, ML, K	30–70	330–1560	110–1100	30–70	15–28	20–33	28–50	Russell and Parker (1979)
	Lias	epeiric sea	K, I, ML	50–65	855–975	110–240	30–150	11–23	20–38	20–37	Chandler (1972); Coulthard (1975)
Triassic	Keuper Marl	playa, desert plain	I, C, ML	10–50		300–1200	100–150	5–15	12–40	10–35	Chandler (1969)
Carboniferous	Coal Measure Shales	mainly fluvio-deltaic	I, K, C	24–87		38–50 mPa	100–180	8–9	6–14	9–19	Taylor and Spears (1970, 1972)

at a slower rate than the more weakly bonded Eocene London Clay. Since the primary agent of chemical weathering is oxygenated water entering the clay mass, the onset of this process is closely related to rebound rate and thus to the strength of original bonding. Below the weathering zone, the combined effect of rebound and moisture content increase is loss of strength (e.g. Burland *et al.* 1977, Figure 4). It is reasonable to suggest that the rate of strength loss will decrease towards the surface as the effect of frictional strength (enhanced by low moisture contents) is reduced relative to purely cohesive strength (i.e. interparticle bonds).

Although some variation in clay mineralogy between pre-Pleistocene mud-rocks is evident (Sellwood & Sladen 1981), the predominant clay mineral is almost always illite. In addition, kaolinite and mixed-layer clays are significant accessory clay minerals in post-Tertiary units with chlorite replacing kaolinite in this role in Triassic and Carboniferous mudrocks. Of these clay minerals, chlorite is the most sensitive to weathering (Droste 1956). However, since these clay minerals, in addition to the detrital quartz silt ubiquitously associated with them, have normally been through at least one cycle of weathering and erosion, they are extremely stable chemically. The same cannot be said of the associated non-detrital minerals such as pyrite which are present in all units except the Keuper Marl. Calcite is common as fossil remains and as a cement in all but the Carboniferous mudrocks where siderite is the carbonate mineral present.

Weathering processes

Significant changes in the physical properties of marine overconsolidated clays are caused by chemical weathering. The oxidation of sulphide minerals, acidification of groundwater, dissolution of calcite and other carbonate minerals and recombination of elements to form gypsum is well documented (Morgenstern 1970, Russell & Parker 1979). Removal of potassium from the edges of illite crystallites, thus weakening interparticle bonds between clay particles, is indicated by distribution of jarosite and illite crystallinity in Oxford Clay weathering profiles (Russell & Parker 1979).

The character of weathering profiles in any material is controlled by fractures, along which oxygenated water gains entry into the rock mass. Sequential phases of weathering intensity can therefore be described and classified by reference to the degree of alteration away from discontinuity surfaces (see Table 21.2, and Chandler 1969, 1972). Only minor modifications to this weathering classification are necessary to account for slightly differing original lithologies (see below).

In the British context, it is commonly assumed that former weathering mantles of pre-Pleistocene age have either been removed or obscured by glacial and periglacial processes. For example, calcite-rich mantles, interpreted to have been of solifluction origin, were found overlying *in situ* weathered clay at four widely separated sites underlain by Oxford Clay (Russell 1977). The process of calcite redeposition at $1-2$ m depth, demonstrated by Catt (1980) for weathered tills, could also occur in weathered Oxford Clay overlain by calcareous materials. Periglacial processes, such as ice lensing which causes brecciation of mudrock

Table 21.2   Classification of weathering zones in Oxford Clay (Russell & Parker 1979). Applicable to marine overconsolidated clays of Jurassic to Tertiary age.

Weathering group	Description
IV	Superficial material of various origins, often flinty or chalky fragments in clay matrix. Usually desiccated relative to underlying clay.
III	Totally oxidised clay, often with gleyed fissures. Frequent, often sugary, gypsum crystals.
IIa	Clay is oxidised beyond immediate area of fissures, making up 30% to 60% of the clay. Large gypsum crystals.
IIb	Clay is oxidised along surfaces of fissures only. Blue-grey colour and laminar fabric visible away from fissures. Some gypsum.
I	Unweathered, dark blue-grey silty firm clay, some fissures horizontally laminar fabric, no gypsum, no oxidation.

fabric, are also significant (Chandler 1973, Jones & Derbyshire 1983). Boulton and Dent (1974) have also documented the extreme rapidity, at modern ice margins, of periglacial subsurface processes.

Geotechnical weathering profiles in Oxford Clay

Index properties.   No consistent variation in Atterberg limits is observable in weathering profiles developed on Oxford Clay. Figure 21.2 shows depth plots for plasticity index from three sites (Minety, near Swindon; Sandy, near Bedford; Fleet, near Weymouth). Changes in plasticity index in the weathered zones at Minety and Sandy are less than the variations apparent in Zones I and II. Zone IV samples (i.e. the surficial material) show wide variation in plasticity index. At Fleet, the enhanced plasticity index in the lowermost weathered material may be due to illuviation of smectitic clay from the overlying Zone IV material. Values of activity (plasticity index divided by clay fraction) shown in Figure 21.2 again show no rational variation ascribable to weathering processes.

Liquidity index (the difference between the field moisture content and the plastic limit, divided by the plasticity index) is a measure of the consistency of a clay soil, since it relates the *in situ* moisture content to the plasticity parameters. For weathering profiles which do not suffer significant changes in plasticity, variation in liquidity index will follow that of moisture content. Figure 21.2 shows depth plots of liquidity index for the three sites discussed above. The profiles sampled were all affected by surface desiccation. However, below this

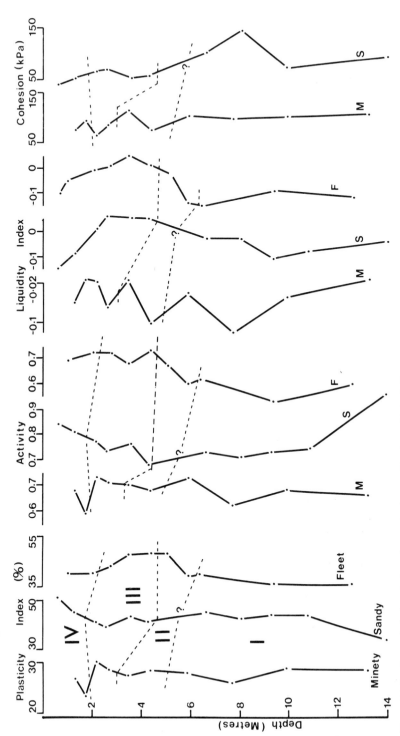

Figure 21.2 Geotechnical variations in weathering profiles from Oxford Clay.

effect the pattern is one of significantly higher liquidity index in Zone IIb than in the underlying clay. This pattern persists even at Fleet, where changes in plasticity should cause a depressed liquidity index. Clearly, the processes causing moisture content increase are dominant in controlling consistency of the clay.

Strength properties. The most important feature of the variation in undrained strength, or cohesion (Fig. 21.2), in Oxford Clay weathering profiles is that greatest strength loss occurs just above the weathering front, and that no statistically significant further loss of strength occurs above this zone (Russell & Parker 1979). Above Zone II, virtually all pyrite has been oxidised, leaving no source for generation of acidic groundwater. Since groundwater movement is usually downward, the acidified solutions will be concentrated on Zone II clays as will the bond-weakening processes (dissolution of calcite cement and clay mineral edges).

The modification of the texture of the Oxford Clay implied by the changes in strength are also reflected in reduction in the anisotropy of various physical properties, such as strength, particle orientation and ultrasonic velocity (Russell 1977). These changes are interpreted as reflecting the disturbance caused by increase in moisture content, allowed by bond weakening and breakdown.

Weathering of Quaternary (glacial) overconsolidated clays

Initial clay properties

Of considerable geological and engineering significance are the overconsolidated pebbly and sandy clays (tills) of relatively recent glacial origin that cover 30% of the total land area of mainland Britain (Fig. 21.1). This cover is not stratigraphically or genetically homogeneous but can be divided for geotechnical purposes into *clast-dominant* and *matrix-dominant* till types (McGown & Derbyshire 1977). Matrix-dominant types are defined as having more than 45% of dry weight smaller than sand-sized; the engineer is primarily concerned with the behaviour and performance of this matrix rather than the individual coarser particles. These tills, equivalent to the 'boulder clays' of much of the older British literature, occupy an eastern and southern outcrop reflecting the glacial erosion of underlying clay and shale bedrock by extensive ice sheet lobes, and subsequent subglacial deposition as drumlinised till plains. In contrast, clast-dominant tills behave essentially as granular aggregates. These are most common in Scotland and areas of high bedrock relief occupied by former valley glaciers, but may also occur locally within the area of matrix-dominant till cover in association with belts of outwash sands and gravels and strongly morainic morphology (e.g. Paul 1983). The effect of weathering on clast-dominant tills is not significant, and the following discussion of variation in geotechnical properties resulting from Postglacial weathering is restricted to matrix-dominant tills of last glaciation (Devensian) age. Comprehensive reviews of the genesis, sedimentology and geotechnical properties of these tills can be found in Eyles

Table 21.3 Classification of weathering zones in till (after Eyles and Sladen 1981).

Weathering state	Zone		Maximum depth (m)
highly weathered	IV	Oxidized till and surficial material. Strong oxidation colours. High rotten boulder content. Leached of most primary carbonate. Prismatic gleyed joint. Pedological profile usually leached brown earth.	3
moderately weathered	III	Oxidized till. Increased clay content. Low rotten boulder content. Little leaching of primary carbonate. Usually dark brown or dark red brown. Base commonly defined by fluvioglacial sediments.	8
slightly weathered	II	Selective oxidation along fissure surfaces where present, otherwise as Zone 1.	10
unweathered	I	Unweathered till. No post-depositionally rotted boulders. No oxidation. No leaching of primary carbonate. Usually dark grey.	

and Menzies (1983) and Sladen and Wrigley (1983). Representative values of compositional and geotechnical data for matrix-dominant tills are shown in Table 21.4.

Weathering processes

The significance and characterisation of deeply weathered till surfaces have long been recognised in North America (Leighton & MacClintock 1962) but, in contrast, reddened Postglacial weathering profiles in British tills have been mapped and interpreted as separate stratigraphic units (e.g. 'upper' tills, 'upper' prismatic clay or boulder clay) until very recently. The foremost example of a soil profile developed in till and formerly considered to be a discrete depositional unit is the Hessle 'Till' mapped over nearly 3000 km² of eastern England and north of the limit of Devensian ice to the Scottish border (Bisat 1940, Carruthers 1947). A wide variety of glacial depositional mechanisms has been invoked for this unit (Carruthers 1953, Lunn 1980). Subsequently it has been shown by mineralogical analyses (Madgett 1975, Madgett & Catt 1978) that this and other 'upper' tills (e.g. Marsh Till, Hunstanton Till) originate through Postglacial (Holocene) weathering (see also Eyles & Sladen 1981).

There are marked stratigraphic controls on the depth and definition of Post-

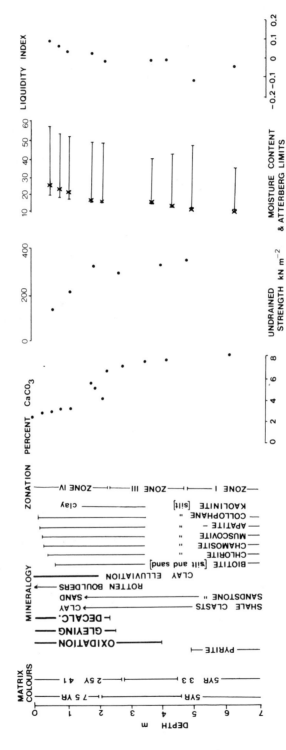

Figure 21.3 Typical geotechnical variation in weathering profiles from matrix-rich tills in eastern England. Mineralogical data after Catt (1980).

glacial weathering of tills. The importance of permeable horizons within lodgement tills (resulting from episodic drainage of subglacial melt waters during till deposition), in acting to 'underdrain' and accelerate drainage of overlying material by virtue of enhancing bulk permeability, is stressed by Madgett and Catt (1978) and Eyles and Sladen (1981). Such horizons may also impart a sharp base to the reddened weathering profile contributing to interpretations of the uppermost material as a separate laterally-correlatable depositional unit (i.e. an 'upper till') particularly in areas of poor or intermittent outcrop (Eyles *et al.* 1982).

A distinct sequence of changes which occurs during deep Postglacial weathering of tills has been defined by Madgett and Catt (1978), who identify the importance of decalcification and the extensive alteration that occurs afterwards (Fig. 21.3). Changes before decalcification include (a) oxidation of pyrite and siderite in coarser fractions, giving a reddish-brown colouration, (b) restricted loss of micaceous minerals from coarse silt and fine sand, (c) breakdown of silt-size kaolinite to clay-size, (d) alteration of mica in the clay fraction, and (e) softening and disaggregation of mudrock clasts. Weathering changes after decalcification, mainly affecting the very topmost horizons (i.e. above the zone of relevance for engineers), are (a) further loss of micaceous minerals, (b) formation of secondary clay chlorite, (c) illuviation of clay from the topsoil, and (d) development of ferruginous concretions. Gilroy (1980), in a study of the properties of different weathered tills in eastern England with varying unweathered lithologies, demonstrated that similar geotechnical weathering profiles develop regardless of initial petrographic differences.

In the United States, Leighton and MacClintock (1962) showed that a mature weathering profile in calcareous till invariably comprises five definable horizons from 5 (unweathered) through 4 (oxidation along joints but initial carbonate content is maintained) to 3 (decreasing carbonate contents), 2 (oxidised, leached and structurally altered) and 1 (surficial soil). The overall depth of the profile is typically $2 - 10$ m. Where only a thin till cover is present the entire column may be oxidised throughout, clay-rich, and difficult to identify as a former till. In mature profiles pedological A and B horizons comprise horizon 1 but in less mature profiles horizons A and B may be equivalent to 1 and 2 respectively.

This weathering scheme is very similar to that developed by geotechnical engineers working in weathered overconsolidated clays (see above), and a four-part scheme for tills is shown in Table 21.3 This scheme has been used in several studies in eastern England (Sladen 1979, Gilroy 1980, Eyles & Sladen 1981). Zone I comprises unweathered till and Zone II is identified by limited oxidation along joint surfaces where the latter are present. A change of matrix colour and increasing clay content owing to illuviation marks Zone III. Much primary carbonate may still be present but rotten disintegrating boulders are evident. These increase in frequency upwards to Zone IV, which is characterised by a prismatic structure, high clay content, development of cutans along joints, and very low carbonate content.

Typical geotechnical properties and their likely variability for the different weathering zones of overconsolidated lodgement tills are shown in Figure 21.3 and Table 21.4. The engineering properties of Zones I and II can be considered the same, and have accordingly been grouped together. These relevant geotechnical properties are reviewed below.

Geotechnical weathering profiles in Devensian lodgement tills

Index properties.   Most chemical and structural weathering changes occur in Zone IV and the increased clay content typical of this zone (Table 21.4) suggests disintegration of larger particles accompanied by alteration of feldspars to clay minerals. In general there is a parallel increase of plasticity index with weathering zonation, although this is also dependent upon initial textural variability prior to weathering, and local relief and drainage conditions. Under conditions of unimpeded drainage, fines produced by weathering are eluviated through the soil column. If this is prohibited by poor drainage, clay particles accumulate in Zone IV, resulting in a clayey plastic horizon referred to in North America as 'gumbotil'. In general, moisture contents increase more quickly than plasticity, resulting in higher liquidity indices. Typical liquidity indices for unweathered till fall within the range −0.1 to −0.35, while those for weathered tills may reach 0.1 (Fig. 21.3). The range of values of activity index reported from tills is considerable, a reflection of very variable clay mineral content and formation. Increases of activity index from 0.64 to 0.68 are reported by Eyles and Sladen (1981) contrasting with the increase from 0.4 to 1.0 documented by Quigley and Ogunbadejo (1976) for weathered clay-rich tills in Ontario, Canada. Sladen and Wrigley (1983) suggest that the degree of alteration in index properties is probably a reflection of initial carbonate content: the higher the latter, the greater the effect of weathering and of increasing clay contents on index properties.

Strength properties.   A systematic relationship is commonly observed between increased plasticity index in more highly weathered till zones and the development of drained brittleness (Vaughan & Walbancke 1978, Vaughan *et al.* 1979). The latter is defined as the drop in shear strength between the peak and residual values divided by the peak strength. Eyles and Sladen (1981) showed that drained brittleness increases sharply at a plasticity index of about 20%. While variations in plasticity index (and therefore drained brittleness) can occur as a result of incorporation of mudrocks, there is a clear relationship between increasing brittleness and increasing plasticity owing to weathering.

Values of undrained shear strength show an inverse correlation with moisture content (Fig. 21.3). Undrained strength is very sensitive to changes in moisture content but it can be seen that undrained shear strength is generally higher for more weathered tills compared with less weathered units at the same moisture contents. A similar phenomenon has been noted by Chandler (1972) with regard to the Jurassic Lias Clays and can be attributed to increasing plasticity. However, this must be qualified by the observation that increases in moisture content resulting from weathering may override this effect, leading to a reduction in un-

Table 21.4 Summary of geotechnical characteristics of weathered tills in eastern England. Modified from Gilroy (1980). Cu is undrained strength, c' is effective cohesion, ϕ' is angle of shearing resistance, and ϕ_r' is residual angle of shearing resistance.

Property	Yorkshire						N. E. England	
	Skipsea Till			Withernsea Till			Northumberland Till	
	I/II	III	IV	I/II	III	IV	I & II	III & IV
natural moisture content (%)	15–17	16–22	25–28	15–18	17–22	25–30	10–15	15–25
bulk density (kg m⁻³)	2150–2300	1900–2200	1900–2000	2150–2300	1900–2200	1900–2000	2150–2300	1900–2200
liquid limit (%)	25–32	38–45	45–50	30–35	35–40	45–50	25–40	35–60
plastic limit (%)	14–15	19–21	20–24	16–18	19–22	24–27	12–20	15–25
plasticity index	14–16	19–23	22–27	15–17	18–21	21–25	0–20	15–40
particle size distribution								
% clay	23	30	32	28	38	40	20–35	30–50
% silt	43	49	62	48	50	50	30–40	30–50
% sand	34	21	6	24	12	10	30–50	10–25
Cu (kNm⁻²)	150–220	115–150	85–110	180–210	120–150	75–110	250–300	100–250
c' (kNm⁻²)	0–20	0–25	0–25	0–20	0–20	0–25	0–15	0–25
ϕ' (degrees)	30–32	22–27	20–22	30–32	26–32	24–26	32–37	27–35
ϕ_r' (degrees)	28–30	18–22	14–18	25–29	18–25	14–20	30–32	15–32

drained shear strength with increased degree of weathering (see discussion below). In addition it is worth emphasising that bulk undrained strengths in the field are influenced by fissures and joint systems that are characteristic of weathered till zones (Vaughan & Walbancke 1978, McGown *et al.* 1978).

Table 21.4 summarises recent work on geotechnical weathering characteristics for lodgement tills in eastern England over a broad zone from the Scottish border to Lincolnshire. Despite petrographic differences in the unweathered tills over this broad area, distinctive changes in strength and index properties with weathering zonation are common. This has considerable significance for geotechnical and engineering geology practice, for once the weathering zone is identified on site by reference to criteria identified in Table 21.3, then geotechnical properties can be estimated (Table 21.4). Inevitably, initial variations in texture and site stratigraphy are also important, but changes in compositional and geotechnical characteristics between weathered zones can be reliably interpreted.

Discussion

The chemical stability of the bulk of both tills and mudrocks is a significant characteristic shared by these two materials. The mineralogies of fresh and weathered clays are remarkably similar to those of, for example, unweathered and weathered mafic crystalline rocks. Those changes that do occur are extremely important, however, in their effect on physical properties, although in contrasting ways for the two types of overconsolidated clay.

One notable contrast is the increasing plasticity of weathered tills and subsequent increase in undrained brittleness − a feature not evident in the case of mudrocks. This is because unweathered till contains a considerable amount of relatively unstable minerals (felspars, mafic minerals) in the form of clasts. Some of these clasts are mudrocks. The initial stable weathering products of these minerals and clasts are clay minerals, with similar properties to those already constituting the matrix of the till. Thus little change is noted in the activity of the till. However, as the relative proportion of the matrix has thereby increased, the consequent increase in plasticity causes important changes in geotechnical characteristics, such as brittleness. The increase in drained brittleness can be ascribed to a greater propensity for alignment of grains along a shear surface in the soils enriched by clay minerals. In contrast, the volumetrically minor mineralogical changes in weathered Oxford Clay are best interpreted in terms of interparticle bond weakening and do not increase plasticity. An increase in brittleness would therefore not be expected for Oxford Clay and similar units. Indeed, Burland *et al.* (1977) demonstrate that the extremely high drained strength brittleness of fresh Oxford Clay accords well with the already high clay content of the unweathered soil.

Contrasting behaviour of the plasticity index between weathered mudrocks and tills is also of importance with regard to the relationship between moisture content and strength. For example, changes in the plasticity of tills such as those caused by weathering have been shown by Eyles and Sladen (1981) to cause enhanced strength at a given moisture content. Sladen and Wrigley (1983) give the theoretical explanation for this phenomenon and have constructed a figure showing predicted moisture content−strength curves for clays with certain plasticity indices (Fig. 21.4). These curves depend on the relationship between the plasticity parameters described by the 'T'−line (Boulton & Paul 1976) and use the lower bound relationship between strength and liquidity index given by Skempton and Northey (1952) assuming the clays are insensitive. These are reasonable assumptions for both clay tills and mudrocks. Generalised points representing moisture content−strength data for the tills described above and the Oxford and Lias Clays (the latter from Chandler 1972) are plotted on Figure 21.4.

It is clear from this diagram that, at low moisture contents (12−18%), the effect of plasticity on strength is highly significant at constant moisture content. Above about 18%, however, the sensitivity of strength to plasticity becomes markedly less. Since the moisture content of the shallowest unweathered Oxford Clay is usually at least 18%, and in any case, plasticity does not systematically increase with weathering, the effect noted by Eyles and Sladen (1979) for tills would not be expected for Oxford Clay. As shown by the location of the points representing Oxford Clay in Figure 21.4, their 'weathering path' is along one of constant plasticity; it is therefore not surprising that Russell and Parker (1979) did not observe systematically enhanced strength in weathered clay at given

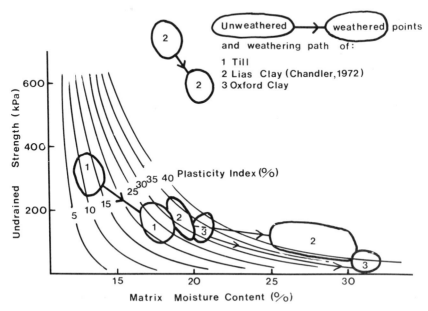

Figure 21.4 Moisture content−strength relationships for clays discussed in text. In part after Sladen and Wrigley (1983).

moisture contents. However, many of the weathered Lias samples tested by Chandler (1972) require an appreciable increase in plasticity to account for their strength variation; this is not seen in index test results. As implied by Chandler (1972), this 'apparent plasticity increase' is probably due to changes in the fabric induced by weathering. However, the exact mechanism by which this process occurs, and why it is apparently not significant for Oxford Clay weathering profiles (despite demonstrated changes in fabric), await investigation.

Similarities in geotechnical weathering characteristics are also evident for mudrocks and tills. An inverse correlation between moisture content and undrained strength (or cohesion) is observed in weathering profiles for all over-consolidated clays. The cause of moisture content increase below the weathered zone is a decrease in vertical load (i.e. overburden thickness). As load is released, stored-in strain energy is released, usually in a visco-elastic fashion (perhaps best envisaged as a mica flake slowly unflexing), and void ratio increases. This causes a drop in strength because of the concomitant reduction in the particle contacts that produce frictional resistance to shear, and in the magnitude of attractive electrostatic forces that are dependent on very small particle separations. The range of moisture content increase depends on the extent and strength of interparticle bonding. Since the opportunity for bonding in tills is relatively restricted, the upper limit in moisture content will be higher for these clays, given similar plasticities. However, since the rebound of the clay fabric is time-dependent, it is unlikely that surface-exposed till has yet reached this maximum moisture content. The upper limit for unweathered Oxford Clay appears to be about 20% moisture content, equating to a cohesion of about 125 kPa. After reaching this point, further strength loss can only be caused by bond dissolution, which apparently occurs suddenly, in Zone II of the weathering profile. In contrast, the rate of strength loss in the till weathering profiles appears to increase towards the surface, suggesting that release of vertical stress is the prime agent of weakening (e.g. Figs 8.12 & 8.19 in Sladen & Wrigley 1983).

This is not the place for a review of the applied engineering implications of geotechnical changes resulting from weathering. We simply emphasise, with regard to mudrocks, the development of pseudo-overconsolidation in the near-surface zones and the consequent risk of differential settlement of foundations where different weathering zones are laterally juxtaposed. Highly weathered tills may be unusable as fill because of their higher moisture content. A ratio of moisture content to plastic limit of 1.2 is comonly used as an upper limit for the use of till as fill (Arrowsmith 1978) and the more weathered till zones commonly exceed this specification (e.g. Cocksedge 1983).

The evolution and form of a weathering profile is controlled not only by material properties but also by complex interaction with stratigraphic and hydrological variables. Following the Devensian deglaciation in the early Holocene (about 10 000 years BP), water tables were presumably at or very close to the surface in the low permeability clays. Weathering would have commenced as conditions changed to allow drawdown, especially during the climatic optimum (about 6000 years BP). Oxidation would have occurred first at the ground surface, penetrating deeper as the phreatic surface was lowered. Thus in Oxford Clay and similar units of consistent internal stratigraphy, the depth of oxidation is controlled by the lowest elevation of the ancestral water table. This level is probably much deeper than at present as a result of the climatic optimum, as has been noted for the London Clay (Chandler 1972). On a longer timescale, evidence for deep interglacial or interstadial weathering prior to the Devensian glaciation remains to be identified.

In tills, the frequent presence of subhorizontal sand or gravel bodies affects the hydrological regime because of their much greater permeability. Thus, lenses of coarse material may form a limiting lower level to weathering in till sequences, above which clay will be intensely weathered. Zones of intermediate weathering intensity may be thin, if not entirely absent.

A failure to recognise the stratigraphic importance of deep Holocene weathering in tills has, as discussed above, caused severe misinterpretations of glacial history in eastern Britain. In addition, recent recognition of thin weathered outliers of Devensian till beyond the traditionally accepted limits of Devensian ice (Madgett & Catt 1978, pp. 99 and 105, Catt 1983a) has shown that the ice margin reconstructions developed from supposed end moraines or ice-marginal drainage channels are often inaccurate.

Future prospects

It is not only the study of the most recently deposited till sheets which can benefit from the observation of weathering profiles. For example, the mineralogical and geotechnical properties of the weathered pre-Devensian matrix-dominant tills outside the limits of the Devensian glaciation, and deposited by previous ice incursions (e.g. paleo-argillic horizons in the Chalky Boulder Clay of Essex, Sturdy *et al.* 1979), still await documentation within a standardised framework of weathering zones. The stratigraphic implications of such documentation are significant and may also shed some light on the origin of clay-rich, decalcified *remanié* sediment mantles found on dip slopes of southern England (e.g. Chiltern Drift) outside the currently recognised limit of Quaternary glaciation (Catt & Hodgson 1976). It may also be that the mantles overlying Oxford Clay weathering profiles, and normally ascribed to solifluction processes, are weathered remnants of pre-Devensian tills. A reliable methodology to identify such deposits would assist in describing the geomorphological evolution of an area, and particularly in engineering evaluations of a landscape where the long term stability of land surfaces needs assessment.

The study of Late Cenozoic weathering profiles and pedogenesis may also provide important clues to Quaternary environmental history for which very little sedimentary evidence survives (e.g. Catt 1983b). The full complexity of such history is evident from the ocean record (e.g. Bowen 1978), and is a reminder that many other variables have influenced the development of weathering profiles, in addition to those few that can be reliably inferred during the course of normal field and laboratory investigations. It is clear that the evolution of weathering profiles in pre-Pleistocene mudrocks and Pleistocene tills cannot be considered in isolation from each other and it is hoped that this preliminary review of an admittedly limited data base will encourage interdisciplinary research into weathering of clays in Quaternary studies. It may be that systematic examination on a regional scale of mineralogical and geotechnical characteristics, and variations in depth of weathering in overconsolidated clays will reveal a wealth of data relevant to the Quaternary history and the geomorphological development of the British land mass.

Acknowledgements

This chapter is published with permission of the Director, Ontario Geological Survey. We are grateful to J. A. Catt and J. A. Sladen for commenting on initial drafts of the manuscript. This work has been supported by grants from the Natural Environment Research Council and the University of Newcastle-upon-Tyne.

References

Arrowsmith, E. J. (ed.) 1978. *Clay fills*. London: Institution of Civil Engineers.
Bickmore, D. P. and M. A. Shaw 1963. *The atlas of Britain and Northern Ireland*. Oxford: Clarendon Press.
Bisat, W. S. 1940. Older and newer drift in East Yorkshire. *Proc. Yorks. Geol Soc.* **24**, 137−51.
Bjerrum, L. 1967. Progressive failure in slopes of overconsolidated plastic clay and clay shales. *J. Soil Mech. Fndns Div. Am. Soc. Civ. Eng.* **93**, 1−44.
Boulton, G. S. and D. L. Dent 1974. The nature and rates of post-depositional changes in recently deposited till from southeast Iceland. *Geogr. Ann.* **56A**, 121−34.
Boulton, G. S. and M. A. Paul 1976. The influence of genetic processes on some geotechnical properties of glacial tills. *Q. J. Geol Soc. Lond.* **9**, 159−94.
Bowen, D. Q. 1978. *Quaternary geology*. Oxford: Pergamon Press.
Burland, J. B., T. I. Longworth and J. F. A. Moore 1977. A study of ground movement and progressive failure caused by a deep excavation in Oxford Clay. *Geotechnique* **27**, 559−91.
Carruthers, R. G. 1947. The secret of the glacial drifts. *Proc. Yorks. Geol Soc.* **27**, 43−57, 129−72.
Carruthers, R. G. 1953. *Glacial drifts and the undermelt theory*. Newcastle-upon-Tyne: Harold Hill.
Catt, J. A. 1980. Till facies associated with the Devensian Glacial maximum in eastern England. *Quat. News.* **30**, 4−10.
Catt, J. A. 1983a. Cenozoic pedogenesis and landform development in south-east England. In *Residual deposits: surface related weathering process and materials*, R. C. L. Wilson (ed.), 251−8. Oxford: Blackwell.
Catt, J. A. 1983b. Soils and Quaternary stratigraphy in the U.K. (Abstract). *Quat. News.* **41**, 44.
Catt, J. A. and J. M. Hodgson 1976. Soils and geomorphology of the Chalk in south-east England. *Earth Surf. Proc.* **1**, 181−93.
Chandler, R. J. 1969. The effect of weathering on the shear strength properties of Keuper Marl. *Geotechnique* **19**, 321−34.
Chandler, R. J. 1972. Lias Clay: weathering processes and their effect on shear strength. *Geotechnique* **22**, 403−31.

Chandler, R. J. 1973. A study of structural discontinuities in stiff clays using a polarising microscope. In *Proceedings of the International Symposium on Soil Structure*, Gothenburg, 78–85.

Cocksedge, J. E. 1983. Road construction in glaciated terrain. In *Glacial geology: an introduction for engineers and earth scientists*, N. Eyles (ed.), 302–17. Oxford: Pergamon Press.

Coulthard, J. M. 1975. *Petrology of weathered Lower Liassic Clays*. Unpublished Ph.D. thesis, University of Aston.

Cripps, J. C. and R. K. Taylor 1981. The engineering properties of mudrocks. *Q. J. Eng. Geol.* **14**, 325–46.

Dearman, W. R. and N. Eyles 1982. An engineering geological map of the soils and rocks of the United Kingdom. *Bull. Int. Assoc. Eng. Geol.* **25**, 3–18.

Droste, J. B. 1956. Alteration of clay minerals in a Wisconsin till. *Geol Soc. Am. Bull.* **67**, 911–18.

Eyles, N. and J. Menzies 1983. The subglacial landsystem. In *Glacial geology: an introduction for engineers and earth scientists*, N. Eyles (ed.), 19–71. Oxford: Pergamon Press.

Eyles, N. and J. A. Sladen 1981. Stratigraphy and geotechnical properties of weathered lodgement till in Northumberland, England. *Q. J. Eng. Geol.* **14**, 129–41.

Eyles, N., J. A. Sladen and S. T. Gilroy 1982. A depositional model for stratigraphical complexes and facies superposition in lodgement tills. *Boreas* **11**, 317–33.

Gilroy, S. T. 1980. *The engineering geological properties of the weathered Devensian tills of Holderness*. Unpublished M.Sc. thesis, University of Newcastle-upon-Tyne.

Jenny, H. 1950. Origin of soils. In *Applied sedimentation*, P. D. Trask (ed.), 41–61. New York: Wiley.

Jones, P. F. and E. Derbyshire 1983. Late Pleistocene periglacial degradation of lowland Britain: implications for civil engineering. *Q. J. Geol Soc.* **16**, 197–210.

Leighton, M. M. and P. MacClintock 1962. The weathered mantle of glacial tills beneath original surfaces in north central United States. *J. Geol.* **70**, 267–93.

Lunn, A. C. 1980. Quaternary. In *The geology of northeast England*, D. A. Robson (ed.), 48–60, Newcastle-upon-Tyne: Northumbria Natural History Society.

McGown, A. and E. Derbyshire 1977. Genetic influences on the properties of tills. *Q. J. Geol Soc.* **10**, 389–410.

McGown, A., W. F. Anderson and A. M. Radwan 1978. Geotechnical properties of tills in west-central Scotland. In *The engineering behaviour of glacial materials*, R. B. Hoole (ed.), 81–91. Norwich: GeoBooks.

Madgett, P. A. 1975. Re-interpretation of Devensian till stratigraphy of eastern England. *Nature* **253**, 105–7.

Madgett, P. A. and J. A. Catt 1978. Petrography, stratigraphy and weathering of late Pleistocene tills in east Yorkshire, Lincolnshire and north Norfolk. *Proc. Yorks. Geol Soc.* **42**, 55–108.

Morgenstern, N. R. 1970. Discussion on black shale heaving in Ottawa, Canada. *Canadian Geotech. J.* **7**, 114–15.

Morris, K. A. 1979. A classification of Jurassic marine shale sequences. *Palaeogeog. Palaeoclim. Palaeoecol.* **26**, 117–26.

Paul, M. A. 1983. The supraglacial landsystem. In *Glacial geology: an introduction for engineers and earth scientists*, N. Eyles (ed.), 71–90. Oxford: Pergamon.

Quigley, R. and T. A. Ogunbadejo 1976. Till geology, mineralogy and geotechnical behaviour, Sarnia, Ontario. In *Glacial till* R. F. Legget (ed.), 336–45, Ottawa: Royal Society of Canada, Special Publn. 12.

Russell, D. J. 1977. *The effect of weathering on physical and chemical properties of some Mesozoic clays*. Unpublished Ph.D. thesis, University of Reading.

Russell, D. J. and A. Parker 1979. Geotechnical, mineralogical and chemical interrelationships in weathering profiles of an overconsolidated clay. *Q. J. Eng. Geol.* **12**, 107–16.

Sellwood, B. W. and C. P. Sladen 1981. Mesozoic and Tertiary argillaceous units: distribution and composition. *Q. J. Eng. Geol.* **14**, 263–75.

Skempton, A. W. and R. D. Northey 1952. The sensitivity of clays. *Geotechnique* **3**, 30–53.

Sladen, J. A. 1979. *Weathering and its effect on the geotechnical properties of tills in southeast Northumberland*. Unpublished M.Sc. thesis, University of Newcastle-upon-Tyne.

Sladen, J. A. and W. F. Wrigley 1983. Geotechnical properties of lodgement till: a review. In *Glacial geology: an introduction for engineers and earth scientists*, N. Eyles (ed.), 184–212. Oxford: Pergamon Press.

Sturdy, R. G., R. H. Allen, P. Bullock, J. A. Catt and S. Greenfield 1979. Palaeosols developed on Chalky Boulder Clay in Essex. *J. Soil Sci.* **30**, 117–37.

Taylor, R. K. and D. A. Spears 1970. The breakdown of British Coal Measure rocks. *Int. J. Rock Mech. Min. Sci.* **7**, 481–501.

Taylor, R. K. and D. A. Spears 1972. The influence of weathering on the composition and engineering properties of *in situ* Coal Measure rocks. *Int. J. Rock Mech. Min. Sci.* **9**, 729–56.

Tucker, M. E. 1978. Triassic lacustrine sediments from south Wales: shore zone clastics, evaporites and carbonates. In *Modern and ancient lake sediments*, A. Mather and M. E. Tucker (eds), 203–24. Oxford: Blackwells.

Vaughan, P. R. and H. J. Walbancke 1978. The stability of cut and fill slopes in boulder clay. In *The engineering behaviour of glacial sediments*, R. B. Hoole (ed.), 209–19. Norwich: GeoBooks.

Vaughan, P. R., D. W. Hight, V. G. Sodha and H. J. Walbancke 1979. Factors controlling the stability of clay fills in Britain. In *Clay fills*, D. Arrowsmith (ed.), 205–18. London: Institution of Civil Engineers.

Index

Numbers in *italics* refer to text illustrations.